SELECTED ACADEMIC STUDIES FROM TURKISH TOURISM SECTOR

Fatih Türkmen (ed.)

SELECTED ACADEMIC STUDIES FROM TURKISH TOURISM SECTOR

Bibliographic Information published by theDeutsche Nationalbibliothek
The Deutsche Nationalbibliothek lists this publication in the Deutsche
Nationalbibliografie; detailed bibliographic data is available online at
http://dnb.d-nb.de.

Library of Congress Cataloging-in-Publication Data
A CIP catalog record for this book has been applied for at the Library of Congress.

Printed by CPI books GmbH, Leck

ISBN 978-3-631-81099-6 (Print)
E-ISBN 978-3-631-81607-3 (E-PDF)
E-ISBN 978-3-631-81608-0 (EPUB)
E-ISBN 978-3-631-81609-7 (MOBI)
DOI 10.3726/b16699

© Peter Lang GmbH
Internationaler Verlag der Wissenschaften
Berlin 2020
All rights reserved.

Peter Lang – Berlin · Bern · Bruxelles · New York ·
Oxford · Warszawa · Wien

All parts of this publication are protected by copyright. Any
utilisation outside the strict limits of the copyright law, without
the permission of the publisher, is forbidden and liable to
prosecution. This applies in particular to reproductions,
translations, microfilming, and storage and processing in
electronic retrieval systems.

This publication has been peer reviewed.

www.peterlang.com

Editor CV

Associate Prof. Dr. Fatih Türkmen was born in Ankara, Turkey, in 1978. He completed his graduation, postgraduation and doctorate in the field of tourism management. He has published many articles in national and international journals apart from books on tourism. Türkmen is associated with the Safranbolu Tourism Faculty at Karabük University, Turkey, and is on a temporary assignment with the tourism department at Ahmet Yesevi University, Turkistan, Kazakhstan.

Book Abstract

A compilation of current academic studies on the tourism sector of Turkey, the 23 chapters of this book have been written by tourism academicians who are experts in their respective fields. The chapters contain qualitative or quantitative research data and current statistical data, making this book useful for academic research. I thank all the contributors and dedicate this book to my family.

Keywords: Tourism, Gastronomy, Academic tourism studies, Turkish tourism sector

Contents

List of Contributors 11

Abdullah USLU
1 Rural Tourism in Durbuy City 15

Ahmet DUYAR and Ahmet AÇIL
2 Effects of Ecotourism Activities in Forests on the Soil 29

Bilal DEVECİ
3 Usage Areas of Salt Mineral and Its Varieties 35

Ebru GÖZEN
4 Hotel Managers' Metaphoric Perceptions for Smart Hotel 43

Ediz GÜRİPEK
5 Innovation in Tourism and Examples in Practice 63

Emre AYKAÇ and Ömer Ceyhun APAK
6 Diaspora Tourism 71

Yasin DÖNMEZ and Sevgi ÖZTÜRK
7 Ecotourism and Geographical Information Systems Applications 89

Emin ARSLAN
8 Gastronomy Tourism and Geographical Indications in Tokat 97

Mehmet TEKELİ and Ezgi KIRICI TEKELİ
9 Sustainable Gastronomic Tourism 113

Mehmet Mert PASLI and Evren GÜÇER
10 To Determine the Recreational Potential of Trabzon 135

Samet GÖKKAYA
11 Digital Transformation & Marketing in Tourism Industry 149

Hakan KENDİR
12 Cultural Heritage Tourism Inventory in Tokat Province 165

Serdar SÜNNETÇİOĞLU
13 Effects of Digitalization in Tourism 177

Handan ÖZÇELİK BOZKURT
14 Climate Change and Tourism 193

Irem BOZKURT and Enes YILDIRIM
15 Overtourism 207

Kansu GENÇER
16 Qualitative Approaches for Tourism Research 225

Mehmet CAN and Çağla ÜST CAN
17 Event Tourism 241

Mustafa Cüneyt ŞAPCILAR and Ahmet BÜYÜKŞALVARCI
18 The Effect of Nepotism 251

Serdar ÇÖP
19 Sharing Economy for Sustainability in Tourism 275

Serdar EREN
20 Sustainable Tourism Criteria and Turkish Restaurants 285

Uğur CEYLAN
21 City Tourism and Kütahya 293

Yeliz PEKERŞEN
22 An Overview of Creative Tourism Concept 309

Yılmaz SEÇİM
23 Gastronomy Festivals in Turkey 323

List of Figures 335

List of Tables 337

List of Contributors

Ahmet AÇIL
Research Assistant, Karabük University, Faculty of Forestry, Department of Forestry Engineering, email: ahmetacil@karabuk.edu.tr

Ömer Ceyhun APAK
Lecturer, Bayburt University, Social Sciences Vocational School, email: ceyhun.apak@hotmail.com.

Emin ARSLAN
Assist. Prof., Dr., Tokat Gaziosmanpaşa University, Zile Dinçerler Tourism and Hotel Management College, Department of Gastronomy and Culinary Arts, Tokat, Turkey, email: emin.arslan@gop.edu.tr.

Emre AYKAÇ
Nevşehir Hacı Bektaş Veli University, Institute of Social Sciences, Department of Tourism Management, email: emreaykac1@hotmail.com.

Ahmet BÜYÜKŞALVARCI
Prof., Dr., Necmettin Erbakan University, Tourism Faculty, Department of Tourism Management, email: ahmetbuyuksalvarci@gmail.com.

İrem BOZKURT
Lecturer, Mardin Artuklu University, Mardin Vocational High School, Department of Hotel, Restaurant and Catering Services, email: irembozkurt@artuklu.edu.tr.

Mehmet CAN
Assist. Prof., Dr., Aksaray University, Güzelyurt Vocational School, Department of Hotel, Restaurant and Catering Services, email: mehmet23can@hotmail.com.

Uğur CEYLAN
Assist. Prof., Dr., Kütahya Dumlupınar University, Tavşanlı Faculty of Applied Sciences, Department of Tourism Management, email: ugur.ceylan@dpu.edu.tr.

Serdar ÇÖP
Assist. Prof., Dr., Istanbul Gelisim University, Faculty of Economics, Administrative and Social Sciences, Department of Tourism Guidance email: scop@gelisim.edu.tr

Bilal DEVECİ
Assist. Prof., Dr., Kırklareli University, Faculty of Tourism, Department of Gastronomy and Culinary Arts, email: bilaldeveci@gmail.com.

Yasin DÖNMEZ
Assoc. Prof. Dr., Karabük Üniversity, Faculty of Forestry, Department of Landscape Architecture, Karabük, email: yasindonmez@karabuk.edu.tr.

Ahmet DUYAR
Assist. Prof., Dr., Karabük University, Faculty of Forestry, Department of Forestry Engineering, email: ahmetduyar@karabuk.edu.tr

Serdar EREN
Assist. Prof., Dr., Kutahya Dumlupinar University, Tavsanli Faculty of Applied Sciences, Department of Gastronomy and Culinary Arts, email: serdar.eren@dpu.edu.tr

Kansu GENÇER
Ph.D, Kütahya Dumlupınar University, Tavşanlı Faculty of Applied Sciences, Department of Gastronomy and Culinary Arts, email: kansu.gencer@dpu.edu.tr

Samet GÖKKAYA
Ph.D., Karabük University, Safranbolu Tourism Faculty, Department of Tourism Management, email: sametgokkaya@karabuk.edu.tr

Ebru GÖZEN
Assist. Prof., Dr., Akdeniz University, Manavgat Tourism Faculty, Department of Recreation Management, email: ebrugozen@akdeniz.edu.tr

Evren GÜÇER
Assoc. Prof., Dr., Haci Bayram Veli University, Faculty of Tourism, Department of Recreation Management, email: evrengucer@gazi.edu.tr

Ediz GÜRİPEK
Assist. Prof., Dr., Tokat Gaziosmanpaşa University, Zile Dinçerler Tourism and Hotel Management College, Tourism and Hotel Management Department, Tokat, Turkey, email: ediz.guripek@gop.edu.tr

Hakan KENDİR
Assist. Prof., Dr., Tokat Gaziosmanpaşa University, Zile Dinçerler Tourism and Hotel Management College, Department of Tourism and Hotel Management, Tokat, Turkey, email: hakan.kendir@gop.edu.tr.

Ezgi KIRICI TEKELİ
Lecturer, Ezgi KIRICI TEKELİ, Iğdır University, Iğdır Vocational School of Higher Education, Department of Travel-Tourism and Entertainment Services, Tourist Guidance Program, email: ezgi.kirici@igdir.edu.tr.

Handan ÖZÇELİK BOZKURT
Assist. Prof. Dr., Sinop University, School of Tourism and Hotel Management, Department of Gastronomy and Culinary Arts, email: handanozcelikbozkurt@gmail.com.

Sevgi ÖZTÜRK
Assoc. Prof. Dr., Kastamonu University, Faculty of Engineering and Architecture, Department of Landscape Architecture, Kastamonu, email: sevgiozturk37@gmail.com.

Mehmet Mert PASLI
Assist. Prof., Giresun University, Bulancak School of Applied Sciences, Tourism and Hotel Management, email: mert.pasli@giresun.edu.tr

Yeliz PEKERŞEN
Assist. Prof. Dr., Necmettin Erbakan University, Tourism Faculty, Gastronomy and Culinary Arts Department, Konya/Turkey, email: yeliz.ulusan@gmail.com.

Mustafa Cüneyt ŞAPCILAR
Assist. Prof., Dr., Necmettin Erbakan University, Tourism Faculty, Department of Tourism Management, email: mustafcuneyt@gmail.com.

Yılmaz SEÇİM
Assist. Prof. Dr., Yılmaz SEÇİM., Necmettin Erbakan University, Tourism Faculty, Gastronomy and Culinary Arts Department, email: yilmazsecim@gmail.com

Serdar SÜNNETÇİOĞLU
Assist. Prof., Dr., Çanakkale Onsekiz Mart University, Tourism Faculty, email: serdarsunnetcioglu@comu.edu.tr

Mehmet TEKELİ
Mehmet TEKELİ, Nevşehir Hacı Bektaş Veli University, Institute of Social Sciences, Department of Tourism Management, email: tekelimehmet@hotmail.com.

Abdullah USLU
Assist. Prof., Dr., Akdeniz University, Manavgat Tourism Faculty, Department of Tourism Management, email: auslu@akdeniz.edu.tr.

Çağla ÜST CAN
Instructor, Aksaray University, Güzelyurt Vocational School, Department of Hotel, Restaurant and Catering Services, email: caglaust@gmail.com.

Enes YILDIRIM
Nevşehir Hacı Bektaş Veli University, Institute of Social Sciences, Department of Tourism Management, email: enesyildirim@nevsehir.edu.tr.

Abdullah USLU

1 Rural Tourism in Durbuy City*

Introduction

Lately, rural tourism became a significant concept in the tourism industry and has developed immensely. World Tourism Organization envisages that from 2020s onwards, tourists will participate in individual tourism movements rather than mass tourism and will head towards alternative and special interest tourism activities with a desire to experience the untried. In most rural areas, tourism is deemed as an important alternative for the purposes of economic growth and regional development. Studies going a long way back reveal that rural tourism has a positive economic impact. Tourism brings variety to the economy of rural communities, which is mainly dependent on agriculture. It provides new opportunities for rural areas. Rural cities such as Durbuy consider tourism as an economic value. Such cities endeavor to develop their tourism industry. However, restricted competitive advantages have an impact on destinations in a way to affect the tourism capital, as well. Building a dreamy or attractive place for tourists is quite difficult in rural tourism areas. For instance, for most tourists the city of Brussels can be a much more interesting place than a rural tourism destination.

Being the biggest village of Medieval Europe, Durbuy embodies various rural tourism activities and is a very important rural tourism destination. From that perspective, the rural tourism potential of Durbuy city will be evaluated and discussed in the current study. This evaluation is believed to serve as a good role model in the sense of developing other rural tourism areas throughout the world. In this chapter, firstly the concept of rural tourism, which is at least as important as the trio of sea, sand and sun, the importance of rural tourism and its impacts will be addressed, and following that, Durbuy city which is exemplary for its active rural tourism activities and was selected the most distinguished rural tourism destination within the scope of European Destinations of Excellence (EDEN) project commenced by the European Commission in 2007 will be introduced. Lastly, the rural tourism activities in Durbuy will be listed.

---

* This chapter of the paper was presented in VII. National 2nd International Rural Tourism and Development Congress, 10–13 May 2018, Bodrum/Muğla/Turkey.

Rural Tourism

With the strong winds of globalization which rose within the last century and as a result of economic, political, cultural and technological changes, countries realized that they need to develop sustainable policies and to revise their existing ones. People living in rural areas should not be restricted only to development from agricultural and livestock activities and should be provided with rural tourism opportunities which will develop their entrepreneurial capital. The quest for a sustainable and livable environment has pushed people to seek different things. New dimensions were introduced to vacation, recreation and entertainment activities, bringing with them new and popular approaches to tourism. Rural tourism is one of such new and modern approaches. Rural tourism, together with its relevant activities, contributed to the balanced development of countries and improvement of the prosperity of rural communities. Likewise, by enabling an active and effective use of rural areas, rural tourism activities also prevent the migration of people living in rural settlements to larger cities.

In the sense of developing upon the concept of rural tourism, it can be seen that academics from different countries have proposed different definitions. Some definitions emphasize places with low population while some others place importance on the tourists' expectations for the natural, traditional and historical features of settlements (EC, 1999). Still other definitions consider rural tourism as a kind of tourism wherein tourists take part in activities such as ranch tours and hikes in the nature, riding horses, trekking, fishing etc. There is another definition which addresses rural tourism, in the broadest sense, as all tourism activities carried out in a rural area (Özkan, 2007).

World Tourism Organization (UNWTO) defines rural tourism and the rural culture as an essential part of the recreational tourist product. According to the definition provided by the EU, rural tourism is what is desired by people who enjoy spending their vacation time in a rural environment, being involved in the rural way of living and seeing the rural heritage (European Commission, 2003). The components of rural tourism are service, transportation, accommodation, activity areas, centers of attraction, events and support services (Veer & Tuunter, 2005). Another definition describes rural tourism as a vacation activity carried out in a safe and peaceful rural settlement, participating in the use of natural resources, and being a part of cultural values and the rural manner of living (Middleton & Rebecca, 1988). Likewise, rural tourism is a complex activity containing multidimensional practices ranging from nature activities, festivals, historical and traditional events, art shows, agricultural tourism, folk theater to farm-based tourism and educational travelling (Kiper, 2006). According to the World Travel & Tourism Council (WTTC), for a tourism activity to be considered within the scope of rural tourism, it needs to be carried out in a rural setting, and during such activity, the natural fabric of that rural setting needs to be protected on a sustainability basis. In line with these conditions, rural tourism needs to place importance on the authentic natural assets of the area and the socio-cultural diversity should be emphasized.

Rural tourism emerged as a different kind of tourism in line with people's desire to get away from the routine of life and concretion of urban areas and to satisfy their need to see different sights and their interest in recreational activities at peaceful and natural areas (Ahipaşaoğlu & Çeltek, 2006: 2). In terms of the sustainability of tourism, rural tourism has an importance that grows with each passing day due to the fact that the activities it contains do not harm the nature (Fuentes-Fuentes et. al., 2006: 48). Another phenomenon which boosts the popularity of rural tourism is that it brings together the local culture and local products, weaving a different touristic fabric (Liu, 2006). The two most important main features of rural tourism are the facts that it serves as a livelihood for the local residents by providing socio-economic benefits in the rural area and that it is an effective instrument for the protection of cultural and natural values (Snieškaa et al., 2014).

The reasons impacting upon the rapid popularization and development of rural tourism throughout the world are the changes in the expectations of tourists, the desire to see new places and cultures, and the wish to move individually instead of being involved in mass tourism (Kutukız et al., 2016). Rural tourism encompasses both a travel from one country to the rural areas of another country and a short visit to the rural areas nearby a person's household and contributes greatly to rural development. The items in the list provided below are the driving forces of rural tourism (Hall et al., 2005);

- Income-generating activities,
- Changes happening in rural and urban areas,
- The multiplied effect in the case of direct investments of a relatively small scale,
- Strengthening of local/regional structures through networking,
- Promotion of developments in the physical infrastructure, and
- Increasing the diversity of economic activities.

In the literature, rural tourism is also called farm-based tourism, agro-tourism, green tourism or eco-tourism (Dubois & Schmitz, 2013; Peláez, 2004). Rural tourism is a type of tourism allowing for activities such as shopping, skiing, horseback nature rides, rafting, sports, hunting and fishing and arts and culture activities (Ryglova et al., 2017).

The rural tourism capital comprises rural areas, rural heritage, rural life and rural activities. Rural areas are mainly made up of the nature, mountains, forests, lakes and rivers. While rural heritage encompasses the traditional and local architecture and historical ruins, the rural life contains local food and activities, handicrafts and traditional music. The most prominent rural activities are hiking, fishing, horse riding, biking and water sports (Özdemir, 2012). Tourism activities in rural areas are shaped by the topography, vegetation, climate, historical and cultural assets, water resources, socio-economic structure and fabric and traditions of the area. Accordingly, the rural tourism identity of a given area is formed by its authentic attributes.

According to OECD (1994), activities such as hiking, photography, biking, climbing, hunting, fishing, horse riding, festivals, village strolls, rural sightseeing, grass skiing, birdwatching, rafting, canoeing, folklore and sporting activities belonging to the local culture can be considered within the scope of rural tourism activities.

The Importance of Rural Tourism

Rural tourism is of great importance for rural areas and national economies as a whole. It has a marked difference in that it has the advantage of being available in all twelve months of the year, contains a rich variety of activities and is suitable for all. Rural tourism is not limited only to coastal areas and bears the potential to be developed and to develop in all rural areas. Rural tourism contributes positively to the social, economic and cultural development of rural areas and also serves to protect the natural, historical and cultural heritage in rural areas. When carried out in a planned manner, rural tourism activities provide the local people, especially women, with new employment opportunities and play an important role in making remarkable contributions to rural economies (Soykan, 2003: 2). Moreover, through the refreshing effect of rural tourism areas, positive results can be achieved in the sense of resolving the economic, social and psychological issues faced by tourists due to their being boxed up in a certain place during their vacations (Gürer, 2003). Again, according to Kesici (2012: 33) the importance of rural tourism can be summarized as follows;

- Rural tourism has no seasonality and can be enjoyed in any season.
- Rural tourism has an important balancing effect in the geographical distribution of tourism.

- Rural tourism can be integrated to numerous other types of tourism.
- Rural tourism contains a great variety of unique recreational activities.
- Rural tourism has a different tourist profile.
- Rural tourism contributes to the protection of the natural environment and the cultural heritage.
- Rural tourism serves the understanding of sustainable tourism.
- Rural tourism is an important tool for the publicity of a country.

Another important attribute of rural tourism is that it has lower costs. It is a tourism activity taking places in rural areas which can grow without relying on external sources through the contribution of local administrations and the participation of small scale enterprises. The installation costs are also rather low and the installation processes are easy. Likewise, rural tourism activities can be developed regardless of the need for other companies' establishing new businesses or making greater investments in the region (Wilson et al., 2001: 132).

Positive and Negative Impacts of Rural Tourism

In the 1990s it was found out that rural tourism has an effect in the development of rural areas. With a study carried out in the United Kingdom, it was revealed that local residents were involved in the tourism industry in order to increase their income (Fleischer & Pizam, 1997). The facts that people living in rural areas have lower income, that their working conditions are not very well, that the livelihood earned from agricultural activities is low, that unemployment rates are higher in the rural areas and seasonal labor is prevalent have rendered rural tourism a more attractive alternative. The purpose here is supporting the orderly distribution of income throughout the year and also resolving the issue of latent unemployment, enabling the distribution seasonal work to all 12 months of the year and, as a result, providing a positive contribution to household and national economies (Torun, 2013: 34). At the same time, rural tourism activities have both positive and negative impacts on the socio-economic situation, rural development, social structure, physical environment and culture.

Rural tourism may also impose both positive and possible negative effects on rural areas. Positive effects can be generally listed as below (Morgül, 2006: 68–70; Uslu et al., 2015);

- Rural tourism prevents the migration of the residents
- It contributes to the development of relations between people residing in rural areas and those living in cities.
- It enables family members to spend more time together.

- It contributes to the development of rural areas.
- It enables the generation of income for the residents from varied activities and boosts the development of small enterprises.
- It improves the living standards for the residents
- It eases the burden on other tourism destinations.
- It revitalizes domestic tourism.

Negative impacts of rural tourism on rural areas can be listed as below (Morgül, 2006: 28–70; Uslu et al., 2015);

- As a result of residents generating income from rural tourism and ignoring the agricultural sector, rural tourism causes a decrease in the number of cultivated lands.
- It causes speculative increases in real estate prices such as those for lands, buildings etc.
- It causes the degradation of natural and cultural areas due to the overuse of rural areas and distorts the ecological balance.
- As a result of overcrowding in rural areas, their attractiveness may decrease over time and they may lose their authenticity.
- This also gives way to a cultural pollution which may even end up with the locals losing their cultural identity.
- It changes the crop patterns and decreases productivity.

Durbuy and Rural Tourism in Durbuy

Durbuy is located in the south of Belgium and 120 km (1 hour) distant from Brussels. It is a small city in Wallonia region with a surface area of 156.61 km². The population of Durbuy is 11.459 (1st of December 2017). Durbuy was an important trade and industrial center in the Middle Ages. In our day and time, the main activity in Durbuy is tourism and recreation. In 1331, John I, Count of Luxembourg contributed to increasing the value of the city. It has been referred to as 'Old Town' since the Middle Ages. The old town of Durbuy is renowned as the smallest city in the world. It is separated from the surrounding 40 villages and is recognized as 'town' since the Middle Ages.

Neighboring to the three geographical areas, Durbuy embodies plenty of countryside areas and also a rich culinary culture. With its colors, flowers, cultural heritage discovered by tourists, it has always been at the forefront throughout history. In 2007, it was dubbed 'European tourist destination of excellence'. For more than 150 years, it is far-famed with its cuisine and dishes. Dishes are prepared with locally produced products. Apart from being a rural tourism area, the main driving force behind Durbuy's economy is the popularity of its cuisine. Furthermore, agriculture and forestry activities as well as tourism are the income sources in Durbuy. Tourists visit Durbuy in December at most. This is due to the Christmas market set up in the city in December. There are a great number of restaurants (45 luxury restaurants), cafes and bars in the city. Durbuy attracts the attention of tourists with its cobblestone pavements, narrow streets and medieval houses ornamented with flowers.

In rural areas characterized by small family businesses, rural tourism activities are the main source of economic development (Fleischer & Felsenstein, 2000). For rural tourism activities, small accommodation facilities such as hotels, motels, and apart hotels etc. are rather preferred (Getz & Carlsen, 2000). Besides rural tourism may have direct economic impacts on this type of accommodation facilities, it provides indirect advantages to other small family businesses such as grocery shops and so on (Wilson et al., 2001: 132). In the scope of accommodation activities in Durbuy, tourists visiting the city usually prefer small family businesses for accommodation. There are a wide range of accommodation alternatives for visitors such as attractive, charming small hotels and facilities, camping and caravan areas, small guest rooms, houses for business travelers, rural cottages, holiday villages, and farmhouses. 22 small but luxurious hotels operate in Durbuy. Female labor force that initially starts working as a hobby is actively engaged in these activities (Getz & Carlsen, 2000).

In cooperation with all stakeholders (local residents, administrators, non-governmental organizations and tourism professionals) in Durbuy, well-organized tourism plans were developed in order to render the tourism activities professional. Thus, seasonal fluctuations were minimized and a diverse range of tourism activities were organized to increase the number of overnight stays in the region. Close cooperation between local administrations and tourism stakeholders was ensured and an administrative structure looking after the rights of local residents was established. All those efforts were aimed to ensure customer satisfaction and sustainable development. The fact that it was dubbed in 2007 'European tourist destination of excellence' has raised the value of tourism in the region. Sustainable tourism activities were developed based on realistic policies. Effective use of Internet and dynamic communication strategy has paved the way for successful management of tourist activities and tourism impacts.

The plans dedicated to render tourism more sustainable in Durbuy are listed as follows (EDEN, 2007):

* Increase the volume of tourist arrivals
* Discover alternative walking routes
* Raise Durbuy's profile
* Make tourist reception areas more visible and more similar to one another
* Encourage tourism stakeholders and improve quality standards
* Establish a system that presents customer views
* Support local economy via tourism
* Render tourist taxes more viable
* Promote out-of-season tourism activities
* Maintain the quality of the natural environment for each and every activity and attraction
* Guarantee water quality in river areas
* Reduce population in rural areas
* Raise tourists' awareness of their responsibilities.

Rural Tourism Activities in Durbuy

Rural Tourism and recreation are its main activities nowadays in Durbuy city. Durbuy is often described, albeit without much justification, as the smallest city/town in the world (www.durbuy.be). Durbuy offers a wide variety of attractions and activities for tourists. Some of them are as follows (EDEN, 2007; www.europeanbestdestinations.com; www.discoveringbelgium.com; www.walloniabelgiumtourism.co.uk):

* One of the most visited places in Durbuy is Modave Castle (The Château of Modave). Modave Castle is a medieval castle from the 13th century. It underwent a major restoration in the 17th century. This stone castle has 20 luxurious rooms reflecting the art of that period. Inside, there is a bathtub made of cut stones.
* Besides being the smallest city in the world, Durbuy embodies the largest natural pruning park, as well. The Durbuy Topiary Garden covering an area of 10 000 m2 embodies more than 250 topiary human and animal figures, which polishes the park with an impressive and blinding look.
* The Durbuy Diamond Museum, a jewelry museum with a wide variety of diamond sets, is another attraction visited frequently by tourists.
* Trip by train that travels within Durbuy and passes by nearly all attractions in the city provides visitors with a magnificent view of Durbuy.
* Durbuy is also home to Belgium's most popular beer (Durboyse). Tourists can find a lot of opportunities throughout their holidays to taste the amber-color Durboyse.

* Durbuy is also renowned with its great variety of jams. The shops with the widest variety of souvenirs in Belgium are also located in Durbuy. Durbuy is far-famed with spices, tea, honey and sweets, as well.
* Tourists can reach the Saint Amor Jam factory by train or climbing up hills, watch the jam-making process and purchase different varieties of jam if they wish.
* Durbuy hosts different events every month of the year: Valentine's Day in February; Chocolate Market in March; Lobster Festival in June-July; Floral Carpets, Beer, Bread and Cheese Market in September; Hunting Events and Concerts in October-November; and Christmas Market in December.
* The Chèvreried'Ozo Goat Farm, which is located nearby Durbuy, is an important attraction for tourists with its high quality cheese and milk.
* In Durbuy, tourists can visit a wide variety of picnic areas and also enjoy more than 60 tourism activities such as rafting, canoeing, kayaking, horseback riding, geo-caching and mountain-biking and so on.
* Durbuy has dozens of hiking trails (Durbuy-Palenge 5.83 miles, Durbuy Fond Vedeur 4.54 miles, etc.).
* Durbuy has the best 18-hole golf course in Belgium.
* In Durbuy, various festivals are organized throughout the year. For instance, Rock Festival, Choco Palace Festival and so on.

Academic Studies on Durbuy for Rural Tourism

Cities come to the forefront with different tourist products, besides being a brand on its own. Places, regions and cities which turn into a touristic value with the help of the design and presentation of tourist products can be considered as a measure of comparison with other places, regions and cities (Yalçın, 2015).

In our day and time, rural cities are very similar to one another in terms of both climate characteristics and surface features. Therefore, rural tourism areas should come to the forefront with unique attributes that differentiate them from others. Benchmarking and marketing the tourism products in the cities with rural tourism characteristics is very important to ensure the contribution of the tourism economy to the local economy.

So far, only few academic researches have been conducted on Durbuy in the context of tourism. The first one of these studies was conducted by Dujardin in 2008. In this study, Durbuy's rural tourism character was put forward and the destination marketing slogan of Durbuy was set as follows: 'Durbuy, the smallest city in the world'.

The second study conducted by Dubois, Cawley and Schmitz and published in Tourism Management Journal in 2017 examined agro-tourism in Durbuy. As a result of the study, it was concluded that agro-tourism differs in line with the experiences of the tourists (1148 tourists were asked to fill in the questionnaire and 68 stakeholders residing in Durbuy were interviewed) and that agro-tourism is an important niche market in Durbuy. In addition, it was highlighted that agro-tourism should have a romantic image beyond its modern agriculture image. Furthermore, it was indicated that there may be inconsistencies between agro-tourism and actual agricultural practices and this can be solved with an eligible marketing approach.

Conclusion

Especially in recent years, the tourism industry has been going through major developments. The developing economy drives changes in individuals' holiday expectations and needs, as well. Due to the growing environmental pollution, distorted urbanization, significant increase in the number of artificial physical spaces and the overcrowding in tourist destinations, tourists are nowadays leaning towards natural, cultural and historical places for their holidays. Rural tourism is a perfect option for those who would like to prefer such kind of vacation. Offering tourists with numerous rural tourism activities, Durbuy sets an example for rural tourism destination all over the world. Durbuy and its surroundings will continue to preserve and maintain its potential as one of the important rural tourism destinations in the coming years, particularly in Belgium and Europe, with its natural beauties and cultural attributes.

Durbuy, 120 km distant from Brussels which is deemed to be center of the Europe, features as an important rural tourism destination. Despite the fact that Durbuy is a small town with a population of 3.5 million in Wallonia, Belgium, it is visited by approximately 8.5 million people annually (Dujardin, 2008). Promotion of rural tourism activities in Durbuy is of great importance to pave the way for other rural areas with similar climatic and geographical characteristics to become a tourism destination and to ensure a sustainable tourism approach all over the world.

It is seen that Durbuy provides the visitors with the opportunity to spend their spare time by enjoying a wide range of social and sportive activities and it has a large number of routes (for trips with ATV), roads and parkour. Cycling, hiking and trekking trails are aimed to offer various services to visitors. There are restaurants and recreational facilities on the routes.

Durbuy is a perfect example of why the elements that may hamper tourists' interest in rural tourism areas should be taken into consideration and of how such challenges can be overcome. In this context, Durbuy is a city that has overcome the challenges of inaccessibility, fragility and diversity resulting from the mountainous characteristics of the region.

Another important finding revealed in the study is that Durbuy is a role model rural tourism destination as it has succeeded in spreading natural, cultural and agricultural activities throughout the year and thus preventing the exceeding of carrying capacity of the destination.

In order to ensure the variety in touristic products, there is even a diamond museum in Durbuy city, the castle, which is a historical building from the Middle Ages, has been restored and it is visited by the tourists. In general, tourists feel high satisfaction about the attractions in the city.

Within the umbrella of tourist products, traditional crafts and folk arts are presented. When we evaluate the niche characteristics, Durbuy beer, which is defined as the best and popular beer in Belgium, and handmade jams specific to the region remarkably shine out the niche characteristics of the region.

The suggestions in the context of all these results are as follows;

All rural tourism cities in the world should diversify their tourism products and activate local non-governmental organizations run by the residents in the region. There are a lot of niche characteristics of many rural cities in Turkey (e.g. Şavşat/Turkey-local architecture, handicraft, natural life etc.) and in the world. If the unique attributes of those cities featured separately and each city was promoted with a unique story and legend (the case of Şirince, Turkey), overcrowding and exceeding of carrying capacity in tourism would be prevented. In addition, in the field of rural tourism, brand cities emerge and remarkable tourism revenues are obtained, preventing local residents migrating from their home places.

References

Ahipaşaoğlu, S. & Çeltek, E. (2006). *Sürdürülebilir Kırsal Turizm*, Gazi Kitapevi, Ankara.

Dubois, C., Cawley, M. & Schmitz, S. (2017). The tourist on the farm: a muddled image, *Tourism Management*, 59, p. 298–311.

Dubois, C. & Schmitz, S. (2013). What is the position of agritourism on the Walloon Tourist Market? *European Countryside*, 5, p. 295–307.

Dujardin, S. (2008). Tourisme et Valorisation Des Ressources Territoriales en Milieu Rural-Analyse de L'offreTouristique de La Commune de Durbuy, *Bulletin de la Société Géographique De Liège*, 50, p. 27–35.

EDEN (2007). *European Destinations of Excellence*, Durbuy, Belgium-Wallonia, https://ec.europa.eu/growth/sectors/tourism/eden/destinations/belgium_en, (Accessed in November 15, 2019).

European Commission (1999). *Towards Quality Rural Tourism: Integrated Quality Management (IQM) of Rural Tourist Destinations*, European Commission, Brussels.

European Commission (2003). Fact Sheet. Rural Development in the European Union, European Commission, Brussels.

Fleischer, A. & Felsenstein, D. (2000). Support for rural tourism does it make a difference, *Annals of Tourism Research*, 27/4, p. 1007–1024.

Fleischer, A. & Pizam, A. (1997). Rural tourism in Israel. *Tourism Management*, 18/6, p. 367–372.

Fuentes-Fuentes, M. F., Albacete-Saez, C. A. & Liorens-Montes, J. (2006). Service quality measurement in rural accomodation, *Annals of Tourism Research*, 34/1, p. 45–65.

Getz, D. & Carlsen, J. (2000). Characteristics and goals of family and owner-operated businesses in the rural tourism and hospitality sectors, *Tourism Management*, 21, p. 547–560.

Gürer, N. (2003). *Kırsal Geleneksel Konut Dokusunun Turizm Bağlamında Değerlendirilmesi Cumalıkızık Örneği*, Gazi Üniversitesi Fen Bilimleri Enstitüsü, Yayımlanmamış Yüksek Lisans Tezi, Ankara.

Hall, D., Roberts, L. & Mitchell, M. (2005). *New Directions in Rural Tourism*, Publishing Company, UK.

Kesici, M. (2012). Kırsal Turizme olan Talepte Yöresel Yiyecek ve İçecek Kültürünün Rolü, *KMÜ Sosyal ve Ekonomik Araştırmalar Dergisi*, 14/23, p. 33–37.

Kiper, T. (2006). *Safranbolu Yörük Köyü Peyzaj Potansiyelinin Kırsal Turizm Açısından Değerlendirilmesi*, Ankara Üniversitesi Fen Bilimleri Enstitüsü, Yayımlanmamış Doktora Tezi, Ankara.

Kutukız, D., Uslu, A., Öztürk, B., Özdek, E. & DerinkökA. E. (2016). Kırsal Yoksulluğun Giderilmesinde Kırsal Turizmin Yeri ve Bir Uygulama, *Uluslararası Sosyal Araştırmalar Dergisi*, 9/47, p. 899–908.

Liu, A. (2006). Tourism in rural areas: Kedah, Malaysia, *Tourism Management*, 27/5, p. 878–889.

Middleton, V. T. C. & Rebecca, H. (1988). *Sustainable Tourism; A Marketing Perspective*, Butter Worth-Heinemann Linance House. Jordan Hill, Oxford.

Morgül, Ş. M. (2006). *Trakya Bölgesinde Kırsal Turizm Potansiyelinin Değerlendirilmesine İlişkin Analiz: Kırklareli Örneği*, Trakya Üniversitesi Sosyal Bilimler Enstitüsü, Yayımlanmamış Yüksek Lisans Tezi, Edirne.

OECD (1994). *Les Strategies Du Tourisme Et Développement Rural*. Paris, www.oecd.org/dataoecd/30/48/2755188.pdf, (Accessed in November 10, 2019).

Özdemir, S. (2012). Kırsal Kalkınmada Kırsal Turizmden Yararlanma Olanakları: Gökçeada Örneği. *KMÜ Sosyal ve Ekonomik Araştırmalar Dergisi*, 14/23, p. 19-21.

Özkan, E. (2007). *Türkiye'de Kırsal Kalkınma Politikaları ve Kırsal Turizm*. Ankara Üniversitesi Sosyal Bilimler Enstitüsü. Yayınlanmamış Yüksek Lisans Tezi, Ankara.

Peláez, L. V. (2004). Rural tourism: a diversifying alternative. Strategic lines of its expansion, *Papeles de Economía Española*, 102, p. 298-315.

Ryglova, K., Rasovska, I. & Sacha, J. (2017). Rural tourism-evaluating the quality of destination, *European Countryside*, 9/4, p. 769-788.

Snieškaa, V., Barkauskiene, K. & Barkauskas, V. (2014). *The Impact of Economic Factors on the Development of Rural Tourism: Lithuanian Case*, 19th International Scientific Conference; Economics and Management 2014, ICEM 2014, 23-25 April 2014, Riga, Latvia.

Soykan, F. (2003). Kırsal Turizm ve Türkiye İçin Önemi, *Ege Coğrafya Dergisi*, 12, p. 1-13.

Torun, E. (2013). Kırsal Turizmin Bölge İnsanına Katkıları, *Karamanoğlu Mehmet Bey Üniversitesi Sosyal ve Ekonomik Araştırmalar Dergisi*, 1, p. 31-37.

Uslu, A., Sancar, M. F., Akay, B. & Kutukız, D. (2015). Siirt İli Kırsal Turizm Potansiyeli ve Turizm Eğitimi Alan Öğrencilerin Kırsal Turizm Algıları Üzerine Bir Araştırma. *Akademik Bakış Dergisi-Uluslararası Hakemli Sosyal Bilimler E- Dergisi*, 49, p. 350-365.

Veer, M. & Tuunter, E. (2005). *Rural Tourism in Europe*, Expert and Innovation Centre, the Hague, the Netherlands.

Wilson, S., Fesenmaier, D. R., Fesenmaier, J. & Van Es, J. C. (2001). Factors for success in rural tourism development. *Journal of Travel Research*, 40, http://jtr.sagepub.com/content/40/2/132.full.pdf+html. (Accessed in November 11, 2019).

Yalçın, B. (2015). Marketing of touristic products: a comparative analysis between City of Innsbruck and Gümüşhane, Gümüshane University Electronic Journal of the Institute of Social Science, 6/14, p. 12-24.

http://discoveringbelgium.com/2012/02/08/discovering-durbuy/ (Accessed in November 10, 2019).

www.durbuy.be (Accessed in November 10, 2019).

https://www.europeanbestdestinations.com/destinations/eden/durbuy/ (Accessed in November 15, 2019).

http://walloniabelgiumtourism.co.uk/en-gb/3/where-to-go-in-wallonia/eden-8-destinations-of-excellence/eden-european-destinations-of-excellence (Accessed in November 15, 2019).

Ahmet DUYAR and Ahmet AÇIL

2 Effects of Ecotourism Activities in Forests on the Soil

Introduction

Nowadays, forestland recreation areas have become a refuge for people who are suffocated by intensive urban life (Cetin & Sevik, 2016: 1). As with all natural resources, the utilization of forest areas subject to recreation activities should be sustainable. While people want these areas to be natural, clean, accessible and safe, ecological degradation and environmental pollution are the most important risks for sustainability (Sezer & Akova, 2016: 109). The sustainability of recreation areas is affected by topography, forest structure, tree species, soil type and vegetation composition. In addition, depending on the type and intensity of ecotourism activities, travel infrastructure development and habitat fragmentation, establishment-related air, water and soil pollution, damage to soil and vegetation associated with the activity, and harassment of wildlife may occur (Leung et al., 2008: 21). All these reasons put sustainable use of recreation areas at risk.

In forest areas, mineral soil is generally covered with a litter layer that forms the organic soil layer (Duyar, 2019: 2362). In addition, the forest floor is covered with an herbaceous and woody vegetation cover that grows depend on stand canopy. These litter and vegetation cover maintain the physical and chemical properties of mineral soil. All kinds of forest activities such as forestry, recreation and human activities can cause damage to the plants that make up the vegetation and deteriorate the dead cover layer. The extent of the damage, which depends on the duration, frequency and severity of the activities, can impact to the deterioration of the characteristics of mineral soil horizons.

In this study, as a result of the ecotourism activities in the forest recreation areas, the effect of soil and the potential of ecotourism activities to harm the soil and restoration of the damaged soils are examined.

Impacts of Ecotourism Activities on the Soil

From the activities in the forestlands, all components of the soil are impacted simultaneously but in different intensity. During ecotourism activities, soil horizons and vegetation cover may be exposed to crushing, trampling, pressure or burning. In an area opened for recreation for the first time, the greatest damage can occur on the vegetation that forms the trampled, crushed and shredded. The second degree damage can occur on the litter or mineral soil, depending on the thickness and composition of the litter layer. If the litter layer is thin and loosely stacked, it can transmit the effect of trampling and pressure directly to the mineral soil. In mineral soil exposed to pressure, the pores may deteriorate and the soil may become compact. Therefore, the infiltration capacity of the soil will decrease and the risk of surface runoff and erosion will increase. In addition, seed germination and root growth in compressed soils will be reduced (Cole, 2004: 46). These physical changes in the soil may also lead to chemical and biological changes. This can cause damage to soil fauna and microbiota. On the area where the fire is burned, vegetation and litter are completely disappear by burn up. Serious physico-chemical degradation also occurs in the mineral soil layer in the burning area. In addition, all of dwelling fauna in the litter and the soil can perish (Parlak, 2018: 33). The impact of burning is more severe than trampling and crushing. The activities than continuously in the recreation area will increase the impact of deterioration. Afterwards, the site may lose its quality due to compacting and erosion.

Potential Impacts of Ecotourism Activities to the Soil

There are total 1304 forestland recreation areas and 145 urban forests that used for recreation purpose in Turkey (Kuvan et al., 2018: 117). These areas are managed by the General Directorate of Forestry (Anonymous, 2014: 3). People in Turkey prefer to use for recreation in the dry season (i.e. summer 88 % and spring 8 %) to the forestlands (Kurdoğlu & Düzgüneş, 2011: 205). The most common ecotourism activities preferred in forest areas are picnics, hiking and camping (Kiniş & Duyar, 2017: 65).

Picnics are usually organized in uneven aged crowded groups such as family and friends. The most common activities in picnics are burning barbecues, playing games, trekking around and relaxing on the floors. Due to the risk of that cooking food and tea on the hearth fire may be caused to forest fires (Parlak, 2018: 30), picnics are generally carried out in planned recreational forest areas. Therefore, the number of visitors and the visiting frequency are quite high in the forestland recreation areas (Şenol, 2018: 101). In these areas, the soil is under the effect of intense of trampling and crushing. Depending on the severity of use, vegetation is almost completely destroyed (Şenol, 2018: 105), the litter layer becomes thinner and mineral soil is compacted and deteriorated (Duyar & Kiniş, 2017: 54). Especially in areas where fires are burnt, the vegetation and litter disappear and physical and chemical deterioration is observed in the soil (Parlak, 2018: 33).

Hiking activities are carried out in planned recreation areas, on forestland traces or randomly preferred routes. Sabri et. al. (2018: 167) investigated soil compaction affected by ecotourism activities in Pahang National Park, Malaysia. In his study, he compared the soil penetration resistance between the camping area and trekking trail with the natural area. The penetration resistance were respectively measured in the trekking trail (2.19 MPa), camping area (1.19 MPa), and natural area (0.95 MPa). In A similar study conducted in the Bolu Aladağ forests, soil compacting was compared to the natural area with the trekking trail used by 100 walking tourists. Trekking trail (926 g/l), which was trampled only once by 100 people, was found to be significantly different from soil bulk density natural area (825 g/l) (Duyar & Kiniş, 2018: 40).

Camping activities in forest areas are concentrated around planned camping areas, streams and rivers edges and water resources close to the highway. Obua, (1997: 221) enounces that the negative impact of camping activities in the Kibale national park on the soil and vegetation are more than the wet season in the dry season. Because in the dry season, the vegetation is more sensitive and fragile, resulting that is caused in greater deterioration. In addition to the seasonal impacts, Cole (1995: 413) tried to explain the extent of the exposure in the camp areas in terms of frequency of using. In a study of the impact of visitors on campgrounds and vegetation in the United States, he states that frequently used areas are more affected than rare used ones and are more susceptible to degradation. The examination in the Bolu Aladağ camp center was occurred that vegetation was completely destroyed and mineral soil was eroded in the areas that were trampled like the front of the bungalow huts. In addition, soil bulk density was found to be significantly different between widely used areas and field edges that people do not use (Duyar & Kiniş, 2017: 54).

Restoration of Recreation Areas

The main destruction caused by ecotourism in forest recreation areas is the destruction of the living cover and erosion of soil horizons. On the restoration of recreation areas may not be possible to fulfill for the erosion caused soil loss, but it is possible to revive the dead cover and vegetation through interventions. When an area is used for recreational purposes, the effect of soil degradation develops more rapidly in the first periods, but in the following periods it slows down relatively. Although the magnitude of the deterioration effect varies according to the type of ecotourism and the characteristics of the site, it is directly related to the duration of exposure. If a degraded site is left in its natural state, the site may be restored after a while. Although the deterioration effect is rapid, the healing process takes quite a long time. This healing process can range from a few years to a few decades (Cole, 2004: 52).

The diversity of the original plant species of the site, community structure and regeneration techniques should be planned in an integrated manner during the regeneration of vegetation. Instead of native species, the use of alien species should be avoided in order to satisfy visitors and improve the beauty of the landscape. Biological restoration should be carried out in the form of adding the lost ones, while preserving the natural plant species. It is appropriate to consider the diversity of plant species, endemism, vegetation structure and composition values in the vegetation necessary to support wildlife. Restoration and ecotourism management plans can be combined to increase success in both activities. Success can be assessed not only for biological and ecological purposes, but also for sustainable ecotourism. It is important to create models for the integration of restoration and tourism development programs and to plan the possible areas to be implemented (Hakim & Miyakawa, 2018: 6).

Conclusions and Recommendations

Recreation is one of the indispensable requirements of societies. During the recreation activities in forest areas, the vegetation covering the forest soil and over is trampled, crushed or burned. As a result of these negative effects, three kinds of deterioration may occur on the forest floor; (i) susceptible species within the vegetation may be damaged and the composition of the vegetation may be degraded or all destroyed, (ii) the litter layer becomes compacted, thinned or destroyed, (iii) the mineral soil layer is compacted, the pore structure deteriorates and may become vulnerable to erosion. Continuous activities in the recreation area will increase the impact of soil degradation. Unless precaution is taken, both soil and recreation areas may lose their qualifications completely.

Some of the precautions for the sustainability of forest recreation areas are listed below:

- Recreation areas should be planned in accordance with the intended ecotourism activities.
- The capacity of the areas should be planned in accordance with the visitor potential and the recreation load should be distributed evenly throughout the area.
- Preventing fire at random places in the area, fire pits or hearths that can be fired without harming the environment should be built.

- Places such as footpaths or tables circumferences that need intensive use should be fortified with gravel, concrete etc. material that suitable for general landscape.
- Sports fields that are suitable for the requested nature sports activities should be arranged and prevent irregular trampling of the ground.
- Vehicle access to the site should not be allowed and should be restricted on specially arranged roads or parking lots.
- Finally, administrative and technical precaution should be arranged to protect the soil and vegetation cover.

References

Anonymous (2014). Communiqué on Recreation Areas Application, General Directorate of Forestry, Communiqué Issue No. 300, Ankara, p: 121.

Cetin, M. & Sevik, H. (2016). *Evaluating the Recreation Potential of Ilgaz Mountain National Park in Turkey*. Environmental Monitoring and Assessment, 188(1), 52, 1–10. Doi:10.1007/s10661-015-5064-7.

Cole, D. N. (1995). *Disturbance of Natural Vegetation by Camping: Experimental Application of Low Level Stress*. Environmental Management, 19(3), 405–416.

Cole, D. N. (2004) *Impacts of Hiking and Camping on Soils and Vegetation*. In: Buckley R (ed), Environmental Impacts of Ecotourism. CABI Publishing, Wallingford, UK, pp. 41–60.

Duyar, A. & Kiniş, S. (2017). *The Effects on the Soil Compaction of Some Ecotourism Activities in Bolu Aladağ Forests*.4th European Ecotourism Conference, Karabuk, Proceedings book, p. 115.

Duyar, A. (2019). *The Relationships between the Litterfall and the Canopy Closure of Uludağ fir (Abies nordmanniana (Stev.) subsp. bornmulleriana (Matff.)) Forests*. Applied Ecology and Environmental Research, 17(2), 2357–2372, Doi: 10.15666/aeer/1702_23572372.

Duyar, A. & Kiniş, S. (2018). *The Effects of Trekking Activities on Physical Soil Properties in the Bolu-Aladağ fir Forests*. Forestist, 68(1), 36–41, Doi: 10.5152/forestist.2018.004.

Hakim, L. & Miyakawa, H. (2018). *Integrating Ecosystem Restoration and Development of Recreation Sites in Degraded Tropical Mountain Areas in East Java, Indonesia*. AIP Conference Proceedings, 2019(1), 040016, AIP Publishing.

Kiniş, S. & Duyar, A. (2017). *Bolu Aladağ Yaylacılarının Ekoturizme Yaklaşımı*. Karabuk University Journal of Institute of Social Sciences, S3, 59–70, Doi: 10.14230/joiss442.

Kurdoğlu, O. & Düzgüneş, E. (2011). *Artvin kent ormanının rekreasyon olanakları ve kullanıcı tercihlerinin irdelenmesi.* Artvin Çoruh Üniversitesi Orman Fakültesi Dergisi, 12(2), 199–210.

Kuvan, Y., Erol, S. Y. & Şahin, G. (2018). *Management of Forest Areas Used for Ecotourism and Recreation in Turkey.* Forestist, 68(2), 114–121.

Leung, Y. F., Marion, J. L. & Farrell, T. A. (2008). Recreation ecology in sustainable tourism and ecotourism; A strengthening role. In: McCool, S. F. and Moisey, R. N. (Eds). Tourism Recreation and Sustainability, 19–37.

Obua, J. (1997) *Environmental Impact of Ecotourism in Kibale National Park, Uganda,* Journal of Sustainable Tourism, 5(3), 213–223, DOI: 10.1080/09669589708667286.

Parlak, M. (2018). *Çanakkale (Eceabat, Akbaş Şehitliği) orman yangınıyla bazı fiziksel ve kimyasal toprak özelliklerinin zamansal değişiminin belirlenmesi.* Toprak Bilimi ve Bitki Besleme Dergisi, 6(1), 29–38.

Sabri, M. D. M., Suratman, M. N., Kassim, A. R., Shari, N. H. Z., Khamis, S. & Daim, M. S. (2018). Light Intensity and Soil Compaction as Influenced by Ecotourism Activities in Pahang National Park, Malaysia. In: Suratman, M. N. (Ed), National Parks: Management and Conservation, 157–171.

Sezer, B. & Akova, O. (2016). *Kent sakinlerinin rekreasyon tercihleri, rekreasyon alanlarının algılanan değeri ve gerçek kullanımı arasındaki ilişki.* Faculty of Economics and Administrative Sciences E-Journal, 5(2), 94–115.

Şenol, E. (2018). *Boraboy Gölü (Amasya) ve çevresinin, rekreasyon amaçlı kullanımdan kaynaklanan başlıca sorunları.* Doğu Coğrafya Dergisi, 23(39), 95–112.

Bilal DEVECİ

3 Usage Areas of Salt Mineral and Its Varieties

Introduction

Salt is defined as a mine and mineral which has an important place and value throughout the history of civilization (Gürsoy, 2012: 15). There is archaeological evidence that people on the shore of Yuncheng Lake collected salt from the lake, which was evaporated by the summer sun, in 6000 AD in China. It is stated that conflicts and wars emerged to have these collected salts (Kurlansky, 2003: 22). It is known that there were salt mines or processing facilities in countries known as Poland today in 5000 AD and as Austria and Spain in 3500 AD (Republic of Turkey Ministry of Health, 2016: 2–3). It is known that Babylonian Civilization, which was founded in Sumerians in 4000 AD and later, used salt (sodium chloride) to store nutrients for a longer time (Gençtoy, 2017: 74). In 2700 AD, salt was used in China as a medicine. There are sources that salt trade was used by Hittites in Anatolia in 1200 AD and in 800 AD in China. Because of its value, it was given to the Roman soldiers in 200 AD in return for their salaries. In addition, it was used as a means of exchange like today's money (Avcı, 2003: 24–25). It is claimed that the word 'salary' used instead of wage or allowance comes from the word 'salt' (Ayaz, 2008: 7; Avcı, 2003: 27; Bloch, 1999; www.history.com).

Salt mineral (sodium chloride) has been used by humans since ancient times to make food more durable and delicious (Akgün, Genç & Arıcı, 2018: 361). Salt is defined as the mineral form of sodium chloride. Salt mineral is obtained in pure form from lakes, seas, oceans and salt caves. The salts that are subsequently refined are used effectively in food preparation and storage. Natural food salt (sodium) is naturally present in the structures of untreated foodstuffs (meat, fish, seafood, various fruits, various vegetables, milk, cereals, etc.). Therefore, the sources of salt minerals (sodium chloride) and natural food salt (sodium) are quite different from each other (Baysal, 2016: 194). When salt (sodium chloride) is pure, it is composed of approximately 60 % chlorine and 40 % sodium (Oğur, et. al., 2008: 138; Ayaz, 2008: 7).

Salt mineral (sodium chloride) contains sodium and chlorine as well as many natural minerals such as selenium, magnesium, lithium, calcium, phosphorus and vanadium. Salt is naturally available in solid and liquid form. The solid rock salt is available in the liquid form in dissolved form in lakes, seas and oceans. It is also available in solid form in dried lake and inland sea beds. Salt has the potential to provide the minerals that humans need such as mineral waters and spring waters and it is very vital for human health (Ünal & Yılmaz, 2014: 188). In other words, salt contains many important minerals for the survival of the living organisms. Therefore, it is stated that the salt mineral is necessary for human survival (Demirkol, Çifçi & Çifçi, 2018: 298).

For millions of years, there has been less than 0.25 grams of salt (sodium) in the daily diet of animals and humans. The main reason for this is the low amount of salt (sodium) in the structure of the foods consumed. However, it is known that salt (sodium chloride) was started to be used for long-term storage of food in China about 5000 years ago and that salt storage method was discovered in China. After this important discovery, salt (sodium chloride) became an important economic value of the empires and contributed to the development of settled life thanks to long-term stored foods (Republic of Turkey, Ministry of Health, 2016). In addition, thanks to salt trade, small cities have become metropolitan areas. Therefore, salt was named as white gold in history (Gençtoy, 2017: 74). The first salt is estimated to be obtained from sea or lake water (Brisay, 1981: 910).

Salt Mineral and Its Production

When the salt mineral is used in cooking, preparing and finishing food, it displays four different characteristics. These are crystal structure, moisture content, mineral content and place of origin. ***Crystal structure*** is one of the most important features of salt mineral. The fineness, thickness, granularity, firmness, smoothness, pyramid-like structure of the crystal structure and the fragility of the leaves determine the effect on the palate. The delicacy and crispy structure in meats and seafood originates from the crystal structure of salt minerals. ***Moisture content*** is very significant for the chewy salts to melt easily in the mouth. This prevents your mouth from drying out during meals. Moreover, it does not lose its form on the nutrients for a long time. ***Mineral content*** varies between 3 % and 30 % in all salt types. It provides flavor to salts with mineral, crystal structure and moist content. Salt types obtained from salt water and underground have their own mineral types. The minerals in the salt determine the true taste of the salt. ***The place of origin*** does not directly affect the sense of taste but affects the enjoyment of food. Salt reflects the cultural values of the place where it comes from. It is therefore important. Salt is closely related to the natural environment, culture, history, the region in which it is produced and the cuisine of that region (Bitterman, 2016: 12).

Salt (sodium chloride) is obtained using three different methods. In the first method; it is obtained by evaporation of the lake, sea and ocean waters around shallow and large areas with the help of the sun. In the second method; salt is extracted and processed from underground and mines with the help of manpower and construction machinery. In the third method; salt is extracted by melting the solidified salt of the underground prehistoric salt lakes and taken to the surface. It is then treated and separated from water and dried. In addition, the method of obtaining salt from salt lakes is by collecting it from the surface by the construction machinery with the evaporation of lake water. It is then washed with water at about 40°C in large boilers and divided into smaller pieces. These small pieces are separated according to their dimensions by means of a machine which resembles a swaying screen. The salt (sodium chloride) is then dried in the drying machines at 180–200° C (Avcı, 2003: 30–36; Ergin, 1988: 11–17; Yalçın & Ertem, 1997: 209; www.history.com; www.mortonsalt.com; www.maden.org.tr).

Salt Mineral Types

Salts are different in color, taste, smell, aroma, sodium, chloride and dozens of different minerals. Salt mineral consists of basic varieties such as table salt (refined salt), Himalayan salt, rock salt and sea salt (Brown, 2008: 91). Kosher salt, Celtic salt, crystal salt, smoked salt, seasoned salts, ocean salt, brine salt and leaf salt are also included in salt types.

Table salt (refined salt): It is available in the nature in rock form. It is crystallized and white in color. It is the most known and most widely used salt type after adding refined process iodine (www.billurtuz.com.tr).

Himalayan salt: It is available in the nature in rock form and its formation takes millions of years. It is light pink and red in color. It is obtained from salt mines located at the foot of the Himalayan mountains of Pakistan (www.yemek.com).

Rock salt: It is defined as the salt obtained from the underground in solid form. It is usually found in natural state in gray, black and red clay color. After refining, the color changes to white (www.mta.gov.tr).

Sea salt: It is defined as the salt obtained by evaporation of sea water with the help of sun and wind (Demirkol, Çifçi & Çifçi, 2018: 302). Although the amount of sea water on our planet is very high, the amount of sea salt production is quite low. The main reason for this is that the pool capacities of sea salt production are very low and the climatic conditions are not suitable (Yalçın & Ertem, 1997: 208). There are more than 80 minerals in calcium including calcium, magnesium, potassium, sodium, etc. In particular, iodine is naturally available in sea salt (www.ruhundoysun.com).

Kosher salt: Iodine added salt is defined as coarse-grained salt generally produced from crystals. It is obtained by evaporation from the lake, sea or oceans (www.thekitchn.com).

Celtic salt: It is defined as a kind of sea salt in light gray color with a small amount of water and high mineral content (www.tabiat.nl).

Crystal salt: It is a kind of salt obtained as a result of crystallization or transparency of rock salt under pressure for millions of years. It is rarer than other salt types (www.yemek.com).

Smoked salt: It is defined as the aromatic salt obtained by the fumigation of sea salt with natural substances (wood types: oak, beech, poplar, etc.) having different aromas (www.anatoliafood.com).

Seasoned salts: It is a kind of salt which is formed by grinding unrefined Himalayan or rock salt with various spice and spice mixtures and grinding in salt mills. Salt is combined with spices such as oregano, tarragon, pepper, turmeric, galangal, basil, ginger, rosemary, anise, cumin, etc. (www.efsina.com). It is also available in garlic, onion, tomato, lemon, orange and wine flavors.

Ocean salt: It is defined as unrefined salt containing tens of useful minerals obtained by evaporation of ocean water. It is supplied from the locations of clean ocean currents (www.yemek.com).

Brine salt: It is a coarse-grained matte white salt obtained by washing the sea salt. It is used to make pickles, tomato paste, cheese, olives and canned food (www.billurtuz.com.tr).

Leaf salt: As a result of the growth of sea salt molecules at different speeds, the surface and edges are of different sizes and shapes. It looks like snowflakes. All leaf salt grains have different flat shapes and are crisp (www.anatoliafood.com).

Usage Areas of Salt Mineral

There are 40 different types of salt in 'Salt' restaurant of Ritz Carlton Hotel. Some of those are prismatic Cyprus salt, bamboo flavored salt, aguni salt, lavender salt, black Hawaiian salt, Shio salt, red alaea Danish Viking salt and Bolivian salt. Different types of salt and flavored salt mixtures are used for all food and beverages produced in the restaurant. It is used especially in the production of coffee varieties, caramel, ice cream, chocolate, sorbet and various desserts (www.history.com; www.ritzcarlton.com). In addition, these kinds of salt add depth to hot dishes; it is used to add charm, vitality, flavor and put the final touch to warm dishes (www.milliyet.com).

Coarse salt (Gros Sel De) is defined as a light gray matt colored French type salt obtained from the sea in coarse grains. It is generally used in soup, sauce, pasta boiled waters and home-made bread (www.leguerandais.fr). The broad and flat grained and snowflake-like leaf salts are generally used in meat dishes, desserts, especially caramel and chocolate making (www.ruhundoysun.com).

Brittle sea salts with delicate, moist, granular crystals and different mineral taste are often used to make fish, game animals more tender and for vegetables. It is also used in the production of toast, omelet, etc. consumed for breakfast. Gray, moist, rich mineral salts are used for beef, lamb, root vegetables and all kinds of grills and fries. Salts, which are pyramid-shaped and fragile, are used for fresh greens, salads and seafood cocktails. Especially when the food is consumed, the feeling of salt is quite short. The most known white and fine viscous salts are used multipurpose for almost all cooked foods. Fine flakes and fine-grained salts are often used in pickling. It is also used in steamed vegetables, fish, soups and pickles. Thick and fine grinding salts, such as hard gravel, have very high adhesion. Therefore, it is used in fried foods and potato chips. Flavored and natural element-added salts are used to give foods a rich content and aroma. It is complementary to taste and flavor and provides depth (Bitterman, 2016: 9–11).

Salt plays an important role in the processing of animal products. Thanks to the protective properties of salt, meat is processed and turned into products that can be stored for a long time such as salami, garlic sausage, sausage and ham (Ruhlman, Polcyn & Keller, 2005). Table salt (sodium chloride) is used as a flavoring agent in the world cuisine. Because while salt can mask undesirable tastes, it can increase the intensity of some flavors. For example; it reduces the sensation of pain and increases the sense of sweetness. Salt is also used to reduce or increase taste contrasts (Akgün, Genç & Arıcı, 2018: 363–364). Salt is used primarily to provide flavor, then to reduce the amount of moisture in foods and to control some bacteria that cause nutrient degradation. It is also used as a preservative in almost all processed and packaged products (Baysal, 2016: 194). Approximately 75 % of the amount of salt (sodium chloride) consumed by people on a daily basis is from processed foods (packaged delicatessen products, various biscuits, olives, pickles, cheese, tomato paste, etc.) (Şahin, 2016: 22) and around approximately 4 % of the natural salt is obtained from unprocessed natural nutrients (Gençtoy, 2017: 74). The remaining part is obtained from the salt (sodium chloride) varieties that we used in the table.

Salt is being used extensively for centuries in the protection, drying and storage of fish, vegetables and meats (Aktan, Yenigün & Kısa, 1995: 389). Leaf salts similar to Kosher salt are used to separate the liquid contained in freshly caught fish and freshly cut meat from the main body. This method helps meats to dry and store them for a long time (Demirkol, Çifçi & Çifçi, 2018: 303–304). Salt types are also used to clearly detect the tastes of foods (Batu, 2017: 31).

Himalayan salt blocks are used as a means of cooking and presenting foods using cooking pan or grill (Bitterman, 2013).

Conclusion

Salt (sodium) is naturally available in nutrients and in very small amounts. It is naturally available in fruits, vegetables, cereals, meat varieties, milk, eggs in nutrients such as milk and eggs. Salt (sodium chloride) consists of 40 % sodium and 60 % chloride. It is obtained from lakes, sea, ocean and underground caves by using different methods. Salt meets the deficiencies of sodium and many minerals that the human body needs. Salt is known to be of critical importance for human survival.

There are many types of salt (sodium chloride). Salt is classified based on its structural shape, color, region or region, crispy structure, mineral variety, moisture content, and its usage during cooking and presentation. In this study, salt types were studied in terms of gastronomy. The types of salt associated with the storage, cooking, preparation and presentation of foods were examined and a basic classification was made.

In terms of gastronomy, the methods and uses of salts (sodium chloride) are quite diverse. Therefore, it is used both in the production of many milky, pasted dessert varieties and in the decoration of products such as sorbet and ice cream. It is used to increase the sweetness and chocolate aroma in almost every process of chocolate making. It is used in salads, seafood cocktails, decorating cooked meats and giving crispiness. Also salt blocks cut into square or rectangular shape are used as kitchen equipment for cooking meat, vegetables and fruits.

References

Akgün, B., Genç, S. & Arıcı, M. (2018). Tuz: Gıdalardaki Algısı, Fonksiyonları ve Kullanımının Azaltılmasına Yönelik Stratejiler, *Akademik Gıda*, 16/3, p. 361–370.

Aktan, H. T., Yenigün, A. & Kısa, Ü. (1995). Askeri Garnizonlarda Tüketilen Sofra ve Mutfak Tuzlarının Kimyasal Özelliklerinin Belirlenmesi, *Ankara Üniversitesi Veteriner Fakültesi Dergisi*, 4/2, p. 389–392.

Avcı, S. (2003). Ekonomik Coğrafya Açısından Önemli Bir Maden: Tuz, *Coğrafya Dergisi*, 11/1, p. 21–45.

Ayaz, A. (2008). *Tuz Tüketimi ve Sağlık*, Ankara: Sağlık Bakanlığı Yayınları.

Batu, A. (2017). Moleküler Gastronomi Bakış Açısıyla Gıdaların Tat ve Aroma Algıları, *Aydın Gastronomy*, 1/1, p. 25–36.

Baysal, A. (2016). Tuz Tüketimi ve Sağlık, *Beslenme Diyetetik Dergisi*, 44/3, p. 194–195.

Bitterman, M. (2013). *Salt Block Cooking*, USA: Andrews McMeel Publishing.

Bitterman, M. (2016). *Bitterman's Craft Salt Cooking*, USA: Andrews McMeel Publishing.

Bloch, D. (1999). Salt and the Evolution of Money, http://www.salt.org.il/money.html, 25.10.2019*Review of the International Commission for the History of Salt*.

Brisay, K. D. (1981). The Ancient Industry of the Salt-Makers, *The Geographical Magazine*, 53/14, p. 910–916.

Brown, A. (2008). *Understanding Food: Principles and Preparation*, (Third Edition), USA: Thomson & Wadsworth.

Demirkol, Ş., Çifçi, İ. & Çifçi, H. (2018). Kaya Tuzunun Gastronomi ve İnanç Açısından Önemi: Hacıbektaş Kaya Tuzu, *1. Uluslararası Turizmde Yeni Jenerasyonlar ve Yeni Trendler Konferansı (01-03 Kasım)*, p. 298–309.

Ergin, Z. (1988). Tuzun üretim Teknolojisi ve İnsan Sağlığındaki Yeri, *Madencilik Dergisi*, 17/1, p. 9–30.

Gençtoy, G. (2017). Tuz ve Böbrek Yetmezliği, *Türkiye Klinikleri J Nephrol-Special Topics*, 10/2, p. 73–83.

Gürsoy, D. (2012). *Baharat ve Güç*, 1. Basım, İstanbul: Oğlak Yayıncılık.

Kurlansky, M. (2003). *Tuz-İnsanlığın Tuzlu Tarihi*, (Çev. A. Çakıroğlu), İstanbul: Aykırı Yayıncılık.

Oğur, R., Saygı, S., Göçgeldi, E., Korkmaz, A., Uçar, M. & Tekbaş, Ö. F. (2008). Piyasada Satılan Tuzların İlgili Mevzuata Uygunluk ve İçerik Yönüyle İncelenmesi, *Türkiye Klinikleri J Med Sci*, 28/1, p. 137–142.

Republic of Turkey Ministry of Health. (2016). Türkiye Aşırı Tuz Tüketiminin Azaltılması Programı 2017-202, Ankara: Sağlık Bakanlığı Yayınları.

Ruhlman, M., Polcyn, B. & Keller, T. (2005). *Charcuterie: The Craft of Salting, Smoking and Curing*, New York: W. W. Norton Company Inc.

Şahin, M. (2016). Tuzla İlgili Son Söz, *Hipertansiyon Haber Bülteni*, 3/5, p. 22–23.

Ünal, K. & Yılmaz, F. M. (2014). Tuzu kısıtlamak mı? Farklı tuz kullanmak mı?, *Spatula DD*, 4/4, p. 187–190.

Yalçın, E. & Ertem, M. E. (1997). Deniz Tuzlalarının Türkiye Tuz Potansiyelindeki Yeri, *2. Endüstriyel Hammaddeler Sempozyumu, (16-17 Ekim)*, 2/1, p. 208–215.

http://www.maden.org.tr/resimler/ekler/019c8091693ef c_ek.pdf.

https://www.mortonsalt.com/heritage/.

https://www.history.com/shows/modern-marvels/season-14/episode-34.

http://www.mta.gov.tr/v3.0/bilgi-merkezi/kaya-tuzu.

http://www.billurtuz.com.tr/tuz_nedir.html.

https://yemek.com/sozluk/himalaya-tuzu/.

https://www.leguerandais.fr/de/produits/das-gros-sel.

https://www.thekitchn.com/kosher-salt-where-it-comes-from-why-its-called-kosher-ingredient-intelligence-219665.

https://www.tabiat.nl/tr/tuz-ueruenleri/109-keltik-deniz-tuzu.html.

https://yemek.com/sozluk/kristal-tuz/.

http://www.anatoliafood.com/tuz-tutsulenmis.php.

https://www.ruhundoysun.com/yazilar/dunyanin-tuzu/.

http://www.efsina.com/baharat_cesnili_himalaya_tuzu_gurme_tuz.html.

http://www.billurtuz.com.tr/product/salamura_tuzu.html.

https://yemek.com/sozluk/okyanus-tuzu/.

http://www.anatoliafood.com/tuz-yaprak.php.

http://www.milliyet.com.tr/pazar/renkli-tuzlar-arasinda-2237000.

https://www.ritzcarlton.com/en/hotels/florida/amelia-island/dining/salt/more.

Ebru GÖZEN

4 Hotel Managers' Metaphoric Perceptions for Smart Hotel

Introduction

In the fourth industrial revolution, which is now called Industry 4.0 in the world, internet, internet of things, artificial intelligence, augmented reality, cognitive technology, sensors, nanotechnology, quantum computing, wearable technologies, intelligent signaling, intelligent networks, intelligent robots, big data and new generation technologies such as 3D are driving many industry applications. This situation leads service companies to use information technologies more intensively. As a result, it is seen that information technologies are widely used in accommodation and travel establishments. These technologies have led to the emergence of smart concepts in the tourism sector. In this context, it is seen that concepts such as smart tourism, smart destinations, smart hotels, smart transportation and smart activities are frequently encountered (Yalçınkaya et al. 2018: 86).

One of the important points about smart applications is that user comfort is important in smart hotel concept. However, in order to ensure customer satisfaction, customers must be aware of the smart applications they will encounter before the stay. More important than that, hotel management and employees must be familiar with these practices before the hotel guests. As a result of the literature view, it was decided to carry out this study due to the lack of studies revealing the perceptions of hotel managers, one of the important take holders of the sector. The aim of their search was to determine the metaphoric perceptions of hotel managers about the concept of 'smart hotel'. From this point of view, the concept of smart hotel is approached metaphorically and the metaphoric perceptions of hotel managers toward the concept of smart hotel are analyzed within the scope of their search.

Smart Hotel Concept

When a smart device is assigned to a particular device, it is easier to use, steer or program. Therefore, its functionality is better for its user. It is also more friendly and intuitive in usage. It is frequently characterized by a higher level of safety and cost effectiveness, which means that such device generates lower costs and results in higher benefits than the solutions deprived of the smart attribute (Jaremen et al. 2016: 66). The most prominent feature of contemporary service enterprises is the more intense use of technological developments (Sarı, 2018). Hotels are also among these contemporary service enterprises.

Technological advances also have some advantages for hotel enterprises. These advantages are; decrease in costs, having green image, hiring fewer staff at reception through online check-in faster service to the customer, shortening the waiting times of the in-house services, the effect on the physical environment such as temperature, smell, music in the hotel rooms, providing a more personalized service by keeping the customer behaviors under record and delivering the customer satisfaction online to the hotel. (Yalçınkaya et al. 2018: 91). In addition, Jaremen et al. (2016) discusses the following advantages of smart hotel applications in their study;

- Automation of the majority of hotel procedures (e.g. check-in, check-out),
- Shortened service realization time
- Reduced risk of making mistakes,
- Higher customer satisfaction,
- Higher efficiency and easier internal communication of staff,
- Automation of documents and information circulation
- Upgraded room standard
- Lower room price – resulting from energy savings and employment reduction
- Maintain positive word-of-mouth (Erdem & Çobanoğlu, 2010).

As a result of all these benefits, international chain hotels adopt the latest Technologies to create a high-tech image (Leung, 2017).

Table 1 provides examples of smart hotel applications that provide these benefits to hotel businesses.

Perhaps the most interesting of the hotels that have implemented the practices in Figure 1 is the Henn-na Hotel, located in Japan, where employees are robots.

A dinosaur-shaped robot welcomes the guest, asks for his passport with all his hospitality, makes the transactions quickly and delivers the room key to the guest at the reception of the Henn-na hotel. There are no buttons in the hotel rooms to turn the lights off. The guests command the tulip-shaped robot Tuly. To give an example of robot-hosting hotels in Europe, the Marriott Hotel Ghent in Belgium can be a good example. Robot Mario can speak 19 different languages. This eliminates the problem of incorrect ordering due to language difficulties. The other one example of smart hotel applications is Hilton Worldwide. This hotel performs energy control and monitoring through smart applications.

Table 1: Smart Applications Used in Smart Hotels

Smart phone
Presented with advanced interface; search for suitable hotels with rich criteria
Hotel Application
Check-in without having to wait at the front desk
Identity recognition with NFC-enabled smartphone and use the phone as a room key
To be able to manage the sources like heat, light, mini-bar and so on.
As a concierge service, you will have Access to the hotel's guide service, including descriptions of attractions as well as the hotel's different facilities such as spa, pool, fitness room
Benefit from multimedia content used to provide more comprehensive visualization
View the hotel environment with augmented reality technology produced by the virtual computer in the form of sound and graphics, and receive information about the surrounding places by holding the phone screen with augmented reality technology
By making check-out at the hotel, you can pay the bill, which includes the drinks they consumed in the mini cup board and the extra payments through the application.
Make in-house payments with contactless payment.
Personalized Services
The smart hotel system keeps customers' past preferences on TV, air conditioning, mini-bar usage and services used in the customer profile database and personalizes their services according to this data when they arrive. (For example, the most preferred drinks in the mini-bar in the hotel room)
Inroom-personalised experiences like watching a holographic music concert (Yıldız & Davutoğlu, 2019).
Smart Card
Ability to shop in hotel with pre-loaded card
Kiosks
Access to all services provided by mobile application
Loading money on a smartcard
Smart waiter/waitress
Ability to recommend menu items to the customer by using the software installed on the handheld device of the hotel staff in the restaurant, using demographic information as well as past preferences of the customers.
Smart Room
When the customer says that he wants to get information about the activities in the hotel by voice command, the smart big screens in the room reflect daily activities, breakfast / lunch times and menus in the hotel. Then, smart large screens look like replaceable wallpapers
Adjust the temperature and pressure of the water in the bathroom with voice commands and open any video with voice commands from smart big screens
Sensors in the room constantly monitor the room conditions and automatically adjust the air conditioner for temperature and humidity even when guests are sleeping

(continued on next page)

Table 1: (continued)

Smart phone
The system informs the doctor when an emergency occurs in the health situation of the guests. To measure the blood and sugar in the urine of the sensors in the toilet and take necessary first aid measures when a problem condition is detected.
Using IPT (Internet Protocol Phone) instead of traditional telephone devices and systems in hotel rooms (Miočić, et al. 2012)
Employee Performance Management System
Receiving real-time service and waiting times via sensors and hand-held device used by the waiter.
Increasing the efficiency of the employee whose performance is managed and paid and by means of this data, it is possible to identify the problematic places in the system in a short time and to develop the necessary solutions and thus to increase the service speed.
Smart Hotel Resource Management
Converting the information received from the RFID readers in the mini fridge to the task information in the form of putting a new drink in the handheld device of the hotel staff concerned.
Retention of real-time data on inventory, so that no customer experiences any problems due to the end of something.
Hotel management's daily business plans for staff falling on the screens (Sarı, 2018)
Security Management (Sarı, 2018)
Detection of non-customers and transfer of location information to security units
Control of customers from face recognition system, creation of location information and measurement of satisfaction levels
Opening of room doors with facial recognition system
The customer's ability to follow the child wearing the RFID wrist bands from the camera and the wristband on his arm in the large and crowded playground
Saving(Sarı, 2018)
Improvement of energy efficiency in unused areas according to customer density

Source: (Yalçınkaya, et al.2018: 91).

Fontainebleau Miami increases hotel revenue by providing guests with early check-in or late check-out for a surcharge, based on room availability (Terry, 2016). Eccleston Square Hotel is located in London. This hotel is equipped with state-of-the-art airconditioning, which filters airpollution and provides separate ventilation for each room (Hürriyet Haber, 2017). Ritz Carlton applies QR codes to art Works and furniture with in the hotel. Thus, it gives its guests the opportunity to learn everything they wonder about the hotel by scanning codes with their phones. In this way, the Ritz Carlton hotels become art galleries (Turizm

Figure 1: Henn-na Hotel Reception Desk. Source: Yamak, 2017

Güncel, 2015). The Upper House in Hong Kong (the guests receive iPod Touch at check-in, loaded with the set of games, music and information about the hotel for their own use); Novotel München Messe (the guests are greeted by both the real and a virtual receptionist, the hotel provides information and communication systems equipped in touch screens, using which guests can easily find tourist information they need) (Jaremen et al. 2016).

In Turkey, it is observed that a limited number of hotels with smart applications. Tekin (2019) examined hotels using cloud technology, which is one of the smart applications in her research. According to the results of the research, it is seen that 10 hotels have this application. One of them is Hilton Garden Inn Hotel which belongs to Hilton group in Isparta. A digitals witch application is available here (Turizm news, 2018). Energy Saver cards used in energy saving lighting systems and room entrances of Hilton Antakya Museum Hotel are among the environmentally friendly smart applications. Another example can be given from Turkey Divan Istanbul. Divan Istanbul guests can access up-to-date information about the hotel and the environment in Turkish, Arabic and English languages with the smart voice assistant device placed in the hotel rooms and meeting rooms. (Yalçınkaya et al. 2018: 44). In addition, Dedeman Group will be organizing 18 hotels, high quality compact bed and breakfast, high-speed internet and state-of-the-art infrastructure that will be built by 2023 by the Smart by Dedeman project (Türkiye Turizm, 2018).

It has been seen that there are limited number of studies on smart hotels in academic literature. One of these studies was conducted by Jaremen, Jędrasiak & Rapacz (2016). In this study, smart hotel concept has been evaluated in terms of competitiveness of hotels. It is stated that smart hotel includes environmentally friendly applications that provide sustainable development and smart hotel applications are more important in practice than theoretical approach (Atay et al. 2019). Majority studies on implementation of smartness on hotel were related to guest experience on in-room technology (Miocic et al., 2012; Leonidis et al., 2013; Neuhofer et al., 2015). One of the managers who participated in Leung's research (2017), described the smart hotel as a remote home that understands you, while a supplier said that it meets the needs of modern people by saving time and money. Petrevska et al. (2016) describe the smart hotel as follows; a smart hotel represents a simple fail-safe model that can be developed double as fast as any traditional hotel. Atay et al. (2019) found that four hotels could be evaluated within the scope of the research by making use of corporate websites and current publications in the media. According to the findings of the study, it is concluded that the cost of the automation infrastructure to be provided for smart hotel applications is high and as a result, a very limited part of the smart hotel criteria can be implemented in the hotels covered by the research. When we look at the usage areas of smart hotel concept, it is used by practitioners more than theorists. It is usually encountered in the subject literature and in the opinions of hospitality industry specialists. Thus, the idea of smart hotels does not stand for a theoretical concept, created as a result of scientific thought development, describing the functioning of a hotel enterprise. Instead, it is rather a practical business model which adapts new information and communication technologies in the hospitality business (Jaremen et al. 2016).

In the past, most studies on hotel management focused on issues such as traditional business performance and service quality. However, the analysis of digital innovation services and modern management provided by the technologies is lacking (Lai & Hung, 2017). Also, it has been observed that individuals' approaches towards the concept of smart hotels, both at home and abroad, are not among the research topics. It is observed that there is no clarity regarding the definition of smart hotel which is a new concept for all stakeholders and the applications that should be present in smart hotels are mentioned when defining it. In such cases, the metaphorical definition of the concept not only reveals the current perceptions of individuals, but also contributes to the awareness of the subject. Metaphors can be a valuable tool for raising awareness (Crompton, 2016: 12) Social sciences often use metaphors or language tools to identify and help to think about them (Brooks et al. 2006: 332). According to the literature, the essence of metaphor is to understand and experience something from the perspective of something else (Akgül et al. 2017: 286). In this type of phenomenological research, the most accurate data can be explained by the person who lives the case. Although hotel managers are the main stakeholders in the smart hotel system, it has been observed that their opinions are not given much attention in these studies. In this respect, the study is thought to contribute to the literature.

Method

Research Model

This research is designed as a phenomenology research in the qualitative research approach. In this study, this method was used to determine the metaphors reflecting the perceptions and opinions of the lower, middle and upper level managers about '*Smart hotel*' in 5-star hotels in Antalya region. Phenomenology focuses on phenomena that cases in which we do not have in-depth and detailed understanding. Phenomena provides a suitable research base for studies that are not completely alien to us but also aim to investigate cases that we do not fully understand. (Göçer, 2013: 29). This is because phenomenological studies are a questioning strategy applied to reveal the participant experiences about the case. (Ceylan & Gündoğdu, 2018).

Sample

Data sources in phenomenological research are individuals or groups who experience and focus on the phenomenon that the research focuses on. For example, in a research about hotel management, people in this sector can form the sample. From this perspective, the sample of this research is 92 lower, middle and upper level managers who work in 5-star hotels operating in Antalya and want to participate in the study.

Data Collection

In order to determine which metaphors characterize the concept of «smart hotel»;

Forms are used to write their opinions easily. The expression is 'Smart hotel is like…………. Because ………….'on a form. The relationship between the subject and the source of the metaphor is tried to be determined with the word '*like*'; '*Because*' was used to evaluate the back ground of the meaning imposed on metaphors. 15 minutes were given to complete the forms and this application was carried out between 31 October –13 November 2019.

Data Analysis

The researcher should convert the qualitative data set into a shape that makes it easier to work on (Baltacı, 2017). 98 hotel managers completed the form at the beginning of the study. However, 92 hotel managers' forms that could establish meaningful metaphors or explain the reason for the metaphor they established was evaluated. The following questions were sought in the research: 'What are the metaphors that hotel managers have about the concept of hotel smart hotel?', 'Under which conceptual categories can the hotel managers' metaphors concerning the smart hotels be combined?'. In this context, the process of analyzing and interpreting the metaphors developed by the participants was realized in five stages; (1) determination of metaphors (2) classification of metaphors (3) category development (4) validity and reliability and (5) Data processing for quantitative data analysis (Kalyoncu, 2012).

Determination and Classification of Metaphors

The metaphor produced by each participant is coded separately. The metaphor set created by the researcher was examined by 2 experts. In this process, how metaphors conceptualize the concept of smart hotel; subject, source and logical basis of metaphor. In order to determine the metaphors belong to which participant the completed forms are numbered from one to ninety-two, and these numbers are coded so that they are placed behind the statements (for example, 'M2' Form 2 belongs to manager 2).

Category Development

Each metaphor produced by hotel managers is associated with a specific theme from the perspective of the smart hotel and 8 different conceptual categories were created.

Validity and Reliability

In order to ensure the validity and reliability of this research process, the following studies
  Were conducted:

- Studies related to the method, process and results of the research are explained clearly and in detail,
- Purposeful sampling technique was preferred when determining the sample group,

- During the research process, managers were not guided about their opinions and care was taken to create an understanding that they were not expected from them.
- Participants' quotations were taken without adding while reporting the research,
- The metaphor set created by the researcher was examined by 2 experts,
- In order to ensure the internal reliability of the research, it was noted that the metaphors under the conceptual categories constitute a meaningful whole. In addition, in order to ensure external credibility, attention was paid to the fact that all categories were different from each other as well as forming a meaningful whole among themselves.

The following Miles-Huberman model, which is frequently used in the analysis of qualitative data, includes a theoretical framework and application model for the implementation of the analytical induction (Baltacı, 2017: 3). The reliability of the findings was calculated with the formula of percentage of fit suggested by Miles and Huberman:

Reliability: number of Consensus subjects/ terms÷ (number of Consensus subjects/ terms+ Number of subjects / terms without consensus)x100

58÷ (58+6)x100: _90,62_

According to the coding control that gives internal consistency, the consensus among the encoders is expected to be at least 80 % (Baltacı, 2017: 8). According to another source, in the qualitative research, 90 % or more concordance between the researchers provides desirable reliability (Saban, 2009). In this study, the obtained value (90.62 %) was found to provide the expected reliability.

Results

Hotel managers who participate to study distribution according to demographic variables are as follows: Gender: 69 %male, 31 %female, generation: 3 % Boomers, 48 % X generation, 48 % Y generation, 1 % Z generation, title: 19,6 %lower level managers, 55,4 %middle level managers and 25 %upper level managers, educational status: 48,9 %managers have tourism education, 51,1 %managers have others disciplines, industry experience: 46,7 % 0–15 years, 47,8 % 16–31 years and 5,5 % 32 years and above.

Content analysis technique was used in the analysis of the data related to the metaphorical perceptions of managers about the concept of smart hotel. According to the findings, managers produced 64 valid metaphors for the concept of smart hotel. These metaphors produced by the managers are given in Table 2 alphabetically.

According to the frequency values in Table 2, the three most frequently repeated metaphors; 'Home' (f: 7), 'Computer' (f: 6) and 'Smart Phone' (f: 4).

Table 2: Metaphors Created by Managers for Smart Hotel Concept

Row	Metaphor name	F	Row	Metaphor name	F
1	Aladdin's magical lamp	1	33	Latest model car	1
2	Alarm clock	1	34	Life	1
3	Application to reduce the feeling	1	35	Life file	1
4	Artificial intelligence	3	36	Life with robots	1
5	Artificiality	1	37	Memory	1
6	Assistant	3	38	Miracle	1
7	Bank branch	1	39	Nature	1
8	Being proactive	1	40	Need of the age	2
9	Book I want to read	1	41	Over-standard service	1
10	Cartoon	1	42	Positive perception tool	1
11	Clock	2	43	Reputation tool	1
12	Cold room	1	44	Right investment	1
13	Comfortable bed	1	45	Robot	2
14	Comfortable flight	1	46	Saving	1
15	Computer	6	47	Scientist	2
16	Crawling child	1	48	Self-driving car	1
17	Dessert	1	49	Shortcut	1
18	Developing system	1	50	Smart house	2
19	Digital banking	1	51	Smart investment	1
20	Digitalizing world	1	52	Smart people	1
21	Diversity	1	53	Smart phone	4
22	Doing everything with a click	1	54	Smart system	2
23	Dream	1	55	Smiling emoji	1
24	Empty story	1	56	Soulless building	2
25	Exceeded expectation	1	57	Star	1
26	Fairytale land	1	58	Successful team	1
27	Friend you want to see	1	59	Summit of comfort	1
28	Future	2	60	Surrealism	1
29	Heaven	1	61	Technological hotel	3
30	Home	7	62	Water	1
31	Innovation in consciousness	1	63	Wind	1
32	Inter-system communication	1	64	World of dreams	1
	Total				92

When the metaphors that managers have for the concept of smart hotels are categorized by considering the analogy aspects, the data in Table 3 is obtained.

As shown in Table 3, when the metaphors created by the managers were categorized according to similarity aspects, 8 categories were obtained. These categories are defined as 'perception creation tool, need, personalized service, soulless, systematics, sustainability, savings and technological comfort'. The first

Table 3: Categories of Metaphors Created by Managers for Smart Hotel Concept

Categories	Metaphors	n	F	%
Perception creation tool	diversity (1), positive perception tool (1), world of dreams (1), innovation in consciousness (1), smart people (1), surrealism(1), dream (1), cartoon (1), land of tale (1), star (1).	10	10	10,86
Need	Smart system (2), Bank branch (1), computer (1), need of the age (2), digitized world (1), crawling child (1), future (2), developing system (1), reputation tool (1), wind (1), water (1), dessert (1).	12	15	16,30
Personalized service	Aladdin's magical lamp (1), assistant (3), computer (1), home (5), friend you want to see (1), life (1), book I want to read (1), being proactive (1), artificial intelligence (2).	9	16	17,40
Soulless	Empty story (1), soulless building (2), life with robots (1), life file (1), Artificiality (1), cold room (1), application to reduce the feeling (1).	7	8	8,70
Systematics	Computer (1), scientist (2), memory (1), self-driving car (1), robot (2), clock (2), inter-system communication (1).	7	10	10,86
Sustainability	Smart investment (1), successful team (1), nature (1), home (1).	4	4	4,35
Savings	Right investment (1), computer (1), shortcut (1), saving (1).	4	4	4,35
Technological Comfort	Smart house (2), smart phone (4), exceeded expectation (1), computer (2), heaven (1), alarm clock (1), digital banking (1), home (1), smiling emoji (1), doing everything with a click (1), comfortable flight(1), comfortable bed (1), summit of comfort (1), miracle (1), latest model car (1), over-standard service (1), technological hotel (3), artificial intelligence (1).	18	25	27,18
TOTAL		71	92	100

3 categories produced the most metaphors; 'Technological comfort (18), need (12) and perception creation tool (10)'.

Category 1 – *Perception creation tool*: There are 10 metaphors in this category. Some of the metaphor expressions in this category are as follows;

M10; Smart hotel is like innovation in consciousness. Because, fixed, mind-free minds cannot create and adapt to smart hotel concept.

M16; The smart hotel is like surrealism. Because it's where dreams come true.

M84; Smart hotel is like a star. Because when guests stay at a smart hotel, they feel themselves like a star.

Category 2 – *Need*: There are 12 metaphors in this category. Some of the metaphor expressions in this category are as follows;

M24; The smart hotel is like the need of the era. Because, both economically, customer satisfaction, and the necessity of the digital age, the transition to these systems is needed.

M61; Smart hotel is like water. Because it has become one of the most necessary needs today.

M90; Smart hotel is like a crawling child. Because, more technological investment and therefore more time is needed to meet every room type, every region, every culture and every guest profile, meet the needs and become widespread.

Category 3 – *Personalized service*: There are 9 metaphors in this category. Some of the metaphor expressions in this category are as follows;

M91; The smart hotel is like the magic lamp of Alaadin. Because, he fulfills his guest's wishes in the fastest and practical way, he knows what he wants next time.

M72; The smart hotel is like a friend you want to see. Because it makes you feel good by getting to know you and producing special solutions for you.

M58; The smart hotel is like an assistant. Because, people do not recognize the hotel and the culture in the region properly during the holiday period, they want to be taken care of. Smart hotel systems are built in such a way that we can access this information at any time.

Category4 – *Soulless*: There are 7 metaphors in this category. Some of the metaphor expressions in this category are as follows;

M40 veM75; The smart hotel is like a soulless building. Because, it is similar to putting the guest in a closed box. There is no empathy.

M77; Smart hotel is like an application that will reduce the feeling. Because, as technology increases, manpower will decrease and electronic systems will reduce the satisfaction interaction in the guest. It will make you feel cold.

M44; Smart hotel is like living with robots. Because people want to be warmly welcomed in hospitality.

Category5 – *Systematics*: There are 7 metaphors in this category. Some of the metaphor expressions in this category are as follows;

M27; Smart hotel is like computer. Because, there are few bugs and timely service.

M28; It is like a smart hotel clock. Because, when you look inside the watch, you see that it has a very nice systematic. The cogs move the clock in harmony with each other. Smart hotels also dominate this systematic inside.

M49; Smart hotel is like a robot. Because it is equipped with automation system.

M43; Smart hotel is like communication between systems. Because we have to accept this as a process. The existing systems in hotels are interconnected and share some information. Establishing a proper system between them ensures error-free communication

Category 6 – *Sustainability*: There are 4 metaphors in this category. Some of the metaphor expressions in this category are as follows;

M59; Smart hotel is like nature. Because, it has recycling, no waste, it has a functioning and sustainable system.

M68; Smart hotel is like smart investment. Management achieves more accurate results with less workload. More economic and sustainable practices are realized.

Category 7 – *Saving*: There are 4 metaphors in this category. Some of the metaphor expressions in this category are as follows;

M37; Smart hotel is like a shortcut. Because it eliminates unnecessary waste of time.

M13; Smart hotel is like right investment. Because it saves time and profitability.

M73; Smart hotel is like saving money. Because technology has an effect on saving.

Category8 – *Technological Comfort*: There are 18 metaphors in this category. Some of the metaphor expressions in this category are as follows;

M74; The smart hotel is like a miracle. Because, it makes your holiday easier with the technological applications by the miracle way.

M66; Smart hotel is like paradise. Because, there is a very enjoyable holiday. It doesn't tire to work or stay there. Because the main theme is to make life easier.

M21; Smart hotel is like a smart phone. Because it has easily accessible application for all tastes.

M34; Smart hotel is like comfortable bed. Because you make so comfortable holiday, like asleep.

Table 4 shows the category distribution of metaphoric perceptions of managers regarding the concept of smart hotels according to their demographic characteristics.

According to Table 4, it is seen that the female managers participating in the research mostly describe the concept of smart hotel in C8 (technological comfort) and male managers in the C2 (need). According to the generations of the participants, the Boomers' mostly describe the concept of smart hotel inC5 (systematicality), X generation is mainly in C8 (technological comfort) and then C3 (personalized), Y generation is primarily C8 (technological comfort) and then C2 (need), Z generation is in C8 (technological comfort). According

Table 4: Category Distribution by Demographic Data

		CATEGORIES							
		C1	C2	C3	C4	C5	C6	C7	C8
Gender	Female	1	3	4	2	1	1	0	16
	Male	9	12	12	6	9	3	4	9
Generation	Boomers	0	0	0	0	2	0	1	0
	X	4	7	10	4	3	1	0	15
	Y	6	8	6	4	5	3	3	9
	Z	0	0	0	0	0	0	0	1
Title	L.L.M	1	3	3	2	2	1	1	5
	M.L.M	5	7	10	5	6	2	3	13
	U.L.M	4	5	3	1	2	1	0	7
Educational Status	Tourism	7	9	6	3	3	3	2	12
	Other	3	6	10	5	7	1	2	13
Experience	0–15	6	8	5	3	5	2	3	11
	16–30	3	7	9	5	5	2	1	11
	31 and above	1	0	2	0	0	0	0	3
TOTAL		10	15	16	8	10	4	4	25

Abbreviations: **C1:** Category 1 (*Perception Creation Tool*), **C2:** Category 2 (*Need*), **C3:** Category 3 (*Personalized Service*), **C4:** Category 4 (*Soulness*), **C5:** Category 5 (*Systematicty*), **C6:** Category 6 (*Sustainability*), **C7:** Category 7 (*Saving*), **C8:** Category 8 (*Technological Comfort*). **L.L.M:** Lower Level Manager, **M.L.M:** Middle Level Manager, **U.L.M:** Upper Level Manager.

to the titles, educational status and experience of the managers participating in the research, it's seen that the managers mostly evaluate the concept of smart hotel in C8 (technological comfort) and then C2 (need) and C3 (personalized). According to these results, it can be said that the concept of smart hotel is mostly related to technological comfort by the managers participating in the research. This result is consistent with the relevant literature review.

Word cloud was formed for all participants' responses based on word frequency numbers and the results are shown in Figure 2.

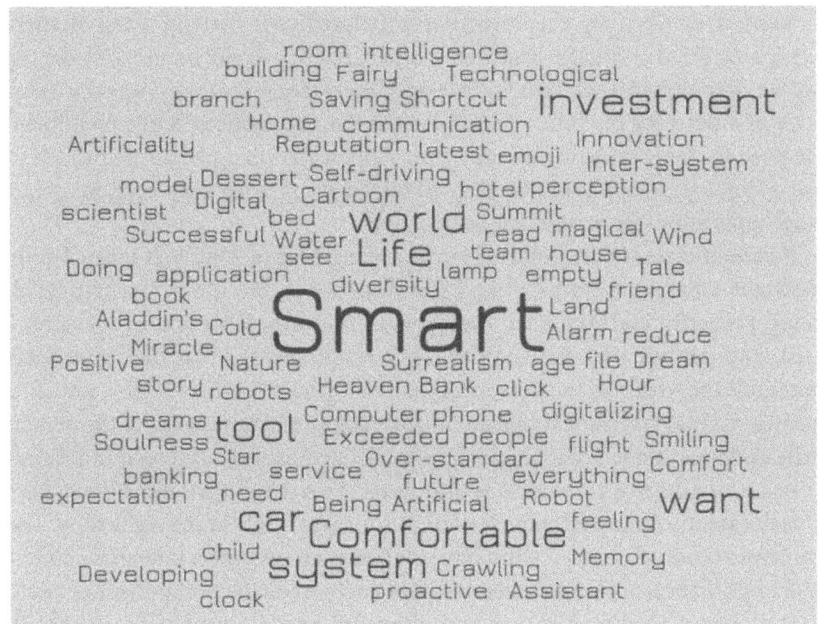

Figure 2: Word Cloud

It is seen that the words 'smart, comfortable, investment, system and tool' which are the elements of smart hotel concept are most emphasized in word cloud.

As a result of this research, 64 metaphors were obtained and these metaphors were collected under 8 categories. When the conceptual categories that metaphors belong to are analyzed, it is observed that managers produce metaphors for the categories of 'perception creation tool', 'need', 'personalized service', 'soulless', 'systematics', 'sustainability', 'savings' and 'technological comfort'. When the results are examined, it is seen that metaphoric perceptions of the concept are mostly combined on positive qualifications and are similar to the expressions used in the definition of smart hotel concept in the literature.

Perception creation tool; In this category, the opinions of participants on the concept of 'smart hotel' has been described by the metaphors of 'Difference, Positive Perception Tool, World of Dreams, Innovation in Consciousness, Intelligent People, Surrealism, Dream, Cartoon, Fairy Tale Land and Star'. The metaphors mentioned in this category are related to the perceptions of people. These applications are related to perception, to make people feel different, to be surreal, like in a dream world or in a cartoon, to see for themselves as a star.

Need; Described by the metaphors of 'Intelligent System, Bank Branch, Computer, Need of the age, Digitized World, Crawling Child, Future, Developing System, Reputation Tool, Wind, Water and Sweet'. Some managers saw the smart hotel as a need for the future of industry in their metaphors. Some mentioned the need for large investments, and some emphasized that people are needed to use technological applications. They all talked about what they need for smart hotel applications from different perspectives.

Personalized service; Described by the metaphors of 'Aladdin's Magic Lamp, Assistant, Computer, Home, Friend You Want to See, Life, Book I want to Read, Being Proactive and Artificial Intelligence'. The common point of metaphors used is to provide personalized service. It is emphasized that the smart hotel can fulfill the wishes of the person instantly like the Aladdin's Magic Lamp. It is mentioned that it includes applications that can predict the needs of the person such as an assistant. Technology for personalized experience creation process to occur, innovative mechanisms and tools are needed that allow businesses to facilitate the right customer service in the right space at the right time. The implementation of smart technology solutions can become a potential catalyst of change that turns standardized services into personalized experiences based on the tenet of 'treating different consumers differently' (Neuhofer et. al. 2015).

Soulless; Described by the metaphors of 'Empty Story, Soulless Building, Life with Robots, File of Life, Artificiality, Cold Room and Application to Reduce The Feeling'. This is the only category that describes the smart hotel as negative. In their explanations, they emphasized that smart hotel applications in a labor intensive sector will create coldness and soulless. Because the people in the hospitality industry will want to be greeted in a warm environment. Though smartness has been discussed in academia for decades, industrial practitioners are still unable to fully understand its functions and definitions

Systematics; Described by the metaphors of 'Computer, Scientist, Memory, Self-Driving Car, Robot, Clock and Inter-system Communication'. The common features of these metaphors are that the system will reduce errors and everything will be programmed. Managers have compared smart hotels with objects such as clocks, robots and computers and compared their functioning within a certain system to that of smart hotels. Smartness refers to the integration of smart features that automate and simplify daily business activities (Leung, 2017).

Sustainability; Described by the metaphors of 'Smart Investment, Successful Team, Nature and Home'. Ensuring sustainability through efficient use of resources in smart hotels is an important issue. Interoperability in the successful team metaphor is the key in smart network

Savings; Described by the metaphors of 'Right Investment, Computer, Shortcut and Saving'. It is emphasized that investing in smart hotel applications will save resources. It is important for the hospitality industry to reach the right conclusion quickly. Smart applications will do this and thus save time and money. Respondents in this category view smart applications as a means of saving. Unfortunately, they are unaware that this investment can not only save cost but also improve operation performance (Li et al. 2013).

Technological Comfort; Described by the metaphors of 'Smart Home, Smart Phone, Exceeded Expectations, Computer, Paradise, Alarm Clock, Digital Banking, Home, Smiling Emoji, Doing Everything with a Click, Comfortable Flight Comfortable Bed, Summit of Comfort, Miracle, Latest Model Car, Overstandard Service, Technological Hotel and Artificial Intelligence'. Technological comfort has been the most metaphor produced category. The common features of these metaphors are that guests are comfortable because the system adjusts everything, comfort is achieved through the combination of technology and service and innovative applications eliminate the ordinary.

The limitation of this study that the research is carried out with the participation of lower, middle and upper level managers of 5-star hotels in Antalya region. It is recommended that a similar study be compared with the opinions of 5-star hotel managers in different regions.

References

Akgül, B. M., Kaya, S., Ayyildiz, T. & Karaküçük, S. (2017). *Rekreasyon UzmanAdaylarınınMesleklerineİlişkinAlgıları: Bir MetaforikÇalışma*, 4. Disiplinlerarası Turizm Araştırmaları Kongresi, 09-12 Kasım, p. 285–303.

Atay, L., Yalçınkaya, P. & Bahar, F. (2019). *İstanbul'daki Akıllı Otel Uygulamalarının Değerlendirilmesi*, Manas Sosyal Araştırmalar Dergisi, 8(1), p. 679–690.

Baltacı, A. (2017). *Nitel Veri Analizinde Miles-Huberman Modeli*, Ahi Evran Üniversitesi Sosyal Bilimler Enstitüsü Dergisi (AEÜSBED), 3(1), p. 1–15.

Brooks, J. J., Wallace, G. N. & Williams, D. R. (2006). "Place as Relationship Partner: An Alternative Metaphor for Understanding the Quality of Visitor Experience in a Backcountry Setting", Leisure Sciences, 28, p. 331–349.

Ceylan, V. K. & Gündoğdu, K. (2018). *Bir OlgubilimÇalışması: Kodlama Eğitiminde Neler Yaşanıyor?*, Eğitim Teknolojisi Kuram ve Uygulama, 8(2), p. 1–34.

Crompton, J. L. (2016). *Evolution of the "Parks as Lungs"Metaphor: Is It Still Relevant?*, World Leisure Journal, 59(2), p. 105–123.

Erdem, M. & Cobanoglu, C. (2010). *The Impact of Consumer-Generated Media and Social Networking in Hospitality: The Implications for Consumers and Hospitality Managers*, Journal of Hospitality Marketing & Management, 19(7), p. 697–699.

Göçer, A. (2013). *Türkçe Öğretmeni Adaylarının Dil Kültür İlişkisi Üzerine Görüşleri: Fenomenolojik Bir Araştırma*, Erzincan Üniversitesi Eğitim Fakültesi Dergisi, 15(2), p. 25–38.

Hürriyet Haber (2017). Dünyanın En Son Teknoloji Otelleri.<http://www.hurriyet.com.tr/seyahat/dunyanin-en-son-teknoloji-otelleri-40692956> (Date of access: 16.11.2019).

Jaremen, D. E., Jedrasıak, M., & Rapacz, A. (2016). *The Concept of Smart Hotels as an Innovation on the Hospitality Industry Market – Case Study of PURO Hotel in Wrocław*, Economic Problems of Tourism, 4, p. 65–75.

Kalyoncu, R. (2012). *Görsel Sanatlar Öğretmeni Adaylarının "Öğretmenlik" Kavramına İlişkin Metaforları*, Mustafa Kemal Üniversitesi Sosyal Bilimler Enstitüsü Dergisi, 9(20), p. 471–484

Lai, W. C. & Hung, W. H. (2017). *Constructing the Smart Hotel Architecture – A Case Study in Taiwan*, In Proceedings of the 17th International Conference on Electronic Business ICEB, Dubai, UAE, December 4–8, p. 67–71.

Leonidis, A., Korozi, M., Margetis, G. & Stephanidis, C. (2013), An intelligent hotel room, in Augusto, J. C. et al. (Eds), Ambient Intelligence, Springer International Publishing (Lecture Notes in Computer Science, 8309), p. 241–246. doi: 10.1007/978-3-319-03647-2_19.

Leung, R. (2017), *Smart Hospitality: Taiwan Hotel Stakeholder Perspectives*, Tourism Review, 74(1), p. 50–62.

Li, Y. P., Xie, Y. Q. & Huang, C. (2013), *Promoting Hotel Managerial and Operational Level with New Technology*, Applied Mechanics and Materials, 380(384), p. 4562–4565.

Miočić, K. B., Korona, L. Z. & Matešić, M. (2012). *Adoption of Smart Technology in Croatian Hotels*, in 2012 Proceedings of the 35th International Convention MIPRO. 2012 Proceedings of the 35th International Convention MIPRO, p. 1440–1445.

Neuhofer, B., Buhalis, D. & Ladkin, A. (2015), *Smart Technologies for Personalized Experiences: A Case Study in the Hospitality Domain*, Electronic Markets, 25(3), p. 243–254.

Petrevska, B., Cingoski, V. & Gelev, S. (2016). From Smart Rooms to Smart Hotels, Zbornikradovasa XXI međunarodnognaučno – stručnogskupaInformacıone Tehnologıje - sadašnjost i budućnostodržanognaŽabljaku od 29. februara do 05. marta 2016. godine, p. 201–204.

Saban, A. (2009). *Öğretmen adaylarının öğrenci kavramına ilişkin sahip oldukları zihinsel imgeler.* Türk Eğitim Bilimleri Dergisi, 7(2), p. 281–326.

Sarı, E. B. (2018). *Reflections of Industry 4.0 To Management of Service Enterprises: Smart Hotels*, International Journal of Contemporary Tourism Research, 2(2), p. 33–40.

Tekin, Z. (2019). *Otel İşletmelerindeki Web/Bulut Tabanlı Teknolojilere Dayalı Yönetim Sistemleri ve İşletme Başarısı İlişkisi*, Journal of International Management and Social Researches, 6(11), p. 130–137.

Terry, L. (2016). *6 Mega-Trends in Hotel Technology*. <http://hospitalitytechnology.edgl.com/news/6-Mega-Trends-in-Hotel-Technology105033> (Date of access: 16.11.2019).

Turizm Güncel (2015). *Oteller Mobil Uygulamalarıyla Öne Çıkmak İçin Neler Yapmalı*. <https://www.turizmguncel.com/haber/oteller-mobil-uygulamalari-ile-one-cikmak-icin-neler-yapmali-h24893.html> (Date of access: 16.11.2019).

Turizm News (2018). *Akıllı Otel Uygulaması İlk Kez Isparta'da*, https://www.turizmnews.com/akilli-otel-uygulamasi-turkiye-de-ilk-kez-isparta-da/12773/> (Date of access: 11.11.2019).

Türkiye Turizm. (2018). *Dedeman'dan Y kuşağına özel Smart by Dedeman*, <http://www.turkiyeturizm.com/dedemandan-y-kusagina-ozel-smart-bydedeman-56092h.htm> (Date of access: 07.08.2018).

Yalçınkaya, P., Atay, L. & Karakaş, E. (2018). *Akıllı Turizm Uygulamaları*, Gastroia: Journal of Gastronomy and Travel Research, 2(2), p. 34–52.

Yamak, M. (2017). *Çalışanlardan Akvaryumdaki Balıklarına Kadar Herşeyin Robot Olduğu Otel: Henn-na Hotel*, <https://www.webtekno.com/calisanlarindan-akvaryumdaki-baliklarina-kadar-her-seyin-robot-oldugu-otel-henn-na-hotel-h26674.html> (Date of access: 16.11.2019).

Yıldız, E. & Davutoğlu, N. A. (2019). Recent Advances in Social Sciences. Efe, R., Koleva, I. & Öztürk, M. (Ed.)., The Transition from Industry 4.0 to Tourism 4.0: Smart Hotels, Artificial Intelligence and Improvements in Robitics, in 229–242. UK: Cambridge Scholars Publishing

Ediz GÜRİPEK

5 Innovation in Tourism and Examples in Practice

Introduction

Due to the increasing competition conditions, innovation is an essential issue in the tourism sector as in all other sectors. It is known that the primary way of increasing the competitiveness of the enterprises operating in the tourism sector by making a difference, responding to the needs and expectations of their guests more effectively, evaluating the opportunities and increasing their competitiveness is known to be through innovation activities (Çakıcı et al., 2016: 12).

Although there is no absolute criterion for achieving success, tourism enterprises that maintain their position in the market and develop continuously, watch for gaps and opportunities, gain new guests by increasing the loyalty of their existing guests, and continue to create value by focusing on innovation and creativity will be more advantageous for providing competitive advantage (Demirkaya & Zengin, 2014: 107).

Besides, the survival of tourism enterprises depends on following the innovative activities and technological developments in the world, exhibiting differences in the services, acting sensitive to the environment, caring for the nourishment, thinking the happiness of their customers, and offering an inspiring atmosphere to their clients (Durna & Babül, 2011: 74).

Concept of Innovation

Innovation, which derives from the word at innovatus' in Latin, is defined as m the introduction of new methods in the social, cultural, and administrative environmental (Elçi, 2006: 1). Innovation is an essential tool for researching, uncovering new ideas, products, processes, production techniques and creating new business areas in the market (Knight, 1997: 213; Covin & Slevin, 1991; Guth & Ginsberg, 1990: 5; Miller & Friesen, 1983; Zahra, 1993; Dess & Lumpkin, 2005).

Innovation aims to contribute to the operating system by introducing products and services by introducing creativity by expressing new product development, service innovation, introducing new production methods and determination of related procedures while emphasizing development and innovation together with technology (Antoncic & Hisrich, 2001: 498; Ming, 2013; Uzun, 2018; Zahra, 1996: 1715).

Businesses that have to keep up with the increasing competition day by day and who want to gain competitive advantage have to realize innovation by performing creative activities in order to increase their power and make their products and services more advantageous against their competitors (Knight, 1997: 2013).

The economist Schumpeter (1934), who proposes innovation as a concept, states that economic innovation creates a creative destruction effect and those old practices create a dynamic effect by replacing products and processes. Innovation is an amalgam of products, new production methods, new markets, new sources of supply, and the emergence of new market structures (Schumpeter, 1934: 55).

From this point of view, innovation is possible by applying any invention or a new idea in the commercial field (Mercan et al., 2011: 31). Therefore, the innovations in the tourism sector should benefit society, economy, customers, and the enterprise.

Types of Innovation

Since innovation can be realized in different ways, sections, and stages the following types of different innovations take place. These are;

Product Innovation: It is defined as the change, innovation, and difference in the existing product or the development of a completely different and new product and its introduction to the market for commercial gain. (Elçi, 2006: 3). For example, Sony's products, such as Walkman and PlayStation, founded in 1946, are essential product innovations.

Service Innovation: Service can be defined as activities that benefit the consumer in terms of place, time, psychology, and form, which consist of abstract activities that do not require ownership of one party to another (Sayım & Aydın, 2011: 246). Service innovation, on the other hand, is a new or significantly altered service approach, creating innovation and difference in service delivery and distribution system and the use of new technologies for this purpose (Simonceska, 2012). For example, the assistance service used in many accommodation establishments is an excellent example of innovation.

Process Innovation: Innovation and significant changes in all processes applied in the enterprise. For example, the use of electronic reservation systems in the distribution process is process innovation.

Marketing Innovation: These are essential changes made by businesses to find new markets or provide better services in order to gain a competitive advantage in existing markets (Johne, 1999: 7). For example, hotels offer 360 ° virtual images on their websites instead of photos of the hotel. Similarly, travel agencies are also conducting similar applications to provide information about destinations to their customers.

Organizational Innovation: First of all, organizational innovation is a prerequisite for the success of other types of innovation (Burmaoğlu & Şeşen, 2011: 3). However, the development of Organizations' necessary capabilities and the creation of an innovation culture is defined as organizational innovation. For example, applications such as expressing new ideas freely in the enterprise, and continuously developing human resources can be shown as an organizational process.

Progressive and Radical Innovation: The progressive development and modification of existing technologies are progressive, depending on the magnitude of innovation; entirely innovation is defined as radical innovation (Manga, 2018: 251). For example, the transition of room keys to the card system can be shown as a radical innovation, and the further development of these cards and their use in payment systems can be shown as progressive innovation.

With the impact of globalization, competition between countries and regions is intensifying and becoming increasingly difficult. Businesses should minimize their costs, make improvements in their products/services or produce new products/services, enter new markets, and increase their consumers through these means in order to adapt to the steep competition and ensure their sustainability. The realization of these situations is only closely related to the motion of innovation activities. The companies performing innovation activities differentiate from other enterprises quickly adapt to the market conditions, gain competitive advantage, decrease their costs, increase their profitability, become the preferred processing because they provide customer satisfaction and reach the leading position in the sector (Kuratko, 2007: 2; Yılmaz & İncekaş, 2018: 157; Ağca & Kandemir, 2008: 215).

It is possible to develop, change, and make more qualified the activities of tourism enterprises with technology and innovation. In this context, tourism enterprises should be able to find a place in the world market and be able to respond to customer expectations and needs. It is an undeniable fact that they need to give importance to innovation strategies in order to be more advantageous than their competitors.

Innovation Examples for Tourism Sector

It is possible to give many examples of types of innovation in the tourism sector. Some examples of innovation for the tourism sector are presented below.

Adam & Eve Hotel, located in Belek, Antalya, offers 11 innovative services full of surprises for couples who wish to get married as a 'romantic concept'. There are holiday assistants under the name of 'Angel Service', who can assist guests in all matters from the beginning to the end of their holiday. All guests are provided with an 'Angel' (www.adamevehotel.com), an expert in human relations and organization, trained in their field, and at least one foreign language.

For those staying in the villas of Rixos hotels, there are 24-hour holiday assistants called Butler. Within the hotel, a Butler is assigned to each floor. Another innovation is the scented pillow service on request (www.rixos.com).

In Denmark, a hotel offers dinner to guests as it contributes to electricity generation in exchange for the electricity generated by a bicycle connected to the generator. Sorrento Hotel in the USA organizes night chats for its guests. Writers, artists, film actors, and experts in other fields offer a culture-oriented environment to their guests at yard midnight symposia ((Durna & Babür, 2011).

The 'Beta Buttons' in the lobby and bars at the Marriott hotel in North Carolina allow customers to gauge every aspect of their accommodation experience, and it is used as a feedback innovation laboratory (www.digitaltalks.org).

An exceptional example is the Rancho Bernardo Inn in San Diego. The hotel prepared a special package called 'Survivor' which has the feature of 'Except Everything'. The hotel has separately priced the room and amenities provided in the standard package. The deluxe double room starts at $ 219 and is all-inclusive. However, guests can reduce this price up to $ 19 by decreasing the services in the room (for example, without pillows, without minibars, without air conditioning). Furthermore, with the application of 'stay more save more', it offers 15 % cheaper accommodation for two nights, 20 % for three nights and 25 % for four and more customers (www.hurriyet.com.tr; www.ranchobernardoinn.com).

The number of organic theme hotel opened for guests who want to eat healthy during the holidays is increasing in the world and Turkey. All materials used in these hotels are organic, including; all food and beverages in minibars, wall paint, flooring, shampoos, soaps, and alcoholic beverages. None of these products are carcinogenic. Also, all paper used in the hotel is recycled.

Ice hotels located in northern countries such as Sweden, Norway, Slovenia, Alaska, and Andorra are another example of innovation. The Ice Hotel, the world's first and most significant, is located in a small town called Jukkasjärvi in Sweden with a population of 1100, everything is almost made of ice. It is being built by the townspeople in October and is open from December to April (www.skyscanner.com.tr).

Ecological hotels, which are in the form of small hotels, far from the city, are becoming increasingly common. For example, the fruits and vegetables used in the ecological hotel operating in Kocaeli Narköy are produced in the hotel's farm. Customers can eat fruits from the garden freshly and can make their vegetables (www.narkoy.com).

Another example of innovation is the Som Dona hotel, which opened in Spain. The hotel only accepts female guests over the age of 14. Most of the employees are women (https://somhotels.es/hotel-som-dona/).

Another application for innovation is the mobile key or critical digital application. Many chain businesses around the world are making serious investments for this application. Customers can also check-in / check-out themselves and even choose their rooms.

The example of a restaurant and bar without a waiter is an example of process innovation. There are successful examples in cities such as New York, Tokyo, and Shanghai. Everything is automatic in the restaurant without a waiter. Customers place their orders on the touch screen on the table. The customer's order from the electronic menu presented via a platform installed on the table. Customers pay with the pos device on the table. Such innovation has reduced labor performance and staff costs.

Process innovation is exemplified by a brainstorming session with the participation of all staff once a week at the first-class holiday village of İberostar Pegasos Palace. In brainstorming meetings, the participation of personnel in the management process and the rewarding of the employees who produce creative ideas can be considered as an indicator of their involvement in the management process. Besides, this is an example of an organizational innovation type (Vatan, 2010: 47).

The rooms developed in the hotel establishments with the consideration of the elderly and disabled are examples of marketing innovation. The differentiating features of the hotel include the preparation of qualified rooms such as concave washbasins and toilets designed for wheelchairs, washbasin mixer with a right-left opening control device, individual grab bars, and flooring for slip risk. Another application is innovations such as changing room lights, illuminated whirlpool, and unique beds, which are applied by many enterprises depending on customer requests.

Recommendations and Conclusions for Innovation in the Tourism Sector

Tourism serves as a sector that contributes significantly to economic development and welfare in terms of providing currency circulation among countries. As a result of the structure of the tourism sector, the travel behavior of people can change rapidly, and the sector can be affected by the political, economic, and socio-cultural changes in the countries. For this reason, country policies should act proactively in order to solve the questions in order to prevent the sector from losing momentum. Besides, to affect the problems that may arise in the country or to minimize the impact level and support the innovation activities that will add value to the tourism sector. In today's competitive conditions, the tourism sector must be constantly aware of developments, be able to change its products, services and production methods continuously and adapt to technological change continuously in order to be different and preferable from the others. However, enterprises must carry out quality, original, and innovative activities that will meet the ever-changing customer expectations and needs. For a successful change, based on innovation, companies need vision strategic thinking and a robust organizational spirit. In this context, tourism enterprises should gain the ability to act effectively for their employees in marketing, production, research, and development. Moreover, they should mobilize employees, offer flexibility to capture market opportunities, and support creative individuals. Tourism enterprises can only develop innovative activities with strong organizational performance and can be strengthened by differentiating them from other enterprises under competitive conditions by eliminating market threats. Therefore, countries and tourism enterprises should adopt the right strategies and policies that support innovation. Also, the realization of innovation as a whole with the local community and the public will provide an opportunity to create a positive image perception.

References

Ağca, V. & Kandemir, T. (2008). *Aile İşletmelerinde İç Girişimcilik Finansal Performans İlişkisi: Afyonkarahisar'da Bir Araştırma*, Sosyal Bilimler Dergisi, 10/3, p. 209–230.

Antoncic, B. & Hisrich. R. D. (2001). *Intrapreneurship: Construct Refinement and Cross-Cultural Validation*, Journal of Business Venturing, 2001/16, p. 495–527.

Burmaoğlu, S. & Şeşen, H. (2011). *Türk Firmalarının Organizasyonelİnovasyon Yeteneğini Etkileyen Faktörler Üzerine Bir Araştırma*, Ankara Üniversitesi, SBF Dergisi, 66/4, p. 1–20.

Covin, J. G. & Slevin, D. P. (1991). *A Conceptual Model of Entrepreneurship as Firm Behavior*, Entrepreneurship Theory & Practice, 16/1, p. 7–26.

Çakıcı, A. C., Çalhan, H. & Karamustafa, K. (2016). *Yiyecek ve İçecek İşletmelerinde İnovasyon ve Sürdürülebilir Rekabet Üstünlüğü İlişkisi*, Kırıkkale Üniversitesi Sosyal Bilimler Dergisi, 6/2, p. 11–33.

Demirkaya, H. & Zengin, R. (2014). *Hizmet İnovasyonu ve Bir Uygulama Örneği*, Elektronik Mesleki Gelişim ve Araştırma Dergisi, 2/1, p. 106–116.

Dess, G. G. & Lumpkin, G. T. (2005). *The Role of Entrepreneurial Orientation in Stimulating Effective Corporate Entrepreneurship*, Academy of Management Executive, 19/1, p. 147–156.

Durna, U. & Babür, S. (2011). *Otel İşletmelerinde Yenilik Uygulamaları*, Alanya İşletme Fakültesi Dergisi, 3/1. p. 73–98.

Elçi, Ş. (2006). *İnovasyon: Kalkınma ve Rekabetin Anahtarı*, İnomer Rekabet ve Kalkınma Yayını, http://inomer.org/wp-content/uploads/2018/05/Inovasyon-SirinElci.pdf, Accessed: 15.10.2019.

Guth, W. D. & Ginsberg, A. (1990). *Guest Editors' Introduction: Corporate Entrepreneurship*, Strategic Management Journal, 1990/11, p. 5–15.

Johne, A. (1999). *Successful Market Innovation*, European Journal of Innovation Management, 2/1, p. 6–11.

Knight, G. A. (1997). *Cross-Cultural Reliability and Validity of a Scale to Measure Firm Entrepreneurial Orientation*, Journal of Business Venturing, 1992/12, p. 213–225.

Kuratko, D. F. (2007). *Corporate Entrepreneurship*, Foundations and Trends in Entrepreneurship, 3/2, p. 1–53.

Manga, M. (2018). *İnovasyon ve Turizm İlişkisi*, Makroekonomik göstergelerle turizm ekonomisi, (Işık, N. & Dineri, E.), Ankara: Nobel Yayıncılık.

Mercan, B., Göktaş, D. & Gömleksiz, M. (2011). *AR-GE Faaliyetleri ve Girişimcilerin İnovasyon Üzerindeki Etkileri: Patent Verileri Üzerine Bir Uygulama*. Paradoks: The Journal of Economics, Sociology & Politics, 7/2, p. 27–44.

Miller, D. & Friesen, P. H. (1983). *Strategy-Making and Environment*, Strategic Management Journal, 1983/4, p. 221–235.

Ming, L. T. (2013). *The Likelihood of Corporate Entrepreneurship in Large Corporations*, British Journal of Economics, Management & Trade, 3/4, p. 442–452.

Sayım, F. & Aydın, V. (2011). *Hizmet Sektörü Özellikleri ve Sistematik Olmayan Risklerin Sektör Menkul Kıymetleri ile Etkileşimine Dair Teorik Bir Çalışma*, Dumlupınar Üniversitesi Sosyal Bilimler Dergisi, 2011/29, p. 245–262.

Schumpeter, J. A. (1934). *The Theory of Economic Development*, London: Transaction Publishers.

Simonceska, L. (2012). *The Changes and Innovation as a Factor of Competitiveness of the Tourist Offer (The Case of Ohrid)*, Procedia-Social and Behavioral Sciences, 2012/44, 32–43.

Uzun, C. (2018). *Alanya Otel İşletmelerinde İç Girişimcilik Algısı*, VII. Ulusal Doğu Akdeniz Turizm Sempozyumu Bildiriler Kitabı, 20–21 Nisan, İskenderun Teknik Üniversitesi, İskenderun, Hatay. p. 14–26.

Vatan, A. (2010). *Turizm İşletmelerinde İnovasyon: İstanbul'daki 5 Yıldızlı Konaklama İşletmelerinde Bir Araştırma*, (Yüksek Lisans Tezi), Balıkesir Üniversitesi Sosyal Bilimler Enstitüsü, Balıkesir.

Yılmaz, Z. & İncekaş, E. (2018). *Türkiye'de İnovasyon ve Bölgesel Kalkınma*, Kırklareli Üniversitesi Sosyal Bilimler Dergisi, 2/1, p. 154–169.

Zahra, S. A. (1993). *Environment, Corporate Entrepreneurship, and Financial Performance: A Taxonomic Approach*, Journal of Business Venturing, 1993/8, p. 319–340.

Zahra, S. A. (1996). *Governance, Ownership, and Corporate Entrepreneurship: The Moderating Impact of Industry Technological Opportunities*, Academy of Management Journal, 39/6, p. 1713–1735.

www.digitaltalks.org/2016/07/30/marriott-inovasyon-oteli-ile-yeni-jenerasyona-hazirlaniyor/, Accessed: 12.11.2019.

www.adamevehotel.com, Accessed: 11.11.2019.

www.rixos.com, Accessed: 11.11.2019.

www.ranchobernardoinn.com, Accessed: 10.11.2019.

www.hurriyet.com.tr/ekonomi/bu-da-hersey-haric-otel-12111361, Accessed: 10.11.2019.

www.skyscanner.com.tr, Accessed: 10.11.2019.

www.narkoy.com/, Accessed: 10.11.2019.

www.somhotels.es/hotel-som-dona, Accessed: 10.11.2019.

Emre AYKAÇ and Ömer Ceyhun APAK

6 Diaspora Tourism

Introduction

Mankind has been in a constant state of mobility since the beginning of time. This mobility is manifested as a desire to travel and the compulsion to travel. In this context, voluntary and compulsory migrations go back as far as human history. Reasons such as natural disasters, war and economic crisis are important factors necessitating forced migration to other countries. However, the outstanding features of migrant societies are that they do not break their spiritual ties with the homeland they have been forced to leave and transfer their culture to the second and third generations. With the development of technology and transportation opportunities and thanks to the transferred culture, easier and faster travelling has increased the opportunities for diaspora to visit their homelands.

Diaspora tourism is defined as the travels made by the members of a nation or faith who have left their homeland and live as a minority in other countries who have not broken their material and moral ties with their homeland and are motivated by various desires (visiting family or friends, discovering their own origins, connecting with their ancestral culture, experiencing ancestral life).

In the current globalizing world questions such as, 'Where is my home and what is my identity?' have gained importance and as a result, the desire of diaspora communities to travel to ancestral homelands has increased. Travels to ancestral homelands are considered within the scope of tourism. Diasporas constitute an important tourism market for their respective homelands. Nowadays with intense digitalization, advertising and promotional messages that can be transmitted to diaspora tourists via social media, word of mouth marketing or database marketing play an important role in diaspora's homeland visits.

In this section, migration, the predecessor of the diaspora, diaspora and the concept of diaspora tourism and the tourist types and tourism market of diaspora and Turkey-based diaspora tourism issues are discussed.

1. Immigration, the Recursor of the Diaspora Phenomenon

One of the most important concepts that is associated with diaspora is migration. According to one hypothesis, the initial origin of humans is Africa. They spread into the world from Africa, for example they migrated to the Near East one hundred thousand years ago, to Southeast Asia and Australia sixty thousand years ago and to America thirty five thousand years ago (Yaldız, 2014: 384). People have migrated from their origins (country, homeland, habitat, hearth, etc.) to another place within or outside their countries either individually or collectively, temporarily or permanently either through necessity or voluntarily due to economic, political, climate-related reasons, natural events, hunger, epidemic diseases, war, terror, social, cultural, educational, health, individual and similar reasons (Beyaz, 2019: 11). The most obvious expression of migration, which starts with the existence of human beings, is the relocation of people for economic, social or political reasons. From a sociological aspect, migration can be expressed as the displacement of individuals or communities in time and place which depend on many factors and the subsequent continuation of its effects within a process (Tören, 2014: 7).

Migrations are classified from different angles (Güllüpınar, 2012: 57):

- In terms of purpose; economic migration and non-economic migration,
- In terms of the factors triggering migration; voluntary migration and involuntary migration,
- In terms of duration; temporary and permanent migration,
- In terms of the last settlement; transitory migration and settled migration,
- In terms of legal status; legal migration and illegal migration,
- In terms of the characteristics of the immigrant; skilled (brain) migration and unskilled migration are the main categories.

Two factors appear to be effective on the basis of the emigration types. These are considered as push and pull factors. 'Push' factors are determined by factors such as unusual events such as wars, hunger, political or religious repression, long periods of recession such as high inflation and low wages while 'pull' factors are determined by factors such as good work, high wages, good education or attractive environment, religious freedom, family or affinity with a given community (Tuzcu & Bademli, 2014: 57).

At this point, it should be noted that these types of historical migration have differentiated with globalization today. There is migration within the migration. This shows that the structural and functional impact of 21st century diaspora communities will be strengthened. Present day migrants are different from the unqualified migrant workers in the 19th century and beyond. They consist of wealthy, talented and qualified people. Therefore, the definition of migration in the 21st century can be redone once more to move in line with future goals rather than the past.

Migrations have created homogeneous societies and removed borders with the displacement of people. Migrations have also brought a multicultural everyday life to local cultures (Morawska, 2011: 1). The person who migrated has naturally started to live with a new culture. Furthermore, many migrants do not forsake their former culture either. Therefore, many migrants have embraced a 'bi-cultural' structure by adopting both cultures and living in a second culture without losing their main culture. As a matter of fact, migrations have displaced the building blocks of culture and introduced new values (Balcıoğlu, 2007: 58). It has challenged, severed, changed, and sometimes strengthened the former ties of people (Faist, 2003: 42). Even if they live far away from the motherland (homeland) as a result of migrations, people feel devotion to their roots, that is to their ancestral land, for many different reasons. This commitment has led to the establishment of today's diasporas. Conceptually they have developed (İlhan & Sözbilen, 2018: 372).

2. Transnational Existence 'Diaspora' and Diaspora Tourism

Diaspora: As is the case in many areas, it is not possible to speak of a single concept regarding diaspora. In this respect, there are two sorts of presupposition. The first one is the classical (ethnographic) approach, which interprets and accepts the diaspora in the oldest contexts, and the second one is the modern approach that has been brought up in the 1980s, particularly around the interaction of modern migration with globalization, transnationalism and post-modernism (Ulusoy, 2017: 140).

Therefore, in order to understand the meaning of the current concept of diaspora and the characteristics of diaspora communities, it is necessary to look at the emergence process of the concept. Currently, the diaspora concept means: the place where any nation or members of faith live as a minority abroad; the branch of any nation separated from its homeland, a fragment (https://sozluk.gov.tr, 2019). However, when the etymological roots of its origins are delved, it is necessary to go back in history in order to perceive the present meaning of the concept of diaspora which is depicted as an adventurous word (Marienstras, 1999: 363) describing the adventure of humanity which is the subject of history and fortune. In etymological terms, the use of the verb diaspeiro formed by the combination of the Greek words dia and sperien is encountered during the 5th century BC in the works of Sophocles, Herodotus and Tucidides (Dufoix, 2011: 17). The concept of diaspora, derived from the Greek word diaspeiro / diasperien, was used in the Septuagint to describe Jews living in exile outside their homeland and pointing to the Hellenic Jewish communities living in Alexandria (Yaldız, 2013).

The diaspora is conceptualized as a transnational population broken off from the homeland, which develops a strong ethnic group consciousness, a sense of alienation or solidarity and various levels of desire to return home. Superficially, each immigrant community consists of the same or similar ethnic groups living outside their homeland. (Li, McKercher & Chan, 2019: 2). According to another definition that draws the lines for being a diaspora is that not every group that is forced to leave their country or territory or does so of their own will automatically becomes a diaspora community upon arrival in another territory. In order for a diaspora consciousness to develop, there must be an affinity with the homeland or difficulties in adapting to the country of migration (Wilkoszewski, 2010).

Edwards, on the other hand, considered the diaspora as a term bearing the traces of transnationalism followed by the shift, emphasizing that the diaspora should not be considered as a failure or negativity of internationalization. However, he also indicates that there is a concept that includes abstraction in the implementation process. (Edwards, 2003: 11).

One important point is that cultural memory and identity play a major role in the formation of diaspora communities. In this respect, Stuart Hall expresses the concept of cultural identity in two different ways. The first is a society that shares the same identity, and the second is a society that agrees with the idea of having a single identity. At the same time, cultural identity includes many artificial and natural identities that people share and embrace. From this point aspect, the feeling of wanting to belong for diasporas living in different locations around the world underlines the strength of their cultural identities and the importance of keeping societies together. However, if people do not learn from their environment some information about their culture and cultural identity, they may forget these values and be assimilated in the societies they live in (Hall, 1990: 230). When the mentioned conditions and the characteristics of diasporas are taken into consideration, it is evident that culture and social identity are very important for diasporas. Therefore, the social, cultural and economic activities of diaspora communities differ according to the countries in which they live and the tasks undertaken by the diasporas. Accordingly, some diaspora communities contribute to their own countries both socio-culturally and economically, while others strive to keep their culture alive. Diasporas have opened many associations, foundations, health institutions, educational institutions and business centers in order to protect them from being assimilated into the new societies that they have migrated to. As this situation persisted, topics such as nation-state identity relations, cultural and social economic relations, economic cooperation, foreign investments, returning, tourism and knowledge transfer started to be covered in conferences and congresses in the 21st century. As a result, efforts were made to improve relations with diasporas. Due to the advantage of current developing technology, diaspora societies, whose access to their own culture is easier, have increased their commitment to their homeland (Tosun, 2016).

Diaspora Tourism: This type of tourism has emerged as a result of the travels of individuals who have settled abroad back to their homelands.

Diaspora tourism means that people in the diaspora travel to their ancestral homeland to seek their origins or feel connected to their ancestors. While most tourists are bound to a destination after repeated visits, the tourist destination relationship in diaspora tourism is unique, because tourists with migrant backgrounds often feel connected to the destination's people, culture and heritage before actually visiting the place. In the light of the information so far, in fact, migration and tourism are the same on the macro level, because both phenomena involve only the movements of people from different geographical regions. The relationship between migration and tourism is two-fold. On the one hand tourism can produce two types of migration: (1) the migration of workforce needed for tourism related services and (2) depleted migration systems consisting of tourists traveling to their favorite places, such as retirement migration and second home improvement. Migration and diaspora, on the other hand, can also lead to four forms of travel.: (1) Migrants travel to their ancestral homelands, (2) People from the homeland can visit their migrant relatives, (3) Diaspora can travel to places other than their homelands, (4) Diasporic communities can develop their own holiday venues by traveling to places where they can meet people of similar ethnic origin (Huang, Haller & Ramshaw, 2013).

Figure 1 shows the conceptual framework of diaspora tourism. A conceptual framework with important themes has been presented to enable a holistic understanding of diaspora tourism. Tourism scientists and practitioners tend to see diaspora tourism as a homogeneous market that can meet needs with generalized product types. This assumption may conclude the phenomenon in the form of unsatisfactory visiting experience. This leaves a gap between the starting point and destination dimensions. A serious review of a wide range of research on diaspora, migration and diaspora (homecoming) tourism has been discussed. A conceptual framework has been proposed by synthesizing important themes in both dimensions identified from the review. The conceptual framework provides researchers and destination managers with a holistic perspective to explore diaspora tourism. The demand dimension of diaspora tourism is related to the structure of diaspora communities, migration background of diasporic individuals, level of acculturation and sense of place, determining reasons for return, destination dimension (supply), why current diasporic destinations and their products do not counter the needs of different types of tourists and

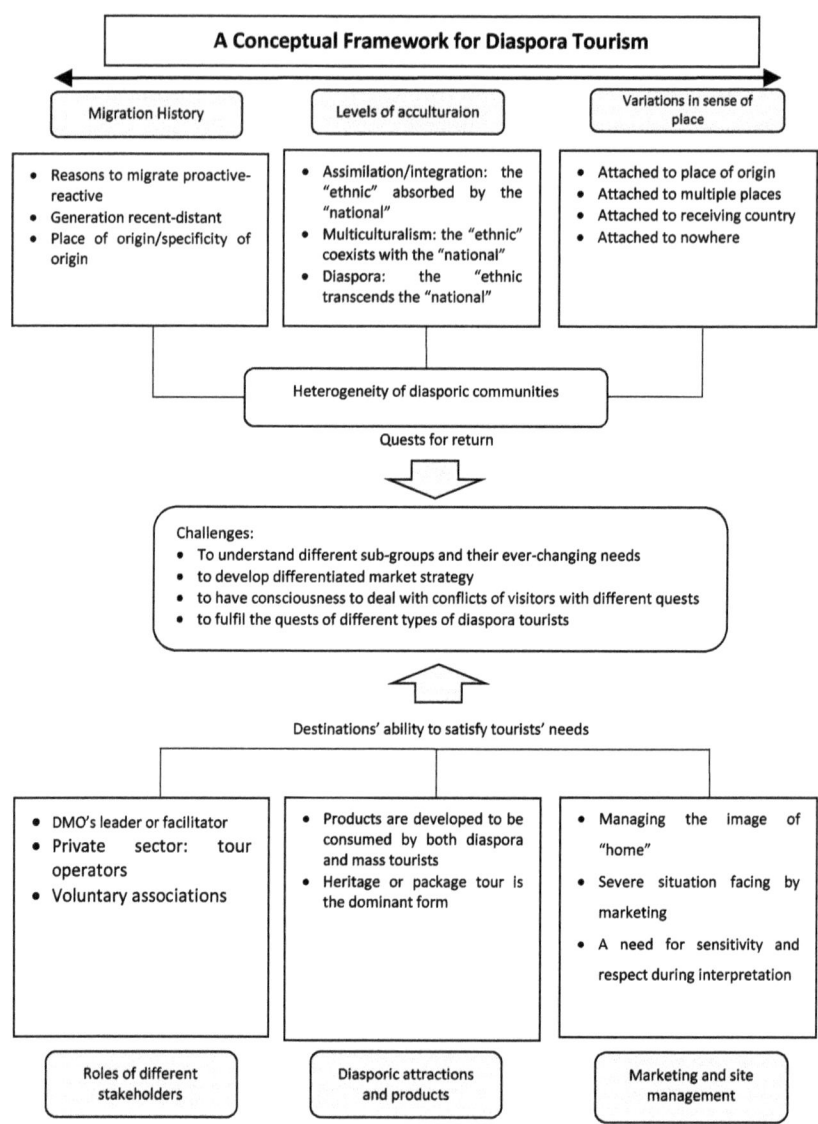

Figure 1: Conceptual Framework of Diaspora Tourism. Source: Li, McKercher & Chan, 2019

relevant matters. Therefore, a comprehensive analysis has been endeavored in order to enable diasporic destinations to focus on a future planning and strategy development that meet the needs of diaspora tourists (Li, Mc Kercher & Chan, 2016: 4). While the concept of diaspora tourism has been handled within the scope of traveling to ancestral lands, its association, similarities and differences with culture-specific special interest types of tourism such as 'cultural heritage tourism, dark tourism, faith tourism and ethnic tourism' have been manifested. Since the definition of diaspora has been defined as the separation of members of a tribe, nation or faith from their homeland and living as a minority in other places for a long time, it has been concluded that diaspora tourism should be considered as a separate type of tourism. The intersection points of diaspora tourism and culture-based special interest types of tourism are explained in the continuation of the relationship between diaspora tourism in Figure 2. (Tören, 2012: 556).

Figure 2: Intersection of diaspora tourism and culture-based special interest tourism types Source: Tören, 2012: 556

For example, if an Orthodox Christian visits the Sümela Monastery in Trabzon it is 'faith tourism' but if a Greek Orthodox who lived there or had ancestors who lived visits and if this visitor has a desire to return to his homeland, this is considered 'diaspora tourism'. A trip to another Circassian locality by another Circassion other than his place of origin to see the ethnic differences of Circassians in that locality is 'ethnic tourism', however, a Circassian visiting this locality (for example, the Caucasus) from where his ancestors had been forced to migrate or due to economic reasons is considered within the context of 'diaspora tourism'. Another example is a visit to Taj Mahal in India by any citizen of a country, which can be considered within the scope of 'cultural heritage tourism', however a visit by an Indian who has emigrated to the USA for economic reasons and has been living there since, to gain knowledge of the culture, history and cultural heritage of his people can be evaluated within diaspora tourism. A comparison between Diaspora tourism and the closely related fourth type of tourism dark sadness tourism can be examined from the aspect of a Jew living in Turkey who visits the Auschwitz Nazi death camp in Poland to revive the events there and feel the sadness and pain. However, if a Polish Jew who lived in the same region (Krakow) and had to leave the region during the war and now lives in the USA visits, this can be considered within the scope of diaspora tourism (Tören, 2012: 556)

3. Diasporic Tourist Types

Travel, a concept associated with moving from one place to another, takes place on connected routes along destinations (Jafari, 2000: 600). Travel is expressed as an activity that requires a person to leave his/her home and move out of their accustomed environments (inside or outside the country). People have traveled for many reasons to date. However, some of these trips have been compulsory (Huang, Haller & Ramshaw, 2013: 286). People have been forced to leave their lands and the places where they live due to natural disasters, wars and economic crises. Today, the most striking example of this situation is manifested by Syrian citizens who have been forced from their homes and take refuge in various countries, mainly Turkey to flee the war. Subsequently, these people are motivated in different ways to return to their homelands or travel to visit (Tanrısever, Yaşarsoy & Pamukçu, 2015: 163).

People who migrate to a different country tend to travel to get information about their family history, connect with their ancestors' origins and cultures, or learn about individual heritage values (McCain & Ray, 2003). Therefore, the people of different types travel within the scope of diaspora tourism (Ari & Mittelberg, 2008: 84).

- Tourists looking for their origin in the homeland,
- Tourists looking for family history or genealogy,
- Tourists traveling to diaspora areas in their homeland,
- Tourists visiting diaspora destinations for tourism purposes only, without being motivated by diaspora purposes,
- Tourists traveling between the homeland and diaspora areas,
- Diasporic people visiting tourism destinations in the host country.

Li & McKercher (2016) have examined diaspora tourist types in five different groups. These are: the positive diaspora tourist, researcher diaspora tourist, reconnected diaspora tourist, distant diaspora tourist and independent diaspora tourist.

Positive Diaspora Tourist: All of these types of tourists are first-generation immigrants who have migrated in adulthood for various reasons. Although they live in another country, they feel that they belong entirely to the country of origin. Visiting the homeland increases the chance for these tourists to connect to their origin. Although they live in another country, they can clearly state that they belong to their homeland. For example, a Turkish citizen who has lived in Germany for the past twenty or thirty for various reasons knows that his homeland is Turkey and will therefore describe himself as Turkish when he describes his identity. In this context, a strong bond to the homeland brings frequent return visits. The experiences of returning to the homeland are referred to as 'going home' and emotions such as 'receiving a warm welcome', 'enjoying visiting friends and relatives' and 'feeling part of a community'.

Researcher Diaspora Tourist: The researcher tourist type can be expressed as the second generation. This generation has not lived in the homeland where the family came from. However, since the influence of the culture and traditions of his family is felt in this generation, his curiosity towards his native land is high. Although he learns his own culture from his family, he also experiences a cultural conflict within himself due to the influence of the culture of his country. Being influenced by the culture of the host country, they feel at an equal distance to their homeland and host country. However, travels to their homelands take the form of learning their culture, strengthening their spiritual ties with their homelands, exploring their essence and exploring the places where their ancestors came from. They see themselves as part of a majority family, not a minority. Nevertheless, this situation cannot change the way they feel about themselves as strangers in their homeland.

Reconnected Diaspora Tourist: This type of tourist, who sees himself as a citizen of his country, feels pleased and comfortable, even though he perceives himself as a stranger when traveling to his homeland. The type of tourist finds the opportunity to discover his origins on his return and experience the tangible and intangible cultural heritage of his ancestors. Even though people in the second or third generation have some communication difficulties in their homeland, the culture they have experienced generates a sense of attachment to their essence again.

Distant Diaspora Tourist: No devotion is generated when the distance tourist visits the homeland. As an example, an individual, who immigrated to another country with his family when he was 5 years old, perceives that country as his own country after growing up and assimilating the culture of that country.

Therefore, he does not feel loyalty for his own country and has no motivation to discover his origins. Rather, they consider their homeland an interesting tourist destination that warrants visiting. The local people in the homeland they visit consider these tourists as foreigners. Especially when they go to their homeland after the first generation, the difficulty in communicating and the possibility of feeling like a foreigner increases due to the common cultural difference.

Independent Diaspora Tourist: After visiting their homeland this type of tourist feels that he has no connection and common ground with his own country. This type of tourist thinks that when he visits his own country they consider him as a stranger. He has adopted the culture of the country he traveled to and concludes that he does not have any connection and common ground with the people he meets when he travels to his homeland. These types of tourists eliminate the possibility of revisiting their homelands because of the inability to communicate effectively with the local community and difficulty in understanding the language when they go to their homeland.

4. Diaspora Tourism Market

Destination managers have started, in various ways, to develop products to address this phenomenon of diaspora tourism, which is widely used to indicate the experience produced, consumed and experienced by diasporic communities with increasing interest and markets (Li, McKercher & Chan, 2016: 1). The United Nations Economic and Social Council (ECOSOC) Population Division published the International Migrant Data for 2019 in the last week of September. The number of international migrants, which was 221 million in 2010 has reached 272 million which displays the extent of the increase. It is also possible to see the top five sending countries of international immigrants. India is the leading sending country with 17.5 million migrants. Mexico follows with 11.8 million, China with 10.7 million and Russia with 10.5 million international migrants. Since most of its citizens fled the civil war in Syria and relocated to other countries in 2011 and mainly to Turkey, Syria ranks fifth with 8.4 million migrants.

Many migrants live in countries such as India, China, Russia, Mexico, the Philippines, the United States and the United Kingdom. These migrants can be an important market segment for their homelands. Migrants often return to their homelands. Nowadays when digitalization is intense, advertising and promotional messages that can be transmitted to diaspora tourists through social media, word of mouth marketing or database marketing play an important role in diaspora's homeland visits (Morgan, Pritchard & Pride, 2003; Huang, Hung & Chen, 2018). Visits by Diaspora tourists are considered as a niche market within the scope of heritage tourism and special interest tourism (Newlan & Taylor, 2010; Huang, Haller & Ramshaw, 2013; Tanrısever, 2016). Diaspora tourism has become an important niche market for many countries, from China to Ghana, Bangladesh to India, the Caribbean Islands and the Philippines. Tourist products are being developed for these people and destinations are involved in efforts to attract tourists.

In the diaspora tourism, which has been changing and developing since the 1970s and is considered a niche market, it is necessary to design products for tourists and to understand the expectations and requirements correctly in order to attract tourists to destinations. It is important to define demanding consumer characteristics, to ensure effective communication through advertising and promotion efforts, and to develop demand increasing strategies. In this context, the creation of demand in diaspora communities is linked to the correct understanding of the purposes for which diasporas are motivated to travel. Travel motivations include relaxation, self-development, socialization, independence, gaining experience in tourism types whereas different aspects can motivate diaspora tourism (Fyall et. al., 2003). According to Salem & Merhi (2019), the diaspora population, which is found in many parts of the world, plays a leading role in the discovery of new cultural products supported by cultural heritage which are specific to their homelands. Unlike international tourists, diaspora tourists take care to stay in local accommodations, wine and dine in local restaurants, buy locally produced products and thus support the local economy. Although diaspora tourism has a tendency to participate more in cultural activities, it emerges as the type of tourism that is the least affected by the seasonal characteristics of tourism. Therefore, it will be beneficial that touristic destinations learn about the travel motivations of diaspora tourists within the scope of diaspora tourism.

These migrants, who were obliged to emigrate to another country or forced to emigrate, travel for different reasons such as visiting their homeland, family or friends, discovering their origin, being able to connect with the culture of their ancestors and experiencing life on their ancestral land. The travel network between the homeland and the country of residence is associated with the bond of the first generation to the homeland. Diaspora tourism can take place if the first generation of migrants can transfer their culture to the second and third generations. Travel motivations (visiting family or friends, discovering their origins, being able to connect with their ancestral culture, experiencing life on ancestral land) play an important role in the development of tourism mobility among diasporas (Arslaner & Erol, 2017: 429).

Travel motivations for diaspora tourists do not apply to all diaspora travels. Diaspora tourists can travel for health, business and commercial, long-term return and cultural heritage. Countries such as the Philippines, India, Taiwan and Cuba provide support to diaspora for health-related diasporic travels (Newlan & Taylor, 2010: 6). German citizens of Turkish origin and second and third generation Turks travelling to Turkey for health services is a major indicator of health related diaspora tourism. Furthermore, there are numerous examples of health related diaspora travels such as people of Indian origin traveling from the U.K. to India and people of Mexican origin traveling to Mexico from the U.S. (www.dosyamerkez.saglik.gov.tr).

Diaspora tourists have great potential in terms of commercial travelling. Together with commercial purposes, the desire to have a holiday and invest in their homeland can be seen as an important market. Long-term returns combine diaspora travel and investment. Long-term returns require investments and new structuring. Visiting cultural heritage has an important place in diaspora tourism. Therefore, it is important that heritage values are not destroyed and are kept alive (Mortley, 2011). Today in Turkey, second and third generation Greek citizens of Turkish origin who have migrated from Gümüşhane in Turkey and are currently living in Athens visit the places where their ancestors used to live. Marketing strategies and promotion of these heritage values should be developed to ensure a balance of conservation-usage of these heritage values in order to increase and sustain these travels.

5. The Position of Turkey in Diaspora Tourism

It is estimated that diaspora tourism, which is seen among the new tourism trends by tourism authorities, has a market size estimated at 2 billion dollars and that Eastern European countries will take the lion's share from this market. Considering Turkey's history and Turks living abroad, Turkey appears to have a significant potential in terms of diaspora tourism as a sending and host country. Scientific studies will draw the attention of tourism authorities in Turkey to this market and enable them to assess the potential of this market correctly. Considering its contribution to national income, it is believed that Turkey's diaspora tourism should be further investigated to study the potential of the market (Unur, Kanca & Ertaş, 2015). In fact, during the 50-year history of migration processes in the Republic of Turkey, the number of citizens who have migrated to other countries has exceeded 6.5 million (Şahin, 2010).

The Population Exchange between Turkey and Greece took place in 1923. Both states resolved to establish a framework for religious integrity and as a result only the religious identity of the peoples was taken as basis to make societies homogeneous in terms of religion while linguistic, ethnic and cultural differences were by-passed and the exchange was carried out. As a result of the exchange, with the exception of those residing in Istanbul, Gökçeada and Bozcaada, all Orthodox Greeks residing in Anatolia and Eastern Thrace were dispatched to Greece while all Muslims living in Greece with the exception of Western Thrace were sent to Turkey (Bayındır Goularas, 2012). Approximately 350,000 Muslim Turks and 200,000 Christian Greeks were subjected to migration. Studies have shown that this forced population exchange affected the parties negatively. Although they seem to have migrated to the land to which they belong, they actually considered the land they left as their motherland (Bozdağlıoğlu, 2014: 10).

Currently, a few million Turks live in Europe as a result of labor migration. This phenomenon created a new social category for Turkey, 'the Turkish expats' (Kastoryano, 2007: 66). The majority of Turks in Europe preferred to settle in the capitals of the countries they migrated to (Akman, 2007: 55). Between 1956 and 1960, during the first years of migration, the authorities did not give much thought to this activity (Kastoryano, 2007: 66). When the number of Turkish workers reached 30 thousand, the State Planning Organization started to focus on this issue and started to prepare detailed reports. During this period, the Turks organized on their own. They founded several solidarity associations to send money to relatives in need of assistance as well as send the bodies deceased workers to Turkey (Abadan et. al., 1976: 51). A very minor part of this community which intended to work and save money and return after a short period of time actually returned to Turkey. The majority brought their families from Turkey and has continued to live in European countries (Akman, 2007: 66).

Turkey is not only a sending country but a receiving country as well. Especially because of the war in Syria in recent years, many people have been obliged to migrate from their homes. Turkey is among the countries that Syrian citizens chose to migrate to. With 5 million Syrian citizens Turkey has become a major diaspora center. Turkey stands out with the diasporas that have been established as a result of outgoing migration (Turkish, Armenian and Greek diasporas) and incoming migration (Circassians, Bulgarian, Tatar, Uigur diasporas) in recent years. Citizens of countries such as Syria, Iraq and Afghanistan have been added to the recent migrations. Many Turkish nationals living in Germany transport their cultural characteristics to other countries and visit both their own homeland (vacation time is 24.70 days on average) and can contribute to foreign citizens visiting Turkey. Furthermore, the Sumela Monastery and the various cities where the Greeks lived (Sinop, Samsun, Gumushane, Trabzon, Bayburt, Ordu, Giresun and some settlements on the Aegean coast) are considered important destinations for Greek diaspora tourism. Thus, the visits of these diasporas to their countries have an important place in the tourism market. (Tören, 2012).

Conclusion

Migrations that started with the existence of human beings incur for many reasons that can be economic, social and political. From a sociological point of view, the migration of people or communities in the form of a relocation activity and the continuation of living thereafter is manifested by driving factors (extraordinary events such as wars, hunger, political or religious pressure, high inflation and low wages) and attractive factors (work, high wages, good education, religious freedom, families or affinity with certain communities).

A significant change incurred when the migration of laborers in the 19th century and onwards was replaced by the migration of wealthy, talented and qualified migrants. This situation emphasizes that migration types also change over time. With this change, developed and developing countries started to see diaspora communities as a tourism market. Major changes in the tourism market depend not only on the change in services or destinations offered, but also on changes in the motivation of travelers. The basis of change experienced by travelers is a function of cultural, social and environmental characteristics (Reisinger, 2013: 30). In this context, diasporas are primarily motivated to travel by factors such as visiting family or friends, discovering their origins, being able to connect with the culture of their ancestors, experiencing life on ancestral land, while travel motivated by health, business and commercial purposes, long-term return and cultural heritage can contribute significantly to the tourism market share of host countries. The difference between the travel motives of diaspora and the other special interest tourism areas of diaspora tourist types manifests the necessity to handle diaspora tourism separately from other types of tourism.

Even if people live away from their homelands as a result of migration, they feel connected to their ancestral lands for different spiritual reasons. Therefore, while traveling to ancestral lands is considered within the scope of tourism, the concept of diaspora tourism allows the examination of culture based special interest tourism types (cultural heritage tourism, dark tourism, faith tourism and ethnic tourism). However, a review of the diaspora tourist types (positive diaspora tourist, researcher diaspora tourist, reconnected diaspora tourist, distant diaspora tourist and independent diaspora tourist) manifests the difference between them and other tourism types and therefore, it can be considered as a separate type of tourism. First generation Diaspora tourists display a positive approach to their homeland, while the attitude of the second generation diaspora tourist is inclined to be researching, reconnecting and distant. Third generation diasporas, on the other hand, appear to be partially independent, that is, disconnected from their homeland and showing no interest. Since these attitudes are approaches that are specific to diaspora tourism, they manifest the difference between diaspora tourism and other types of special interest tourism. Another characteristic of diasporas found in many parts of the world and diaspora tourists is that they are willing to make investments in their own countries and unlike international tourists, to stay in local accommodation establishments. In addition, they take care to wine and dine in local restaurants, buy locally produced products and support the local economy this way. It is important to develop this type of tourism, which gives importance to locality on the basis of all pecuniary and non-pecuniary factors that make up the homeland and its culture, to develop tourism products for diaspora tourists and ensure sustainability in touristic destinations.

References

Abadan U., N., Keleş, R., Penninx, R., Renselaar, H. V., Velzen, L. V. & Yenisey, L. (1976). Göç ve Gelişme, Ankara: Ajans Türk Matbaacılık Sanayi.

Akman, V. (2007). *Küresel göç hikâyeleri*, çev: T. Gezer, İstanbul: Yakamoz Yayınları.

Ari, L. L. & Mittelberg, D. (2008). *Between Authenticity and Ethnicity: Heritage Tourism and Re-ethnification among Diaspora Jewish Youth*, Journal of Heritage Tourism, 3(2), 79–103.

Arslaner, E. & Erol, G. (2017). *Alternatif Turizmin Bazı Türleri Üzerine Bir Değerlendirme*, Journal of Tourism and Gastronomy Studies, 5(4), 422–438.

Balcıoğlu, İ. (2007). *Sosyal ve psikolojik açıdan göç*, İstanbul: Elit Kültür Yayınları.

Bayındır Goularas, G. (2012). *1923 Türk-Yunan Nüfus Mübadelesi ve Günümüzde Mübadil Kimlik ve Kültürlerinin Yaşatılması*, Alternatif Politika, 4(2), 129–146.

Beyaz, C. (2019) *Kırsal Göçmenin Kimlik, Aidiyet ve Bağlılık Mücadelesi: Rizeli Göçmenler Üzerine Bir Çalışma*, Hacettepe Üniversitesi Sosyal Bilimler Enstitüsü Sosyoloji Anabilim Dalı, Yayımlanmamış Doktora Tezi, Ankara.

Bozdağlıoğlu, Y. (2014). *Türk-Yunan Nüfus Mübadelesi ve Sonuçları*, Türkiye Sosyal Araştırmalar Dergisi, 18(3), 9–32.

Dufoix, S. (2011). *Diasporalar*, İstanbul: Hrant Dink Vakfı Yayınları.

Edwards, B. H. (2003). *The Practice of Diaspora-Literature, Translation and Rise of Black Internationalism*, Cambridge: Harvard University Press.

Faist, T. (2003). *Uluslararası Göç ve Ulusaşırı Toplumsal Alanlar* (Translated: A. Gündoğan and C. Nacar), İstanbul: Bağlam Yayıncılık.

Fyall, A., Callod, C. & Edwards, B. (2003). Relationship marketing: The challenge for destinations. Annals of Tourism Research, 30(3): 644–659.

Güllüpınar, F. (2012). *Göç Olgusunun Ekonomi-Politiği ve Uluslararası Göç Kuramları Üzerine Bir Değerlendirme*, Yalova Sosyal Bilimler Dergisi, 4: 53–86.

Hall, S. (1990). Cultural İdentity and Diaspora, Jonathan Rutherford (ed.) Identity: community, culture, difference, London: Larence & Wishart Publisher.

https://dosyamerkez.saglik.gov.tr/Eklenti/10945,03pdf.pdf?0> Accessed: 13 November 2019.

https://sozluk.gov.tr Accessed: 15 November 2019.

Huang, W. J., Haller, W. J. & Ramshaw, G. P. (2013). *Diaspora Tourism and Homeland Attachment: An Exploratory Analysis*, Tourism Analysis, 18(3), 285–296.

Huang, W. J., Hung, K. & Chen, C. C. (2018). *Attachment to the Home Country or Hometown? Examining Diaspora Tourism across Migrant Generations*, Tourism Management, 68, 52–65.

İlhan, İ. & Sözbilen, G. (2018). *Turizmde Güncel Konu ve Eğilimler*. Ankara: Detay Yayıncılık.

Jafari, J. (2000). *Encyclopedia of Tourism*. London–New York: Routledge.

Kastoryano, R. (2007). *Ulusaşırı Türk milliyetçiliği: Yurtdışında yaşayan Türklerin milliyetçilik tanımı. Kökler ve Yollar-Türkiye'de Göç Süreçleri* (Ed: A. Kaya & B. Şahin), İstanbul: Bilgi İletişim Grubu Yayıncılık.

Li, T. E. & McKercher, B. (2016). *Developing a Typology of Diaspora Tourists: Return Travel by Chinese Immigrants in North America*, Tourism Management, 56, 106–113.

Li, T. E., McKercher, B. & Chan, E. T. H. (2019). Towards a conceptual framework for diaspora tourism, Current Issues in Tourism.

Marienstras, R. (1999). On the Notion of Diaspora. (Ed: S. Vertovec, & R. Cohen içinde), *Migration, Diaspora and Transnationalism*. Massachusetts: Edward ElgarPub.

McCain, G. & Ray, N. M. (2003). *Legacy Tourism: The Search for Personal Meaning in Heritage Travel*, Tourism Management, 24, 713–717.

Morawska, E. (2011). *A Sociology of Immigration*, New York: Palgrave Macmillan.

Morgan, N. J., Pritchard, A. & Pride, R. (2003). *Marketing to the Welsh Diaspora: The Appeal to Hiraeth and Home Coming*, Journal of Vacation Marketing, 9(1), 69–80.

Mortley, N. K. (2011). *Strategic Opportunities from Diaspora Tourism: The Jamaican Perspective*, Canadian Foreign Policy Journal, 17(2), 171–185.

Newland, K. & Taylor, C. (2010). *Heritage Tourism and Nostalgia Trade: A Diaspora Niche in the Development Landscape*, Washington, DC: Migration Policy Institute.

Salem, G. & Merhi, E. (2019). *Diaspora Tourism in Lebanon: A Strategy to Maintain Tourism Efficiency during Crises*, European Journal of Hospitality and Tourism Research, 7(1), 31–52.

Şahin, B. (2010). *Almanya'daki Türkler*, Ankara: Phoenix Yayınevi.

Tanrısever, C. (2016). *Diaspora Turizmi: Türkiye-Azerbaycan Örneği*, Turizm ve Araştırma Dergisi, 5(2), 56–64.

Tanrısever, C., Yaşarsoy, E. & Pamukçu, H. (2015). *Diaspora Turizmi: Türkiye-Azerbaycan Örneğ*, 1. Uluslararası Türk Dünyası Turizm Sempozyumu, 19–21 Kasım, Kastamonu.

Tosun, İ. (2016). *Ermenistan ve Azerbaycan Diaspora Politikalarının Karşılaş tırılması*, The Journal of Europe-Middle East Social Science Studies, 2(2), p. 217–251.

Tören, E. (2012). Ata Toprağı Ziyaretlerinin Turizm Kapsamında Değerlendirilmesi. VI. Lisansüstü Turizm Öğrencileri Araştırma Kongresi (Yayına hazırlayan: N. Kozak & M. Yeşiltaş). Ankara: Detay Yayıncılık, p. 550–566.

Tören, E. (2014). Diasporaların Ana vatan Ziyaretleri: Almanya Türk Federasyon Türkiye Kültür Gezisi 2013 Üzerine Bir Alan Araştırması, Eskişehir Anadolu Üniversitesi Sosyal Bilimler Enstitüsü, Turizm İşletmeciliği Anabilim Dalı, Eskişehir.

Tuzcu, A. & Bademli, K. (2014). *Göçün Psikososyal Boyutu*, Psikiyatride Güncel Yaklaşımlar-Current Approaches in Psychiatry, 6(1), 56–66.

Ulusoy, E. (2017). *Diaspora Kavramı ve Türkiye'nin Diaspora Politikalarının Modern Teori Çerçevesinde Sosyo-Politik Bir Analizi*, The Journal of Humanity and Society, 7(1), 139–160.

Unur, K., Kanca, B. & Ertaş, Ç. (2015). *Türkiye'nin Diaspora Turizmi Potansiyeline İlişkin Bir Değerlendirme*, I. Euraisa Internaitonal Tourism Congress: Current Issues, Trends, and Indicators (EITOC-2015), 350–362.

Wilkoszewski, T. (2010). *Türkiye'ye Uluslararası Göç*, (Ed: B. Pusch ve T. Wilkoszewski), İstanbul: Kitap Yayınevi.

Yaldız, F. (2013). Diaspora Kavramı: Tarihçe, Gelişme ve Tartışmalar, Hacettepe Üniversitesi Türkiyat Araştırmaları Dergisi, 2013Bahar, 18, 289–318.

Yaldız, F. (2014). *Uluslararası Göç ve Diaspora ile İlişkili Kavramlar*, Journal of the Human and Social Science Researches, 3(2), p. 382-403.

Yasin DÖNMEZ and Sevgi ÖZTÜRK

7 Ecotourism and Geographical Information Systems Applications

Introduction

Ecotourism, whose origins go back to natural and socio-cultural values, is an ecologically and socio-culturally responsible tourism based on the natural environment. Ecotourism practices serve the sustainability of natural resources, which provide economic benefits to local people by providing rational use of natural resources (Sayın, 2019).

The main purpose of eco-tourism is to reduce the environmental impacts from mass tourism and its infrastructure and to ensure the active activities of the local people. The work to be done within ecotourism is to reduce resource use, support infrastructure designs and adapt it to the local community and traditions. The income from this area contributes to the development of the local people along with the cultural heritage. However, if ecotourism is not well planned, it harms natural and cultural resources and wild environments (Sezen et. al., 2011: 55; Fennel, 2001). In this study, the importance between ecotourism and geographical information systems is emphasized.

Ecotourism

Based on their rich flora-fauna and climatic comfort opportunities, mountainous areas are the preferred locations for the recreation and tourism activities. Recently, increased urbanization leads to the tendency of dwellers to escape from the urban problems. Therefore, urban people who go towards rural and mountainous areas, need to visit their natural and cultural characteristics, learn and experience distinct social conditions there. Ecotourism is a kind of tourism which has been preferred as recreation activity within natural and rural areas (Gökyer et al., 2015; Kendir et al., 2019).

Ecotourism since the early 1980s, different terms (nature tourism, alternative tourism, adventure tourism, trekking, sustainable tourism, non-consumer tourism, endemic tourism, geotourism, responsible tourism, as a natural resource-based type of tourism where the impact on resource values is minimal, focuses on learning and acquiring experience about natural life and contributes to the protection of natural areas (Weaver, 2001; Fennel, 2001; Buckley, 2009; Türkmen & Dönmez, 2015).

Ecotourism supports sustainable development as an alternative type of tourism. The emergence of alternative tourism environmental factors due to touristic activities, increase in the number of tourists and general tendency in tourism towards alternative types of tourism (Düzgüneş & Demirel, 2013)

Today, the world population is increasing. With the rise of socio-economic welfare, living conditions improve and people change and develop.

As a result of increasing education and communication opportunities, human and group behaviors are differentiated and the structure of the population has a different appearance in terms of age. As a result of these changing factors, interest in eco-tourism types is increasing. Eco-tourism includes cave tourism, agriculture and farm tourism, plateau tourism, bicycle tourism, river tourism, hunting tourism, camping tourism, mountain tourism, hiking, bird watching, wildlife watching, botany tourism (Gündüz & Dönmez, 2018).

Eco-tourism which envisages the protection of natural and cultural resources, is a type of tourism that focuses on the theme of studying organisms and their relations with the environment. On the basis of eco-tourism, freedom, naturalness, distance from artificiality, being intertwined with nature and distancing from stress are the basic expectations. Scientific studies within the scope of eco-tourism, which has emerged as a wide area of interest in the tourism market in recent times and where there is frequent debate about what has happened, it has become important (Gündüz & Dönmez, 2018).

Geographic Information Systems

Geographic Information systems are systems that provide graphical and non-graphical data to the user as a whole based on a specific location prepared for the purpose of analysis, querying, storing, and organizing information (Ozyavuz & Dönmez, 2014; Dönmez at. al., 2018; Kahraman, 2019; Yomralıoğlu, 2010).

Geographical Information Systems (GIS) must meet these four items effectively and accurately, including data collection, data management, data processing and data presentation (Bayzan, 2009).

A geographic information system (GIS) is a computer system for storing, managing, analyzing, and displaying geospatial data. Since the 1970s GIS has been important for professionals in natural resource management, land use planning, natural hazards, transportation, health care, public services, market area analysis, and urban planning. It has also become a necessary tool for government agencies of all the levels for routine operations. More recent integration of

GIS with the Internet, global positioning system (GPS), wireless technology, and Web service has found applications in location-based services, Web mapping, in-vehicle navigation systems, collaborative Web mapping, and volunteered geographic information. It is therefore no surprise that geospatial technology was chosen by the U.S. Department of Labor as a high-growth industry. Geospatial technology centers on GIS and uses GIS to integrate data from remote sensing, GPS, cartography, and surveying to produce useful geographic information (Chang, 2008).

Therefore, because Geographic Information System is preferred by different disciplines, it is generally referred to with three main concepts: vehicle, management and system. As a result, the Geographical Information System allows the collection of qualified graphical and non-graphical data obtained by location-based transactions in an editable and updateable manner, It is an information system that performs its functions in a manner that enables it to be stored, analyzed and presented to the user (Yomralıoğlu, 2010; Kahraman, 2019).

Application of Geographic Information Systems in Ecotourism

Since the success of any tourism business is determined by tourism planning, tourism development and research and tourism marketing, the first thing we review in this article is GIS application for tourism planning. Geographic Information Systems (GIS) is a rapidly expanding field enabling the development of applications that manage and use geographic information in combination with other media. In the tourism industry, GIS is used to provide (Chang, K. T., 2008; Jovanović,2016):

- A digital map base for printed maps
- Digital files for Internet mapping
- Digital files for mobile mapping
- Attractions map
- Website with interactive mapping

GIS technology offers great opportunities for the development of modern tourism applications using maps (Table 1). This technology integrates common database operations such as query with the unique visualization and geographic analysis benefits offered by maps. The integration of tourism data and GIS data is a big challenge for the tourism industry, today (Chang, K. T., 2008; Jovanović, 2016).

GIS operates on two data elements: spatial and attribute data. Spatial or geographical data refers to a known location on the Earth's surface. Usually this is expressed as a grid coordinate or in degrees of latitude and longitude. Most

Table 1: Definitions of GIS (Jovanović, 2016)

Properties of GIS		GIS Analytical Functions
A process	A system for capturing, storing, checking, manipulating, analyzing and displaying data, which are spatially referenced to the earth.	Presentation and thematic mapping Data query Spatial query Database integration Route finding
A toolbox	Containing tools for collecting, storing, retrieving, transforming and displaying spatial data.	
A database	Spatially referenced entities	Point in polygon analysis
An application	Cadastral information system, marketing information system, planning information system, etc.	Overlays Buffering Visualisation and 3-D modelling
A decision support system	Integrating spatial data within a problem-solving environment	

Table 2: GIS data elements (Jovanović, 2016)

GIS data elements	Description
Geographical or spatial data	Location aspects: ƒ Explicitly - using a standard geographical frame of reference such as latitude and longitude ƒ Implicitly – using surrogate spatial references such as addresses/postcode
The attribute data	Statistical and non-location data associated with a spatial entity

organizations make use of implicit geographical references as place names, addresses, postcodes, road numbers and so on; implicit spatial references can usually be geocoded into explicit spatial references (Table 2). Technological advance, particularly in software and hardware, has resulted in the development of systems which provide a range of searching, querying, presentation and analytical functions in a more user-friendly manner (Chang, K. T., 2008; Jovanović, 2016).

Results

Ecotourism generally addresses an elite audience. However, unlike this, it is a type of tourism that requires more local people to be intertwined and should benefit the local people. Therefore, if some ecotourism activities are to be carried out, these activities should have an economic contribution to the local people. In the concept of ecotourism, local people also need to play an active role in making decisions affecting themselves and their regions. This action is also a requirement of pluralistic understanding and democracy. Therefore, it is necessary to ensure the active participation of the people in an ecotourism activity to be initiated in the field of research.

It was revealed that there were not enough preliminary researches and investigations in the announcement of Culture and Tourism Development and Conservation Regions. Therefore, in the subsequent detection and announcement processes, it is recommended that adequate researches and investigations should be carried out with the participation of experts in tourism types. For example, if the subject is ecotourism, the assessments of experts from science fields such as biology, ecology, landscape architecture, forest engineering participation. Similarly, the subject of golf tourism, especially those who do golf the participation of relevant experts should be ensured. With this approach, time and money loss can be prevented (Jovanović, 2016; Altıngüzgün, K. G. H., 2019).

It also demonstrated the importance of the use of scientific methods in determining suitability for tourism types. The use of geographic information systems is recommended for determining areas for tourism in general, and for ecotourism in particular. This approach has been proven to be an appropriate and reliable approach to decision-making among alternatives.

As a result, environmentally friendly practices are needed for the sustainability of tourism in the world. These applications range from site selection to physical planning, from implementation to management.

Geographical information systems used in the selection and planning of Culture and Tourism Conservation and Development Regions and approaches that protect the ecological structure and natural and cultural landscapes have an important role in identifying and transferring them to future generations (Altıngüzgün, K. G. H., 2019).

References

Altıngüzgün, K. G. H., (2019). Gebiz kültür ve turizm koruma ve gelişim bölgesinde ekoturizm ve golf turizmine uygun alanların belirlenmesi, Yüksek Lisans Tezi, Akdeniz Üniversitesi, Fen Bilimleri Enstitüsü, Peyzaj Mimarlığı Anabilim Dalı, 91s., Antalya.

Bayzan, Ş., (2009). Gprs Verileri Yardımıyla Araç Rotalarının Belirlenmesi Problemine Farklı Bir Yaklaşım, Akademik Bilişim'09 - Xı. Akademik Bilişim Konferansı, Şubat 11-13, Şanlıurfa, 243-249.

Buckley, R. (2009). Ecotourism: Principles and Practices, CABI International Publishing, Cambridge University Press, Cambridge UK.

Chang, K. T. (2008). Introduction to Geographic Information Systems (Vol. 4). McGraw-Hill, Boston.

Dönmez, Y., Özyavuz, M., Çabuk, S. & Çorbaci, Ö. (2018). Determination of Bioclimatic Comfort Zones by Geographic Information Systems: Karabük Province, Turkey. *Journal of International Environmental Application and Science*, 13 (1), 41-49.

Düzgüneş, E. ve Demirel, Ö. (2013). Maçka Bölgesi'nin Alternatif Turizm Potansiyeli Açısından Değerlendirilmesi, İnönü Üniversitesi Sanat ve Tasarım Dergisi ISSN: 1309-9876 E-ISSN: 1309-9884 cilt/vol.3, sayı/no.7: 1-11, Malatya.

Fennel, D. A. (2001). A Content Analysis of Ecotourism Definitions, *Current Issues in Tourism*, 4(5).

Gökyer, E., Öztürk, M., Dönmez, Y., & Çabuk, S. (2015). Bartın ili dağlık alanlarında coğrafi bilgi sistemleri kullanılarak ekoturizm faaliyetlerinin değerlendirilmesi. *İnönü Üniversitesi Sanat ve Tasarım Dergisi*, 5(12), 25-35.

Gündüz, B. & Dönmez, Y. (2018). Üniversite Çalışanlarının Ekoturizm Algısı. *Bartın Orman Fakültesi Dergisi*, 20 (2), 152-162. Retrieved from https://dergipark.org.tr/tr/pub/barofd/issue/36468/440108.

Jovanović, V. (2016). The Application of GIS and Its Components in Tourism. *Yugoslav Journal of Operations Research*, 18(2).

Kahraman, S. (2019). CBS kullanılarak analiz grupları ve mekansal ağların belirlenmesi, ulaşım ve sosyal etkileşim modellerinin oluşturulması: Kayseri Erciyes Üniversitesi Kampüsü örneği, Yüksek Lisans Tezi, Erciyes Üniversitesi, Fen Bilimleri Enstitüsü, Harita Mühendisliği Anabilim Dalı, 216s, Kayseri.

Kendir, H., Arslan, E., & Asan, H. (2019). Determination of Winter Trekking Routes within the Scope of Winter Tourism Potential in Sivas Province. *Journal of Recreation and Tourism Research*, 6(3), 294-305.

Ozyavuz, M., & Donmez, Y. (2014). Monitoring the Changing Position of Coastlines Using Information Technologies, an Example of Tekirdag. *Journal of Environmental Protection and Ecology*, 15(3), 1051-1058.

Sayın, G. (2019). Kayseri ili doğal peyzajlarında ekoturizm ve görsel peyzaj kalitesi üzerine bir araştırma, Kahramanmaraş Sütçü İmam Üniversitesi, Fen

Bilimleri Enstitüsü, Peyzaj Mimarlığı Anabilim Dalı, Yüksek Lisans Tezi, Kahramanmaraş.

Sezen, I., Yılmaz, S., & Külekçi, A. E. (2011). Ekoturizm için öneri alanlarıyla Bayburt. Kahramanmaraş: I. Ulusal Akdeniz Çevre ve Orman Sempozyumu

Türkmen, F., & Dönmez, Y. (2015). Korunan alanların turizme açılmasına ilişkin yerel halkın görüşleri (Yenice örneği). Karabük Üniversitesi Sosyal Bilimler Enstitüsü Dergisi, 5(2), 189-204.

Weaver, D. B. (2001). Ecotourism in the Context Other Tourism Types, The Encyclopedia of Ecotourism, CABI Publishing, New York, USA

Yomralıoğlu, T. (2010). Coğrafi Bilgi Teknolojileri. *Bilim ve Teknik*, 617, 48-51.

Emin ARSLAN

8 Gastronomy Tourism and Geographical Indications in Tokat

Introduction

One of the physiological needs which is the need for food and drink ranks number one in Maslow's famous hierarchy of needs. As technological developments accelerate, changes incur in people's demands and expectations. As a result, the perception of physiological need has changed and turned into an ostentatious and psychological satisfaction rather than a physical satisfaction. Therefore, eating and drinking behavior has triggered the motivation of people to travel and has been influential in their choice of tourist destination. For example, it is noteworthy that many individuals with this idea and with disposable income travel from place to place only to eat. The popularity of this topic, which has also begun to be studied in national and international literature (Kivela & Crotts, 2006; Kim & Eves, 2012; Phillips et al., 2013; Şengül & Türkay, 2016; Meneguel et al., 2019; Özkoç et al., 2019) is increasing.

The number of people taking part in tourism movements around the world increases from year to year with the development and diversification of transportation opportunities. In addition to the concentration of people in mass tourism (sea-sand-sun), the concentration created in alternative tourism types has become visible (Dönmez & Türkmen, 2018). Gastronomy tourism is one of the most remarkable species within the scope of alternative tourism. In gastronomic tourism, where tourist experiences are at the forefront, modern or regional food and beverages are presented to visitors as an attraction (Richards, 2002; Hall & Sharples, 2003). The widespread use of social media in recent years and tourist experiences reaching the masses through social media has transformed gastronomic tourism into a trend.

Gastronomic tourism is defined as the journeys of tourists to discover food and beverages that are unique to a region, to enjoy by tasting and to experience different experiences. In addition, it is also considered an activity where different tastes and visual shows are displayed in the hotels or independent restaurants located in the destination, in addition to providing catering services to tourists (Rand & Heath, 2006; Kyriakaki et al., 2013). Therefore, destinations that have touristic value want to register the local products to protect them. This is called 'Geographical Indication' and is one of the important issues prioritized by destinations that want to develop gastronomic tourism (Orhan, 2010). At the same time, geographical indication, which contributes economically and culturally to touristic destinations, has been made to 7 products in Tokat province, all of which have gastronomic value. In the light of this information, the aim of the study is to evaluate the geographically indicated products of Tokat, which draws attention with its local cuisine as well as its historical, cultural and natural resources in terms of gastronomic tourism. In addition to this, the aim of this study is to contribute to the promotion, protection and creation of economic value of the geographically indicated products of Tokat and to raise awareness for the decision makers on this subject.

Geographical Indications and Gastronomy Tourism

Products which are considered local contribute to a region through local development, preservation of the traditional structure, increasing the welfare of the local people and sustaining its naturalness. The registration of these products with a 'geographical indication' and becoming a brand creates added value and increases the benefits to be obtained from these products (Suna & Uçuk, 2018). According to the Turkish Patent and Trademark Office, the geographical indication is defined as the 'registration of a product which is identified with a locality, area, region or country of origin in terms of its distinctive quality, reputation or other characteristics'. For consumers, it is a quality mark that guarantees the origin of the product, its characteristics and the connection between these features and the relevant geographical area. Thanks to the geographical indication registration, the protection of local products that have gained fame with their traditionality, raw material and quality is ensured (Turkish Patent and Trademark Office, 2019).

It is noted that gastronomic elements are among the products registered with geographical indications. As of 2019, 339 of the 412 products that have received geographical indications are gastronomic products. On the other hand, there are 3 gastronomic products from Turkey that have received geographical indications in the European Union. Furthermore, 14 gastronomic products from Turkey have been included in the geographical indication evaluation process in the European Union (Turkish Patent and Trademark Office, 2019). Therefore, these products have a very important place in terms of gastronomic tourism.

There are many studies evaluating geographical indications, gastronomic tourism and experience together in tourism literature (Orhan, 2010; Murgado, 2013; Özkaya et al. 2013; Yıkmış & Ünal, 2016; Rinaldi, 2017; Suna & Uçuk, 2018; Erik & Pekerşen, 2018; Hazarhun & Tepeci, 2018). These studies generally focus on topics such as sustainability, local development and destination marketing. Furthermore, it is suggested that the increase in the number of tourists participating in gastronomic tourism will contribute to local development and the sustainability of local products. On the other hand, it is emphasized that geographically indicated products will have a positive impact on destination marketing. Gastronomic tourism is a star that shines throughout Turkey, it is important for the development of destinations with local flavors. Gastronomic tourism is an area that is also emphasized by UNESCO. Gastronomy cities are listed in a separate category in the 'Creative Cities Network List' which was initiated by UNESCO in 2004. Gaziantep from Turkey was included in the 'Gastronomy Cities' list which is updated regularly in 2015 and Hatay was included in 2017 (UNESCO, 2019). It is anticipated that different cities from Turkey which have a rich culinary culture will be included in this list in the coming years.

Gastronomy Tourism in Tokat

Geographically Tokat is located in the Central Black Sea region of Turkey. It is bordered to the north by Samsun and Ordu provinces, Yozgat and Sivas in the south and east and Amasya in the west. Covering an area of 10.072 km2, Tokat has 11 districts (Map 1). As of year-end 2018, the total population of the province was 612.646 (TURKSTAT, 2019). The altitude of Tokat center is 623 meters on average. However, the altitude varies between 200 and 2200 m depending on the geography. All kinds of agricultural products suitable for the climate are grown in the fertile plains of Tokat. These products include vegetables (tomatoes, eggplants, peppers, cucumbers, etc.), fruits (cherries, sour cherries, peaches, apples, grapes, rosehip etc.), cereals (wheat, barley, corn), legumes (beans, chickpeas, lentils), herbs and radical plants (knotweed, sugar beets, potatoes, onions) (Tokat Governorship, 2019). These products are also the gastronomic elements of Tokat cuisine.

Tokat has many attractions in terms of tourism with its historical assets and natural beauties. The number of domestic and foreign tourists visiting Tokat was 246,634 as of 2018. This figure increases from year to year, the increase in recent years is particularly remarkable (Tokat Provincial Directorate of Culture and Tourism, 2019). Tokat, which has just started to mobilize its tourism potential, is making a breakthrough in terms of touristic operations. It is noticed that the capacity has increased thanks to the new hotel and restaurant investments. As of 2018, there were 1 5-star, 2 4-star, 7 3-star, 2 2-star and 4 tourism certified apart-hotels in Tokat. In addition, 16 tourism certified

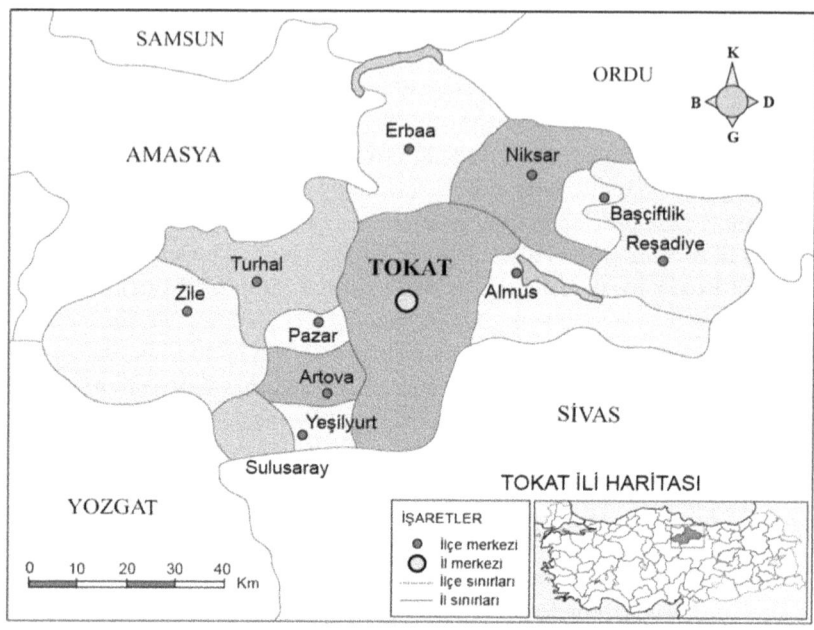

Map 1: Location of the Study Area (Coğrafya Harita, 2019)

hotels have a total bed capacity of 1513. There are 3702 beds in 64 hotels certified by the Municipality. On the other hand, there are 2 tourism-certified restaurants and numerous independent food and beverage establishments in Tokat (KTB, 2019).

The most remarkable types of tourism in Tokat are nature-based tourism, thermal tourism, faith tourism and cultural tourism. However, gastronomic tourism has also become a rising trend in Tokat like in the rest of Turkey. Some gastronomic flavors of Tokat are shown in Figure 1.

Regional gastronomic dishes of Tokat cuisine: Tokat Kebab, pita with dry cottage cheese, knotweed, bat (a dish prepared with green lentils and cracked wheat), stuffed vine leaves with bean filling, stuffed vine leaves with meat, helle soup, bacaklı soup, wedding soup, Tokat keshkek, Tokat fry-up, wrapped rice, chicken with dried okra and wedding pilaf. On the other hand, apart from those dishes, the other gastronomic flavors of Tokat region include Yogurtmach, Zile Churchkhela, Zile Molasse, grape leaves, cloth sausage, Niksar Walnut, Tokat Fenugrek, sweet tarhana, tarhana with dried yogurt, katmer (flaky pastry), oily pastry, Tokat Bagel, walnut muffin and rosehip jam (Tokat Provincial Directorate

Figure 1: Some Gastronomic Flavors of Tokat (Tokat Provincial Directorate of Culture and Tourism, 2019)

of Culture and Tourism, 2019). These gastronomic delicacies are produced in restaurants and offered at local events (weddings, celebrations and festivals) in Tokat province.

Evaluation of Gastronomic Products with Geographical Indications in Tokat Province

The successful geographical indication process in Tokat that started with Zile Molasse in 2006 and the registration of the product in 2009 paved the way for the inclusion of other products in this process. Table 1 shows the gastronomic products of Tokat that have received geographical indications and are still in the process of being evaluated.

According to Table 1, which contains geographically indicated gastronomic products of Tokat province, the geographical indications of 7 products have been registered as of 2019 and the registration evaluation process of one product is in progress. On the other hand, the first registered product was 'Zile Molasse' in 2009 while the last registered product was 'Tokat Narince Pickled Vine Leaf' in 2019. The application for the registration of Tokat Special Sausage was made in

Table 1: Tokat's Geographical Indication Registration and Gastronomic Products in Registration Process

Product name	File no.	Application Date	Registration Date	Registration no.	District
Erbaa Narince Vineyard Leaf	C2016/052	23-06-2016	05.12.2017	258	Erbaa
Niksar Walnut	C2011/033	17-05-2011	17.12.2013	177	Niksar
Cloth Sausage	C2018/142	27-06-2018	Evaluation phase	-	-
Tokat Kebab	C2013/077	26-08-2013	31.07.2015	188	Tokat Center
Tokat Narince Pickled Vine Leaves	C2017/054	10-07-2017	28.02.2019	420	Tokat Center
Turhal Yoghurtmach	C2012/117	31-07-2012	28.01.2014	181	Turhal
Zile Churchkhela	C2014/024	14-03-2014	01.11.2017	224	Zile
Zile Molasse	C2006/029	20-10-2006	17.11.2009	118	Zile

Source: Turkish Patent and Trademark Office, 2019

2018 and the evaluation process is ongoing. The evaluations of the products are given in the table below.

Table 2: Erbaa Narince Vineyard Leaf

Product name: Erbaa Narince **Applicant:** Vineyard Leaf **Usage form:** Erbaa Chamber of Trade and Industry Labeling	

Source: Turkish Patent and Trademark Office, 2019

Narince grape leaves are among the best edible vine leaves in Turkey in terms of quality and production. Most of the product is produced in Tokat's Erbaa district in Turkey and it makes a significant contribution to the region in economic terms. The distinctive features of the Erbaa Narince Vine Leaves, as shown in Table 2 are manifested by their pleasant taste and the surface structure which is smooth, fine and has minimal veins (Turkish Patent and Trademark Office, 2019). Furthermore, it differs from other types of vine leaves because it cooks rapidly and contains dietary fiber. This product, which is also subject to academic studies (Kızılaslan & Somak, 2013; Cangi & Yağcı, 2017), has a high gastronomic potential. It is a product that is preferred for local Tokat cuisine such as bat, stuffed vine leaves with beans, stuffed vine leaves with meat and also for foods throughout Turkey known as stuffed vine leaves. Therefore, it is the raw material of some dishes that tourists coming to the region are curious about and want to taste. This feature makes it valuable in terms of gastronomy tourism.

The most important difference of Niksar Walnut seen in Table 3 from other walnuts is that it consists of only regional qualities and is a preferred species in the food industry. Niksar Walnut, which is grown in the region's unique climatic conditions and suitable irrigation methods, has a more oily structure than other walnuts in terms of taste and quality. Niksar Walnut, which contains an average of 65 % oil, is at the forefront in this regard (Turkish Patent and Trademark Office, 2019). According to a study carried out about Niksar Walnut, it is emphasized that producers prefer this product because of the high income it generates, ease of production and marketing superiority (Kilci &

Table 3: Niksar Walnut

Product name: Applicant: Usage form:	Niksar Walnut Niksar Chamber of Trade and Industry Labeling	

Source: Turkish Patent and Trademark Office, 2019

Kızılaslan, 2016). Therefore Niksar Walnut is a product with advantages. On the other hand, Niksar Walnut is known to add flavor to many local dishes such as desserts and pastries. This increases the importance of the Niksar Walnut in terms of gastronomic tourism.

Table 4: Tokat Kebab

Product name: Applicant: Usage Form:	Tokat Kebab Tokat Chamber of Trade and Industry Branding		

Source: Turkish Patent and Trademark Office, 2019

Tokat Kebab is one of the most special dishes of Tokat. It is quite laborious to prepare, however the flavor is just gorgeous. Tokat Kebab is made using lamb, tail fat, pepper, eggplant, tomato, potato, onion, garlic and thin pita bread. The kebab is cooked in special cookstoves which are unique to the region and the first samples of which date back 300 years. Meat and vegetable skewers are hung vertically on the horizontal iron in the center of the cookstove. They are cooked for 20–25 minutes with the fire of the firewood burning at the edges of the cookstove. There is a tray at the bottom of the cookstove and the fat dripping from the cooking meat is collected in this tray. Specially cooked thin pita bread is spread on a different tray. The meat and vegetables cooked on the cookstove are placed on the pita bread. Cooked tomatoes are placed in the center of the tray. The fat collected in the tray under the cooker is spread thoroughly over the kebab and the kebab is served (Sağır, 2012: 2689).

The main distinguishing features of Tokat Kebab which is shown in Table 4 are as follows: (Turkish Patent and Trademark Office, 2019):

- The meat of 6–9 male lambs raised in the highlands of Tokat is used
- Tokat's local green pepper is used,
- A unique cookstove and cooking technique is used,
- Visuality and presentation of the served kebab

Produced and presented in many restaurants throughout the province, Tokat Kebab is the dish of choice for tourists coming to Tokat in terms of gastronomic tourism.

Table 5: Tokat Narince Pickled Vine Leaves

Product name:	Tokat Narince
Applicant:	Pickled
Usage Form:	Vine Leaves Tokat Chamber of Agriculture Branding, Labeling

Source: Turkish Patent and Trademark Office, 2019

In Tokat, vineyard culture and grape production dates back to the Roman period. The leaves of Narince grapes, which are unique to Tokat, have been used in the local area since ancient times. After harvesting, Tokat Narince vine leaves are fermented in brine of a certain concentration for a while. The product thus obtained is packaged and offered for sale. 15 companies in Tokat produce this product. The most important characteristics of the leaf seen in Table 5 are that it cooks rapidly, it has a fine, soft, smooth structure which is less fragmented, less veined and yellow in color (Turkish Patent and Trademark Office, 2019). This product is used in the production of many dishes specific to Tokat. Therefore, it has an important place in terms of gastronomic tourism.

Yoğurtmach is made only in Tokat-Turhal region in Turkey. The name of the product comes from the words kneading and folding. Yogurtmach is a traditional product that was offered to guests on holidays, celebrations and special occasions before it became a commercial product. The product is made by combining and

Table 6: Turhal Yoghurtmach

Product name:	Turhal Yoghurtmach	
Applicant:	Turhal Chamber of Trade	
Usage Form:	and Industry	
	Branding	

Source: Turkish Patent and Trademark Office, 2019

shaping the filling materials such as poppy and walnut together with the dough with its unique preparation method and then baking it in a stone oven. Turhal Yogurtmach, whose geographical indication is shown in Table 6, applied for registration in 2012 and was registered in 2014 (Geographical Indication Portal, 2019; Gürel, et al. 2016). Due to its unique flavor, yoghurtmach is a gastronomic product that must be experienced by local and foreign tourists visiting the region.

Table 7: Zile Churchkhela

Product name:	Zile	 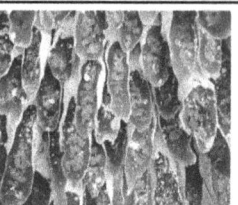
Applicant:	Churchkhela	
Usage Form:	Zile Chamber	
	of Trade and	
	Industry	
	Labeling	

Source: Turkish Patent and Trademark Office, 2019

Zile Churchkhela is a sweet and local food product prepared with the fermented juice of Narince grapes, wheat starch and flour containing walnuts on a string. The walnuts beaded on a 40–50 cm string are grown in Tokat Province. Since the region is between the Black Sea and the continental climate, it has a unique transitional climate. This climate structure leads to a unique taste, aroma, color and texture in walnuts. The thick and sticky gel into which the walnuts are dipped for Zile Churchkhela is called 'hasuda'. 'Hasuda' is prepared with Narince fermented grape juice and starch. Zile Churchkhela, whose geographical indication and shape is shown in Table 7, was registered in 2017 (Geographical

Indication Portal, 2019). As it is a durable and sweet food, it is a souvenir product in terms of gastronomic tourism.

Table 8: Zile Molasse

Product name:	Zile Molasse
Applicant:	Zile Trade
Usage Form:	Exchange Branding

Source: Turkish Patent and Trademark Office, 2019; Tokat Province Culture and Tourism Directorate, 2019

Zile Molasse is the first gastronomic product of Tokat registered with a geographical indication. As shown in Table 8, Zile Molasse is stored in a wooden container and has a rather solid consistency. Molasse produced from local grapes are whitened using egg whites. Zile Molasse differs completely from the other molasse in Turkey because it is white. Production mode, solidity and its white color are the most important distinguishing features of the product (Geographical Indication Portal, 2019). Because of this feature, Zile Molasse has been the subject of articles published in prestigious academic journals. For example, the white color of Zile Molasse was mentioned in an article published in Food Chemistry Magazine in 2003 (Tosun & Ustun, 2003). Therefore, Zile Molasse is one of the most popular gastronomic products of Tokat as well as one of the most demanded products by visitors to the region.

SWOT Analysis of Tokat's Geographically Indicated Products and Gastronomic Tourism

A SWOT analysis was carried out in order to evaluate Tokat's geographically indicated products more effectively in terms of gastronomic tourism. SWOT analysis is also known as a situation analysis. The scope of the SWOT analysis is to identify the 'Strengths and Weaknesses' of a situation. On the other hand, it is also possible to define the 'Opportunity and Threat' elements that may arise within the same situation. This method of analysis is preferred in order to develop strategies and provide suggestions from an academic angle (Dyson, 2004: 632).

Table 9: SWOT Matrix for Tokat's Geographically Indicated Products and Gastronomy Tourism

STRENGTHS	WEAKNESSES
- Registration of 7 gastronomic products of Tokat by geographical indication - Geographically indicated gastronomic products are still being produced in Tokat by traditional methods - Activities such as hotels and recreation facilities to improve tourism in Tokat - Completion of the restoration works of historical and touristic assets to a great extent - Tokat has a tourist attraction like Ballıca Cave	- Tokat is not promoted sufficiently in terms of tourism - Tokat Airport is not yet operational - Tokat lacks a Destination Management Organization for tourism
OPPORTUNITIES	**THREATS**
- The number of geographically indicated products may increase due to the numerous gastronomic products specific to Tokat Cuisine - Potential use of historical buildings for which restoration activities have been completed or are ongoing in gastronomy tourism - Planning for the opening of the airport in mid-2020 - Creating an awareness of the geographical indication in the local administration of Tokat - Expansion of the use of social media in the promotion of gastronomy tourism	- Competitive destinations investing and promoting activities for gastronomic tourism - Negative effects of an economic crisis can be experienced since gastronomy tourism targets spending tourists - The deterioration of the original structure of some local products in Tokat through industrial production - The negative impact of global warming and climate changes on the world agenda on gastronomic products

The SWOT matrix in Table 9 contains the strengths and weaknesses, opportunities and threats for Tokat's gastronomic tourism and geographically indicated products. While the most prominent strengths were Tokat's tourism assets and gastronomic values, the weaknesses were lack of publicity and transportation problems. Opportunities for tourism in the future of Tokat are listed as planned tourism related investments and the potential for numerous geographically indicated products, while competing destinations, environmental and economic problems come to the fore in terms of threats.

Conclusion and Recommendations

As a result of its historical accumulation and geopolitical importance, Turkey has hosted many civilizations. This feature has given Turkey a very rich cultural heritage with both tangible and intangible assets. Therefore, this rich cultural heritage is the main reason for the diversity of cuisine and local products in many regions. The wealth of culinary culture in Turkey has pioneered the process of registering local gastronomic products with geographical indications. In recent years, local products, especially gastronomic delicacies in different regions are being registered with the geographical indication by the Turkish Patent and Trademark Office.

It is frequently stated in academic studies that the registration of gastronomic products with a geographical indication has a positive effect in terms of gastronomic tourism (Murgado, 2013; Özkaya, et al. 2013; Hazarhun & Tepeci, 2018; Meneguel, et al. 2019). Thanks to the geographical indication system, gastronomic values are protected, local agricultural activities are supported, destinations are promoted and tourists come to the destinations. Thus, gastronomy tourism movements can be sustained in a more planned and orderly manner.

Gastronomic tourism is a rising star in Tokat like it is throughout Turkey. Tokat Province, which is very rich in cuisine and regional delicacies, wants to have a share in gastronomic tourism. Tokat has registered 7 gastronomic products so far and by increasing the number of these products, Tokat may be included in the scope of 'UNESCO Gastronomic Cities' in the future. It can be said that the destinations included in the UNESCO Gastronomic Cities attract much more attention in terms of gastronomic tourism, their image is positively affected, and the level of development and prosperity increases (Cunningham, 2002; Smith & Warfield, 2008). At this point, promotion and sustainability activities are rather important. Therefore, decision-makers in Tokat have great responsibilities in this regard.

Improving the transportation infrastructure is very important in order to facilitate the gastronomic tourism movements in Tokat. Ongoing investments in air and highway transportation should be completed and commissioned immediately. Furthermore, a destination management organization must be established urgently in Tokat. This way, tourism and especially gastronomy and cultural tourism can take advantage of all the attractions throughout the province. On the other hand, the authenticity of local gastronomic values should be maintained. Measures should be taken to ensure that the authenticity of these products does not deteriorate due to industrialization. Therefore, local decision makers and stakeholders (governors, municipalities, district governorships, universities and NGOs) should come together and create a gastronomic tourism master plan and road map for Tokat in order to protect geographically indicated products. Finally, increasing the number of academic studies on the tourism potential of Tokat will contribute to a scientific approach and analysis of the subject. Studies should be encouraged and supported by relevant institutions.

References

Cangi, R. & Yağcı, A. (2017). Bağdan sofraya yemeklik asma yaprak üretimi. *Nevşehir Bilim ve Teknoloji Dergisi*, 6, 137–148.

Cunningham, S. D. (2002). From cultural to creative industries: Theory, industry, and policy implications. Media international Australia incorporating culture and policy. *Quarterly Journal of Media Research and Resources*, 102(1), 54–65.

Dönmez, Y. & Türkmen, F. (2018). The relation between the landscape design and brand image in purchase preferences of tourists: The case of Safranbolu and Nevşehir, in Turkey. *Applied Ecology and Environmental Research*, 16(1), 629–643.

Dyson, R. G. (2004). Strategic development and SWOT analysis at the University of Warwick. *European Journal of Operational Research*, (152), 631–640.

Erik, U. & Pekerşen, Y. (2018). Konya ilinin coğrafi işaretli gastronomik ürünlerinin bölge turizminin gelişimi açısından değerlendirilmesi. *Journal of Social and Humanities Sciences Research*, 5(31), 4866–4877.

Geographical Indication Portal. (2019). *Türkiye'nin Coğrafi İşaretleri*. https://www.ci.gov.tr/cografi-isaretler/liste?il=60. D. A. 10.11.2019.

Gürel, E., Gürler, A. Z., Nabalı, B. & Ayyıldız, B. (2016). *Coğrafi işaretlerin kırsal kalkınma açısından değerlendirilmesi: Tokat ili örneği*. 12. Ulusal Tarım Ekonomisi Kongresi. Süleyman Demirel Üniversitesi, Ziraat Fakültesi. 25–27 May 2016, Isparta.

Hall, C. M. & Sharples, L. (2003). *The consumption of experiences or the experience of consumption? An introduction to the tourism of taste*. In: Hall, C. M., Sharples, L., Mitchell, R., Macionis, N., Cambourne, B. (Eds.), Food Tourism around the World. Elsevier Butterworth-Heinemann, Oxford, pp. 1–24.

Hazarhun, E. & Tepeci, M. (2010). Coğrafi işarete sahip olan yöresel ürün ve yemeklerin Manisa'nın gastronomi turizminin gelişimine katkısı. *Güncel Turizm Araştırmaları Dergisi*, 2(ek.1), 371–389.

Kızılaslan, N. & Somak, E. (2013). Tokat ili Erbaa ilçesinde bağcılık işletmelerinde tarımsal ilaç kullanımında üreticilerin bilinç düzeyi. *Gaziosmanpaşa Bilimsel Araştırma Dergisi*, 4, 79–93.

Kilci, M. & Kızılaslan, H. (2016). *Tokat ili Niksar ilçesi ceviz üretim ve pazarlama yapısı*. 12. Ulusal Tarım Ekonomisi Kongresi. Süleyman Demirel Üniversitesi, Ziraat Fakültesi. 25–27 Mayıs 2016, Isparta.

Kim, Y. G. & Eves, A. (2012). Construction and validation of a scale to measure tourist motivation to consume local food. *Tourism Management*, 33(6), 1458–1467.

Kivela, J. & Crotts, J. C. (2006). Tourism and gastronomy: Gastronomy's influence on how tourists experience a destination. *Journal of Hospitality & Tourism Research*, 30(3), 354–377.

KTB (2019). *Yatırım ve İşletmeler Genel Müdürlüğü Turizm İstatistikleri*, https://yigm.ktb.gov.tr/TR-9851/turizm-istatistikleri.html. D. A. 05.11.2019.

Kyriakaki, A., Zagkotsi, S. & Trihas, N. (2013). *Creating authentic gastronomic experiences for tourist through local agricultural products: The "Greek breakfast" project*. 5th International Scientific Conference- Tourism Trends and Advances in the 21st Century. Rhodes: University of the Aegean.

Meneguel, C. R. A., Mundet, L. & Aulet, S. (2019). The role of a high-quality restaurant in stimulating the creation and development of gastronomy tourism. *International Journal of Hospitality Management*, 83,220–228.

Murgado, E. M. (2013). Turning food into a gastronomic experience: Olive oil tourism*Options Mèditerranèennes*, 106, 97–109.

Orhan, A. (2010). Yerel değerlerin turizm ürününe dönüştürülmesinde "Coğrafi İşaretlerin" kullanımı: İzmit Pişmaniyesi Örneği. *Anatolia: Turizm Araştırmaları Dergisi*, 21(2), 243–254.

Özkaya, F. D., Sünnetçioğlu, S. & Can, A. (2013). Sürdürülebilir gastronomi turizmi hareketliliğinde coğrafi işaretlemenin rolü. *Journal of Tourism and Gastronomy Studies*, 1(1) 13–20.

Özkoç, A. G., Arslan, E., Kendir, H. & Erdoğan, T. (2019). Otel İşletmelerinde Yeşil Mutfak Kalitesinin (Y-Mutkal) Ölçülmesi: Nevşehir İlinde Bir Araştırma. *Journal of Tourism and Gastronomy Studies*, 7(3), 2294–2309.

Phillips, W. J., Asperin, A. & Wolfe, K. (2013). Investigating the effect of country image and subjective knowledge on attitudes and behaviors: US upper midwesterners' intentions to consume Korean food and visit Korea. *International Journal of Hospitality Management*, 32, 49–58.

Rand, G. E. & Heath, E. (2006). Towards a framework for food tourism as an element of destination marketing. *Current Issues in Tourism*, 9(3), 206–234.

Richards, G. (2002). *Gastronomy: an essential ingredient in tourism production and consumption?* In: Hjalager, A.–M., Richards, G. (Eds.), Tourism and Gastronomy. Routledge, London, 3–21.

Rand, G. E. & Heath, E. (2006). Towards a framework for food tourism as an element of destination marketing. Current Issues in Tourism, 9(3), 206–234.

Rinaldi, C. (2017). Food and gastronomy for sustainable place development: a multidisciplinary analysis of different theoretical approaches. *Sustainabilty*, 9(10). 1–25.

Sağır, A. (2012). Bir Yemek Sosyolojisi Denemesi Örneği Olarak Tokat Mutfağı. *Turkish Studies - International Periodical for The Languages, Literature and History of Turkish or Turkic*, 7(4), 2675–2695.

Smith, R. & Warfield, K. (2008). The creative city: a matter of values. In: P. Cooke & L. Lazzeretti (Eds.), Creative Cities, Cultural Clusters and Local Economic Development, (287–312), Cheltenham: Edward Elgar Publishing.

Suna, B. & Uçuk, C. (2018). Coğrafi işaret ile tescil edilmiş ürüne sahip olmanın destinasyon pazarlamasına etkisi. *Journal of Tourism and Gastronomy Studies*, 6(3), 100–118.

Şengül, S. & Türkay, O. (2016). Yerel mutfak unsurlarının turizm destinasyonu seçimindeki rolü (Mudurnu Örneği). *Uluslararası Yönetim İktisat ve İşletme Dergisi*, 12(29), 63–87.

Tokat Governorship. (2019). *Tokat'ta Toprak, Tarım, Su, Coğrafya, Turizm ve Dahası...*http://www.tokat.gov.tr/tokatta-tarim-toprak-ve-turizm D. A. 05.11.2019.

Tokat Provincial Directorate of Culture and Tourism. (2019). *Genel Bilgiler.* https://tokat.ktb.gov.tr/TR-60574/genel-bilgiler.html D. A. 05.11.2019.

Tosun, I. & Ustun, N. S. (2003). Nonenzymic browning during storage of white hard grape pekmez. *Food Chemistry*, (80), 441–443.

TURKSTAT. (2019). *Adrese Dayalı Nüfus Kayıt Sistemi.* https://biruni.tuik.gov.tr/medas/?kn=95&locale=tr D. A. 04.11.2019.

Turkish Patent and Trademark Office. (2019). Coğrafi İşaret ve Geleneksel Ürün Adı İstatistikleri. https://www.turkpatent.gov.tr/TURKPATENT/geographicalRegisteredList/ D. A. 04.11.2019.

UNESCO. (2019). UNESCO creative cities network. https://en.unesco.org/creative-cities/creative-cities-map D. A. 11.10.2019.

Yıkmış, S. & Ünal, A. (2016). The importance of geographical indication in gastronomy tourism: Turkey. *International Journal of Agricultural and Life Sciences*, 2(4), 73–79.

Mehmet TEKELİ and Ezgi KIRICI TEKELİ

9 Sustainable Gastronomic Tourism

Introduction

The main goal of gastronomy is to establish a sustainable gastronomy that incorporates the traditional principles of sustainable development. Sustainable gastronomy ensures the social and economic development of communities while making an eco-nurturing commitment to the optimal health of community members by ensuring the environmental sustainability of communities. In this sense, sustainable gastronomy and eco-gastronomy have the same meaning and can interact with all other components of a developing system as seen in sustainable tourism. On a more specific level, sustainable gastronomy is about producing environmentally sensitive food and preparing and nourishing both the mind and the body. In this sense, food quality increases with sustainable gastronomy, it becomes more favorable to health and this contributes positively to individuals and societies. Therefore, access to quality food has become an important issue in current conditions. In this context, sustainable gastronomy has a close association with tourism. Tourists have the chance to experience the culinary culture of the destinations they visit and therefore food safety is important. For this reason, gastronomy studies have become very important in terms of ensuring future food safety for related industries such as the tourism sector. In this section, firstly the concepts of sustainable tourism and gastronomic tourism have been explained, the outlines of sustainable gastronomic tourism and its components have been examined and the 'slow food movement' related to sustainable gastronomic tourism has been included.

1. The Concept of Sustainable Tourism

Together with globalization, sustainability, which is one of the important issues discussed in the world in recent years, has spread to almost all fields including tourism. Although these fields differ from each other, they actually have one thing in common. This common point is that sustainability is based on the fact of 'Human Future' (Bayram, 2016). Sustainability has derived from the Latin word 'sustinere' and passed into our language meaning 'continuing, sustaining, providing, supporting' and used to indicate 'the ability to sustain a situation or process' (Işıldar, 2016). Sustainability is defined as 'The ability of a society, ecosystem, or any other system with continuity to be maintained without interruption, disruption, overuse, or overloading vital resources' (Bayram, 2016). In other words, sustainability is also described as 'a participatory process that ensures the prudent use of all resources (social, cultural, natural and human) and establishes a social outlook on the basis of sustaining it' (Gladwin, Kennelly & Krause, 1995).

With the Industrial Revolution, which started in the late 19th century, economic, social, environmental, cultural and technological developments have taken place in England. The rapid industrialization efforts that accompanied these developments have caused irreparable damage. As a result of these developments in the 1970s, various meetings were held under the leadership of international organizations such as the 'United Nations and the World Bank' which resulted in the emergence of concepts such as 'Sustainability and Sustainable Development' to prevent such damage (Bayram, 2016). Subsequently, the 'United Nations World Commission on Environment and Development' (1987) published a report titled 'Our Common Future', stating that the ecological, cultural and socio-economic resources that people possess and need are unique. After the 'United Nations World Commission on Environment and Development', the idea of sustainable development was further advanced and disseminated with the 'Brundtland Report' (Brundtland, 1987). The concept of 'Sustainable Development' emerged and was adopted based on these statements.

Sustainable development; aims to protect and sustainable use of resources for future generations to use resources; it is defined as 'without destroying opportunities to meet the needs of future generations; the development process in which the needs of the current generation can be met' (Brundtland, 1987).

Three generally accepted dimensions have been used since sustainable development was adopted, discussed and used. These dimensions are classified as economic, social and environmental (Holmberg & Sandbrook, 1992):

Economic: An economically sustainable system must produce products on an ongoing basis, maintain governmental and external debt manageability and avoid sectoral imbalances that damage agricultural and industrial production.

Social: In a socially sustainable system social services such as equality distribution, equality in health and education, gender equality, political responsibility and participation must be delivered at an adequate level.

Environmental: An environmentally sustainable system must avoid the exploitation of renewable resource systems or environmental investment functions without changing the resource base, and consume non-renewable resources only after they have been replaced adequately with new investments. This process should include conservation of biodiversity, atmospheric balance and other ecosystem functions that are not classified as economic resources.

Based on the above definitions and explanations of sustainability, it is accepted that human beings are not the sole focal point in the present day, that social and environmental variables are of great importance in the continuation of vital activities and that participation in supporting sustainable development in all areas is important. The effective use of various principles of sustainability such as honesty, democracy and rights in areas such as economy, ecosystem management, urbanization, industry, agriculture, pollution is not enough; they must also be used effectively in areas such as business, management, law, politics, psychology, and tourism; and strategies covering these areas must be developed (Işıldar, 2016).

The basic principles of sustainability apply in almost all areas of tourism. The main framework of sustainable tourism consists of the preservation of the environment, which can be considered as the main source of tourism, and the use of all socio-economic resources included in tourism in such a way that they are transferable to future generations (Swarbrooke, 1999).

Tourism is one of the sectors that need sustainability the most with its intensive environmental elements. The basis of sustainable tourism is to improve tourism and product quality without adversely affecting the physical and human environment of people (Cronin, 1990). Therefore, sustainable tourism emerges as 'a tourism approach that fully takes into account the current and future economic, social and environmental impacts that counter the needs of visitors, the industry, the environment and host communities' (UNWTO, 2005).

The concept of sustainable tourism is defined by the 'World Tourism Organization' as 'a form of development that preserves the essence of the environment where human interaction occurs without destroying it as well as its cultural integrity, ecological processes, biological diversity and natural life, while all resources are managed in a way to counter the economic, social and aesthetic needs of the people and tourists in the visited region and to meet the same needs of future generations' (United Nations Environment Program and World Tourism Organization, 2005).

Sustainable tourism is a positive approach aimed at reducing the tension and friction caused by the complex relations between the tourism sector and tourists, the environment and the local population. Furthermore, sustainable tourism needs to be actively adopted to ensure the sustainability of both natural and cultural resources (Bramwell & Lane, 1993).

The continuity of tourism movements in a country or region generally depends on some natural, socio and cultural factors. If the sustainable development of these resources is provided and the continuity of tourism movements is desired, it is necessary to manage the economic needs of the sector and the experiential needs of the tourists, as well as to ensure the continuity of the cultural structure, preservation of biodiversity and life support systems (Harris & Leiper, 1995).

Sustainable tourism is the preservation, development and sustainability of natural, socio-cultural, historical and artistic values which are the main attraction sources of tourism. Sustainable tourism, which is not seen as a type of tourism, which should be dealt with in the context of all tourism activities, is a tourism approach that reduces tourism costs, increases its benefits and can be applied indefinitely for the environment and society. In addition, sustainable tourism is regarded as a process that protects local products, supports producers and promotes greater ownership of tourism (Ayyıldız & Kargıglıoğlu, 2018).

The United Nations Environment Program and World Tourism Organization (2005) have set out the objectives of sustainable tourism. These objectives are economic viability, economic welfare, employment quality, community welfare, social justice, visitor satisfaction, cultural wealth, productivity of natural resources, biodiversity and environmental cleanliness. In order to achieve these objectives, it is necessary to adopt some policies in order to increase tourism capacity and the quality of touristic products.

For this purpose, alternative tourism types have an important role in protecting and increasing the quality of touristic resources. Among such types of tourism gastronomic tourism, where local, authentic and organic food and food varieties are the main determinants, has a structure that supports tourism especially in terms of sustainability and has a special importance in this regard.

2. The Concept of Gastronomic Tourism

The nutritional needs of individuals are among the indispensable requirements for human physiology just like breathing. However, the phenomenon of eating has important and different meanings in addition to vital activities. Food has been considered as an indicator of wealth in many societies, played an important role in many rituals and has attested to the efforts of humans to civilize (Özgen, 2013). While eating was perceived as a visual expression of friendship and sharing among individuals, eating activities in the Middle Ages and post-industrial societies were accepted as an important indicator of socialization.

With the increase in social development and welfare level, the idea of eating is now considered with the concept of 'gastronomy' and has led to the emergence of an alternative type of tourism which is now referred to as 'gastronomic tourism' (Akdağ et al., 2016).

The first use of the concept of gastronomy dates back to the ancient Greek period. In the 4th century, Sicilian Greek Archestratus wrote a book on food and wine in the Mediterranean region and called the book 'Gastronomia'. The concept of gastronomy is included in several chapters of this book (Santich, 2004).

About 200 years after the discovery of the concept of gastronomy, the Greek author Athenee wrote several articles on food and drink. However, during the Greek and Roman Empire, some writers' research on food and beverages was not entirely gastronomic (Scarpato, 2002).

Some researchers have argued that the starting point of gastronomy is 19th century 'French Cuisine'. According to Larousse Gastronomique, the concept of gastronomy started to be used in the literature with Joseph Bercholux's study from 1801 titled 'Gastronomie ou L'Homme des Champs a Table (Gastronomy or Humans from the Field to the Table)' (Göker, 2011).

'Grimod de la Reyniere', the founder of the gastronomy newspaper, published 'Almanachs des Gourmands', one of the best-selling publications in 1804. This study describes what the best foods and drinks are, when and how to prepare them (Santich, 2004).

'Manuel Des Amphitryons' introduced the concept of gastronomy as a field of study in 1808. In addition, by addressing the development of the science of gastronomy, he stated that gastronomy departments will be opened soon in universities (Santich, 2004; Göker, 2011).

Gastronomy was defined by 'Charles Monselet' as 'the art of eating good food' and was added to the thesaurus and French culinary literature by the French Academy in 1835 (Lang, 1988; Karim, 2006).

A French lawyer and judge, 'Brillat-Savarin' enriched the concept of gastronomy and made the word available to the public. The work 'Physiologie du Gout (A Handbook of Gastronomy)' written by Brillat-Savarin in 1826, was not only translated into English, German and Spanish but the subject of journals such as 'Le Gastronome (1830–1831), La Gastronomie (1839–1841), Il Gastronomo İtaliano (1866) and Le Gastronome (1872–1873)'. This book includes recommendations on menu planning, food selection, wine selection as well as general guest hospitality. In other words, 'Physiologie du Gout (A Handbook of Gastronomy)' addressed both the concept of gastronomy and evaluated the concept of gastronomy in theoretical and practical terms (Santich, 2004; Göker, 2011).

In terms of word origin, according to Scarpato (2000), the concept of gastronomy derives from the words 'gastros' (stomach) and 'nomos' (laws, rules) in ancient Greece and includes rules about eating and drinking. In the past the concept of gastronomy was expressed as 'fancy dishes for show' (Barkat & Vermignon, 2006). Nowadays, gastronomy is 'a concept that reflects the preparation, production, presentation and eating characteristics of food' (Scarpato, 2000). According to another definition gastronomy is 'the discipline that examines the art of eating' (Kivela & Crotts, 2006). While Richards (2002) defined gastronomy as 'the reflection of the preparation, cooking, presentation and consumption of food, Gillespie & Cousins (2001) describe this concept as 'a science that explores how to enjoy food more and how to exceed the limit of this pleasure'.

In many sources, gastronomy has been considered as 'the interaction of components related to eating and drinking' (Hegarty & O'Mahony, 2001; Santich, 2004; Hegarty, 2005; Hegarty & Antun, 2007). Hegarty & O'Mahony (2001) describe gastronomy as 'a concept that includes the basic food ingredients used in the preparation of the food, the methods of food storage, preparation and cooking, the types and quantities of food and beverages, popular and unpopular flavors, the traditions and customs of food and beverage presentation, the tableware used and the beliefs about food and beverage'. Santich (2004) defines gastronomy as 'providing recommendations and guiding on what, where, when, how and with what ingredients to eat and drink based on historical, cultural and environmental influences'. Hegarty (2005) and Hegarty & Antun (2007) consider gastronomy as a process. Gastronomyis 'a process for individuals to enjoy nutritious and microbiologically safe food and drink with satisfaction' (Hegarty, 2005). In other words, gastronomy is 'the selection, assembly, preparation, processing and service of food and beverages for human consumption' (Hegarty & Antun, 2007).

The concept of gastronomy includes similar definitions and different definitions depending on the countries or authors. In Greek it is defined as 'a pleasure', in the United States it is defined as 'the art and science of eating', in German it is defined as 'the art of cooking', in Italian it means 'applications related to food' and in Spanish it is defined as 'the preparation of good and luxurious food' (Anthelme & Savarin, 2016).

Considering the common characteristics of the definitions related to gastronomy it is possible to define gastronomy 'an art and science branch, which is a reflection of certain cultures on the basis of food preparation, cooking, presentation and eating and drinking experience'. Since gastronomy is a branch of science, it has certain rules. When it is evaluated in terms of art, it is associated with pleasure and aesthetics (Sarıışık & Özbay, 2015).

Gastronomy has its place in economic, social, cultural, ideological and political structures. Le'vi-Strauss (1966) likens the way a society cooks to unconsciously translating it into a language. He also states that there is no community of people who do not use a language in the world and neither is there a society without a culinary culture. Anthelme & Savarin (2016) emphasizes the importance of food culture in society with the phrase 'Tell me what you eat, and I will tell you who you' in his book 'Physiology of Flavor'.

It is evident that gastronomy is associated with many disciplines. Therefore, it is considered interdisciplinary, multidisciplinary and even transdisciplinary. In addition, gastronomy has both scientific and artistic aspects (Gürsoy, 2017).

As a branch of science, gastronomy makes use of both science as well as social sciences. At the same time these areas offer various research topics. The qualifications of foodstuffs, nutrients, taste, hygiene, sanitation and similar issues are evaluated in the context of their relationship with science. The psychology of why some foods or beverages are liked is studied in the context of social sciences, sociology looks into the sharing of food and beverages, economy deals with the motivation of eating and drinking out, history and archeology study the evolution of kitchen utensils, marketing is about menu presentations and customers' satisfaction with the service provided while theology has the last word regarding which foods can or cannot be consumed for religious reasons (Shenoy, 2005). It is possible to increase these disciplines and research subjects, and one of them is tourism.

In the literature, gastronomic tourism used to express food and beverage oriented tourism activities is also expressed with various concepts such as 'Culinary tourism' and 'food tourism'. The concept of 'gastronomic tourism' which is preferred by many researchers has been used in this study (Üner & Şahin, 2016).

The concept of gastronomic tourism has been defined by Long as the idea of tourists experiencing different culinary cultures under the heading of 'Culinary Tourism' (Long, 2005). Gastronomic tourism is the presentation of 'food and beverages as an element of attraction to people visiting a region' (Wolf, 2006). Gastronomic tourism is generally defined as 'a concept defining people's experiences about food and drinks specific to a region' (Karim & Chi, 2010).

Hall et al. (2003) describe gastronomic tourism as 'People visiting food festivals, restaurants and special areas to savor a special type of food or to observe the preparation of a meal or experience food'. In addition, the experience of seeing the different production processes of dishes, tasting a special dish or eating from the hands of a famous chef can also be included in gastronomic tourism. In other words, gastronomic tourism is defined as 'trips made to regions with rich gastronomy resources and enjoy leisure time or have fun while visiting primary and secondary producers of gastronomy products, gastronomy festivals, fairs, events, cooking demonstrations, food flavor testing or any food-related activity' (Lee, Packer & Scott, 2015).

Based on the definitions regarding gastronomic tourism, the 'Gastro tourist' is defined as 'a special interest tourist whose main travel motivation is to have gastronomic experiences and participate in food and beverage activities at a destination' (Üner, 2014) or 'those who have the opportunity to experience these activities' (Uyar & Zengin, 2015).

Hjalager (2002) considers gastronomic tourism as a type of tourism related to new efforts in tourism to achieve year round continuity of tourism and thus sustainability. In fact, gastronomic tourism refers to gastronomic mobility. This includes travels by tourists to taste local food and beverages, to see and experience the local food and beverage culture (Kivela & Crotts, 2005; Kivela & Crotts, 2006). Gastronomic tourism, which is considered as a type of tourism, offers tourists the opportunity to familiarize themselves and experience cultural and regional flavors, smells, structures on their holiday and enjoy a unique eating and drinking experience. In this respect, gastronomic tourism is an important tool in learning about the culture of a society (Yüncü, 2010). In addition, it is a growing sector in which tourists can gain experience in the preparation of food and beverages belonging to different regions and take part in culinary traditions (Richards, 2002). Therefore, gastronomic tourism is a kind of alternative type of tourism realized with the aim of 'observing the production process of foods belonging to different cultures, tasting different foods and beverages, discovering the culinary traditions of different cultures and their eating styles' (Long, 2005; Karim & Chi, 2010).

Gastronomic tourism includes four types of motivation; 'physical, cultural, social and prestige'. According to the first type of motivation which is physical motivation, food is required for the continuity of life and the individual has to eat to live or travel. The second is cultural motivation, and tourists need to eat to learn a certain culture. In this case, simple local food and beverages can motivate tourists. In this regard, restaurants or festivals where local dishes are served are very important. The third is the social motivation of the tourist. The tourist is involved in the gastronomic activity and this provides social motivation. The fourth motivation tool is prestige. The tourist experiences the local gastronomic elements and transfers the knowledge and experiences to different people and thus ensures prestige (Guzman & Canizares, 2011).

Gastronomic tourism is an integrated set of goods and services which are a part of local culture that are important in terms of competition and are indicative of globalization and localization, which help economic, social and environmental development, are important in promoting regional tourism and are consumed by tourists who have specific criteria and preferences (Hall et al., 2003). It also contributes to the promotion and branding of destinations, the maintenance and preservation of local traditions and diversity, and the use and rewarding of originalities. Some destinations use their unique cuisine to brand their targets (UNWTO, 2016).

Gastronomic tourism which has developed in the last decades, is further diversified into 'Wine tourism, beer tourism, chocolate tourism, cheese tourism'. There are many countries in the world that successfully carry out gastronomic tourism. For example Germany, Australia, France, Spain and Italy are among the most successful countries in wine tourism. France, Italy, Switzerland, the Netherlands are world-renowned countries in cheese tourism (Güzel Şahin & Ünver, 2015).

Considering the definitions and explanations made within this scope, the concept of gastronomic tourism can be a natural and authentic source for a sustainable destination in tourism (Yurtseven & Karakaş, 2013). In addition, gastronomic tourism is a rapidly developing tourism type and has an important role in economic, social and environmental development as well as in intercultural communication. More and more tourists claim that destination dishes are an important part of their travel experience and that it is impossible to know a culture without savoring its food and beverages (Yun, Hennessey & MacDonald, 2011). In other words, gastronomic tourism has become a touristic movement with more participants each year. However, as a result of globalization, the concept of 'sustainable gastronomic tourism' has become more and more popular in the literature in terms of the availability of scarce resources and the protection of local cultures (Şimşek & Akdağ, 2017).

3. Sustainable Gastronomic Tourism

When sustainability, which ensures the prudent use of all natural, socio-cultural and economic resources of the society (Atak, 2016) is examined in terms of tourism, it means preserving, developing and maintaining the attractiveness of natural, socio-cultural, historical and artistic values which are the main source of tourism (Kurnaz & Arman, 2018). Therefore, ensuring the sustainability of gastronomic tourism which is an alternative tourism type that has developed rapidly in recent years and is assessed with different scientific approaches is important in the long term (Akdağ et al., 2016).

It is notable that there is no clear definition of sustainable gastronomic tourism in the literature. There are definitions generated from different perspectives.

Everett & Slocum (2013) stated that the establishment of sustainable gastronomic tourism is a very difficult situation, while Gössling & Hall (2013) mentioned that sustainable gastronomic tourism is a system in itself and emphasized that sustainable gastronomic tourism is a study area that needs to be emphasized in terms of food and beverage management.

Londono (2011) argues that sustainable gastronomic tourism is 'a tourism activity that attaches importance to ensuring food safety during the production / preparation of food and the use of environmentally friendly production techniques'.

Sustainable gastronomic tourism is 'a kind of tourism that supports the continuity of the production of local, organic and authentic foods, the maintenance of a home-cooked food culture, the preservation of traditional palatal delights, knowledge of culinary culture and traditional cooking techniques, the preservation of food diversity and the preservation of original gastronomic culture'. In other words, sustainable gastronomy is based on the principles of sustainable development and can be defined as 'an understanding that improves and protects the social and environmental quality of the society through the use of eco-foods while protecting environmental sustainability and the health of community members at the most appropriate level'. When combined with tourism, this understanding of community members' commitment to eco-food does not only aim to protect their environmental quality and health in terms of merely local community members but also by ensuring that tourists are introduced to eco-food. Therefore, sustainable gastronomy and eco-gastronomy are defined in the same way (Scarpato, 2002).

In this context, sustainable gastronomy includes 'producing and selling local foods, revitalizing the culture of home cooking, transferring cooking techniques to future generations, educating individuals to develop new tastes, ensuring social welfare, preserving food diversity, increasing the sense of pleasure and sustaining original gastronomic cultures and transferring it to future generations' (Işıldar, 2016).

Yüncü (2010) argues that the importance of gastronomy in sustainable tourism can be evaluated in terms of three main themes: economic, socio-cultural and environmental.

Sustainable gastronomic tourism should be considered first in economic terms. Local producers, production activities and distribution networks are defined as 'local food systems' in which the relationship between producers and consumers is close. Local economies benefit from local food systems. Through the local food systems, the agricultural sector generates job opportunities, contributes to local taxes and increases the amount of social investment. Local food production has benefits such as employment, job sustainability, supporting local services and increasing local social income (Hall & Wilson, 2008).

Sustainable gastronomic tourism means that local products are bought, local farmers and the local economy are supported, the relations between agriculture and tourism are reinforced, jobs are created, local brands and products are supported and the local agricultural and food sectors are strengthened (Yurtseven & Kaya, 2010).

A local food system also has a positive impact on the social and cultural development of society. The local food system supports small and medium-sized farms or businesses. This situation established an awareness of place, culture and history. Location awareness is as important to tourism as it is to the local community and its sense of identity. Local food systems also connect people in a community. These systems bring producers and consumers and even visitors closer together. Farmers' markets can be an area where visitors can spend time and enhance their socio-cultural development (Delind, 2006; Kivela & Crotts, 2006; Yiakoumaki, 2006). Furthermore, Yurtseven (2011) states that sustainable gastronomic tourism is a tourism activity that contributes to local people and supports agricultural activities in the region. Similarly, Durlu Özkaya, Sünnetçioğlu & Can (2013) emphasize that local gastronomic heritage is preserved through sustainable gastronomic tourism and that socio-cultural development is supported by supporting local people.

Sustainable gastronomic tourism offers many economic and socio-cultural benefits as well as environmental contributions. Sustainable agriculture and food systems are based on relatively small farms. Sustainable agriculture and food systems combine animal and plant production where appropriate, preserve biodiversity and switch to convertible energy. In contrast, large-scale farms lead to biodiversity reduction and depend on locally unavailable substances such as pesticides and fertilizers and generate production waste with negative impacts on the ecosystem or society. Therefore, encouraging the purchase of products from small local farms accelerates assistance to environmental sustainability. Local food production also increases the validity of the traditional farming system, which is perceived as environmentally friendly. However, depending on the techniques and resources used in farm production, the conversion to the traditional farming production system sometimes has environmental consequences. Selective harvesting means that farms can be more sustainable. In this respect, small farms that grow different plants and crops gain importance. Local markets create market gaps by steering farmers to grow wild plants and animal species (Hall & Wilson, 2008).

Durlu Özkaya, Sünnetçioğlu & Can (2013) purport that the production/preparation and presentation of foods that are beneficial to human health, high in nutritional value and sensitive to the environment is an important travel motivation in sustainable gastronomic tourism. In this context, Yurtseven (2011) considers local, authentic and organic food, special restaurants, local food festivals, local life culture, special food production systems, special food organizations, natural shopping centers, wine routes and eco-museums as components that generate touristic attraction for sustainable gastronomic tourism.

When the constituents of sustainable gastronomic tourism are examined, it is accepted that the basic determinants of foods and food varieties that are preferred by tourists are local, authentic and organic (Işıldar, 2016).

3.1 Local Foods

Local foods are one of the basic elements of sustainable gastronomic tourism. Supporting environmentally friendly agricultural activities and local production, providing regional socio-economic benefits, preserving originality and cultural values, local foods also create a significant attraction element by appealing to the sense of pleasure felt by tourists and increase brand value and are among the basic elements of sustainable gastronomic tourism (Işıldar, 2016).

It is not possible to say that there is a clear definition of local food. Local food for some authors is 'food produced, sold and consumed in a limited geographical area' (Pearson et al., 2011; Bosona & Gebresenbet, 2011; Mirosa & Lawson, 2012; Bianchi & Mortimer, 2015). According to another definition, it is defined as 'foods produced in a certain local area in a traditional manner and with certain sensory characteristics' (Stolzenbach, Bredie & Byrne, 2013).

Countries such as France and the UK implemented different policies in the field of food and agriculture in the 16th and 19thcenturies in order to ensure continuity in the production of local foods. In England, a decree of privilege for industrial development was made, while France developed a protective food policy (Brown, Dury & Holdsworth, 2009).

3.2 Authentic Foods

The concept of authenticity is expressed as 'original, real, true and unique' (Pratt, 2008). Heitmann (2011), on the other hand, defines authenticity in the simplest way as 'original, real and hailing from the past'. The concept of authenticity is widely used in language, festivals, rituals, antiques, artworks, architecture, clothes, vintage cars and tourism experiences, and now it is also used as a feature attributed to foods and cuisines (Pratt, 2008; Heitmann, 2011).

In terms of food terminology, authenticity is defined as 'a characteristic that manifests the locality and culture of food by blending the traditional and the local' (Sims, 2009). Authentic character that offers a feature that connects and associates foods and cuisines is expressed as 'foods identified with a given region which are created as a result of a craft process'. In this context, authentic foods are expressed as 'foods reflecting the local culture that has emerged as a result of continuous social accumulation during the historical process of a certain region' (Pratt, 2008).

Sims (2009) argues that authentic foods, which are the main travel motivation for most gastro-tourists, are considered as an element that stimulates gastronomic activities. Furthermore, authentic foods are among the main elements of sustainable gastronomic tourism by developing societies and ensuring the development of economic, socio-cultural and environmental structures (Işıldar, 2016).

3.3 Organic Foods

Organic foods are defined as 'foodstuffs which have been grown and processed without genetic engineering, artificial and similar fertilizers, pesticides, weed killers and fungicides, growth regulators, hormones, antibiotics, preservatives, colorants, additives, chemical coating and brightening agents and chemical packaging materials' (Can Kırgız, 2014). In other words, 'organic foods are produced naturally and do not contain any chemical products, flavors or sweeteners' (UNWTO, 2012).

The concept of organic food is associated with many types of tourism. Types of tourism such as rural tourism, agricultural tourism, farm tourism, eco-tourism and gastronomic tourism are examples of the types of tourism that are associated with the concept of organic food. The fact that the origin of gastronomic tourism is based on agriculture increases the importance of organic food and this is important for the sustainability of organic food (Barış, 2015). Therefore, the consumption of organic foods increases throughout the world. The fact that tourists prefer healthier, environmentally friendly products that are sensitive to social equality principles increases the preferability of organic foods. Therefore organic foods; contributes to the protection of the environment, the development of the local economy and the principles of social justice, and thus emerges as an important element for sustainable gastronomic tourism. (Ahmadova & Akova, 2016).

4. Slow Food Movement

Bratec (2008) argues that sustainable gastronomic tourism is an important approach that should be examined within the scope of sustainable tourism, and that the 'slow food' approach can be a valuable supporter from this perspective.

Slow Food described as 'a non-profit organization for sustainable gastronomy established against the disruption of local nutrition cultures by fast food, loss of local foods, and more and more people not knowing what they are eating and where the food comes from, becoming alienated from cultural flavors and lack of interest in how preferred foods influence the world' (slowfood.com, 2019).

Slow Food was founded in 1986 following the protest regarding the opening of McDonalds, an international fast food chain in Rome, Italy. In 1986, activists led by Carlo Petrini took action to protest against such negative situations as globalization, rapid consumption and the development of industrial agriculture and the standardization of food habits. The 'Slow Food Manifesto' was signed three years after this action (9 November 1989), and Slow Food was officially established (slowfood.com, 2019). Petrini, the founder of the Union, stated in his studies that local cultures should be protected from the homogenizing effect of globalization and industrialization. He stated that this can be achieved by using the concepts of enjoyment and taste. This statement explains the mission of the organization (Schneider, 2008).

The subject of foods is the main starting point of Slow Food. The basis of the Slow Food movement includes targets such as seasonal non-industrial production of food, prevention of genetically modified products, promotion of local products and conservation of biodiversity. Furthermore, Slow Food is not only concerned with the production of food but also with its consumption and has some principles in food culture. The aim is to use local products in dishes, preserve the traditional recipes and convey dishes to visitors with the traditional presentation style. At this point, Slow Food, which can be associated with culture, maintains and preserves cultural values (Sezgin & Ünüvar, 2012).

While the main focus of the Slow Food movement is food, the notion of good, clean and fair concepts that are related to each other to ensure food quality are proposed. With its good principle, it is stated that fresh and delicious seasonal products are a part of the local culture and personally satisfy the senses of individuals. It states that food production and consumption that is clean is carried out in a manner that does not harm people, animals and the environment. Lastly, the principle of fairness means the determination of favorable prices and conditions for consumers and producers (slowfood.com, 2019).

The Slow Food Movement ensures the support of regional production and producers and the promotion of using local quality products. Accordingly, it enables sustainable development (Sezgin & Ünüvar, 2012). In addition, it ensures the satisfaction of local people and tourists on a socio-cultural and economic scale, the protection of environmental and cultural values, the proclamation of the gastronomic culture throughout the world and its transfer to future generations and demonstrates the importance of this movement in terms of sustainable gastronomic tourism (Işıldar, 2016).

Results

Sustainable gastronomic tourism focuses particularly on preserving local gastronomic heritages and providing economic and environmental development by supporting local people. Sustainable gastronomic tourism is not an attraction element created within the scope of gastronomic tourism, it refers to the orientation towards agricultural products, local and traditional gastronomic elements created within the scope of gastronomic tourism. In this respect, sustainable gastronomic tourism is considered as an approach to the protection of regional gastronomic elements. Sustainable gastronomic tourism ensures that consumed local and authentic foods and beverages, which are an important travel motivation factor, are organic, beneficial to human health, environmentally sensitive and produced/prepared with local methods. Therefore, the idea of sustainable gastronomy which aims for long-term development and progress is an important issue for tourism. The idea of sustainability, which can play an important role in increasing the variety of gastronomic experiences and discovering new flavors, has an important role in ensuring the continuity of tourism and gastronomy. In this context, the correct evaluation of the passion of gastro tourists for eating and drinking is extremely important for sustainable gastronomic tourism. Tourists who are satisfied with local, authentic and organic foods and beverages will ensure development and sustainability in terms of gastronomic tourism. At the same time, preserving local, authentic and organic foods and beverages and transferring them to future generations will contribute to the sustainability of both the cultural and the gastronomic heritage.

References

Ahmadova, S. & Akova, O. (2016). Türkiye'de Organik Eko Turizm Çiftlikleri Üzerine Bir Araştırma, *Karabük Üniversitesi Sosyal Bilimler Enstitüsü Dergisi*, 6/1, p. 14–29.

Akdağ, G., Özata, E., Sormaz, Ü. & Çetinsöz, B. C. (2016). Sürdürülebilir Gastronomi Turizmi İçin Yeni Bir Alternatif: Surf & Turf, *Journal of Tourism and Gastronomy Studies*, 4/1, p. 270–281.

Anthelme, J. & Savarin, B. (2016). Lezzetin Fizyolojisi ya da Yüce Mutfak Üzerine Düşünceler. İstanbul: Oğlak Yayıncılık.

Atak, O. (2016). Sürdürülebilir Kalkınmanın Temel İlkeleri, (Editör: Hüseyin Ç.). *Sürdürülebilir Turizm* içinde (p. 1-164). Ankara: Detay Yayıncılık.

Ayyıldız, S. & Kargigilioğlu, Ş. (2018). Konaklarda Sunulan Yöresel Yemeklerin Sürdürülebilir Gastronomi Turizmi Bakımından İncelenmesi; Safranbolu Konakları Örneği, *Akademik Sosyal Araştırmalar Dergisi*, 6/79, p. 367-381.

Barış, Z. (2015). *Turizm İşletmelerinde Organik Gıda Kullanımı Algı ve Tutumlarının Araştırılması: Gaziantep İli Örneği*, İzmir Katip Çelebi Üniversitesi Sosyal Bilimler Enstitüsü, Yayımlanmamış Yüksek Lisans Tezi, İzmir.

Barkat, M. S. & Vermignon, V. (2006). Gastronomy Tourism: A Comparative Study of Two French Regions: Brittany and La Martinique, *Sustainable Tourism with Special Reference to Islands and Small States Conference*. Malta, 25-27 Mayıs 2006.

Bayram, A. T. (2016). Sürdürülebilirlik ve Turizm, (Ed.: Ali Y., Özlem S.). *Özel İlgi Turizmi* içinde (p. 1-18). Ankara: Detay Yayıncılık.

Bianchi, C., & Mortimer, G. (2015). Drivers of Local Food Consumption: A Comparative Study, *British Food Journal*, 117/9, p. 2282-2299.

Bosona, T. G. & Gebresenbet, G. (2011). Cluster Building and Logistics Network Integration of Local Food Supply Chain, *Biosystems Engineering*, 108/4, p. 293-302.

Bramwell, B. & Lane, B. (1993). Sustainable Tourism: An Evolving Global Approach, *Journal of Sustaniable Tourism*, 1/1, p. 1-5.

Bratec, M. (2008). Aiming Towards Sustainable (Tourism) Development: The Case of the Slow Food Movement and Its Impacts in Slovenia. Assignment for the course of Sustainable Tourism Development. University of Southern Denmark.

Brown, E., Dury, S., & Holdsworth, M. (2009). Motivations of Consumers That Use Local, Organic Fruit and Vegetable Box Schemes in Central England and Southern France, *Appetite*, 53, p. 183-188.

Brundtland, G. H. (1987). Report of the World Commission on Environment and Development: Our Common Future. Oxford: Oxford University Press.

Can Kırgız, A. (2014). Organik Gıda Sertifikasyonlarının ve Etiketlemelerinin Türkiye Gıda Sektörü İşletmelerinin İtibarı Üzerindeki Etkisi, *Sosyal Bilimler Metinleri*, 1, p. 1-12.

Cronin, L. (1990). A Strategy for Toursim and Sustainable Developments, *World Leisure & Recreation*, 32/3, p. 12-18.

Delind, L. B. (2006). Of Bodies, Place, and Culture: Re-Situating Local Food, *Journal of Agricultural and Environmental Ethics*, 19, p. 121-146.

Durlu Özkaya, F., Sünnetçioğlu, S. & Can, A. (2013). Sürdürülebilir Gastronomi Turizmi Hareketliliğinde Coğrafi İşaretlemenin Rolü, *Journal of Tourism and Gastronomy Studies*, 1/1, p. 13-20.

Everett, S. & Slocum, S. L. (2013). Collaboration in Food Tourism: Developing Cross-Industry Partnerships, (Editöler: Hall M. C. & Gössling S.). Sustainable Culinary Systems, Local Foods, Innovation, Tourism and Hospitality *içinde* (p. 205-222). New York: Routledge.

Gillespie, C. & Cousins, J. A. (2001). *European Gastronomy into the 21st Century*. Burlington: Butterworth-Heinemann.

Gladwin, T. N., Kennelly, J. J. & Krause, T. S. (1995). Shifting Paradigms for Sustainable Development: Implications for Management Theory and Research, *Academy of Management Review*, 20/4, p. 874-907.

Göker, G. (2011). *Destinasyon Çekicilik Unsuru Olarak Gastronomi Turizmi: Balıkesir İli Örneği*. Balıkesir Üniversitesi Sosyal Bilimler Enstitüsü, Yayımlanmamış Yüksek Lisans Tezi, Balıkesir.

Gössling, S. & Hall, M. C. (2013). Sustainable Culinary Systems: An Introduction, (Ed.: Hall M. C. & Gössling S.). *Sustainable Culinary Systems, Local Foods, Innovation, Tourism and Hospitality* içinde (p. 3-44). New York: Routledge.

Guzman, L. T. & Canizares, S. S. (2011). Gastronomy Tourism and Destination Differentiation: A Case Study in Spain, *Review of Economics & Finance*, p. 63-72.

Gürsoy, Y. (2017). Giresun Merkez Yöresinde Gastronomi Turizmi Üzerine Genel Bir Değerlendirme, *Journal of International Social Research*, 10/51, p. 1296-1304.

Güzel Ş. G. & Ünver, G. (2015). Destinasyon Pazarlama Aracı Olarak Gastronomi Turizmi: İstanbul'un Gastronomi Turizmi Potansiyeli Üzerine Bir Araştırma. Journal of Tourism and Gastronomy Studies, 3/2, p. 63-73.

Hall, C. M. & Wilson, S. (2008). Scoping Paper: Local Food, Tourism and Sustainability, *Observado el*, 9.

Hall, M. C., Sharples, L., Mitchell, R., Macionis, N. & Cambourne, B. (2003). *Food Tourism around the World: Development, Management and Markets*. Londra: Rochester.

Harris, R. & Leiper, N. (1995). *Sustainable Tourism: An Australian Perspective*. Chatswood: Butterworth-Heinemann.

Hegarty, J. A. (2005). Developing "Subject Fields" in Culinary Arts, Science and Gastronomy, *Journal of Culinary Science and Technology*, 4/1, p. 5-13.

Hegarty, J. A. & Antun, J. M. (2007). Celebrate Culinary Science and Gastronomic Knowledge, *Journal of Culinary Science and Technology*, 5/4, p. 1-7.

Hegarty, J. A. & O'Mahony, G. B. (2001). Gastronomy: A Phenomenon of Cultural Expressionism and An Aesthetic for Living, *Hospitality Management*, 20, p. 3-13.

Heitmann, S. (2011). Authenticity in Tourism (Ed.: Robinson P. Heitmann S. & Dieke P.). *Research Themes for Tourism* içinde (p. 45-58). Wallingford: CABI.

Hjalager, A. (2002), A Typology of Gastronomy Tourism, (Ed.: Hjalager A. M. & Richards G.). *Tourism and Gastronomy* İçinde (p. 21-35). Londra: Routledge.

Holmberg, J. & Sandbrook, R. (1992). Sustainable Development: What Is to Be Done? (Ed.: Holmberg J.). *Making Development Sustainable: Redefining Institutions, Policy, and Economics* (p. 19-38), Washington D. C.: Island Press.

Işıldar, P (2016). Sürdürülebilirlik ve Gastronomi, (Ed.: Kurgun H & Bağıran Özşeker D.). *Gastronomi veTurizm* içinde (p. 45-63). Ankara: Detay Yayıncılık.

Karim, S. (2006). *Culinary Tourism As A Destination Attraction: An Empirical Examination of The Destination's Food Image And Information Sources*, Oklahoma State University, Yayımlanmamış Doktora Tezi, Oklahoma.

Karim, S. A. & Chi, C. G.-Q. (2010). Culinary Tourism As A Destination Attraction: An Empirical Examination of Destinations' Food Image, *Journal of Hospitality Marketing & Management*, 19/6, p. 531-555.

Kivela, J. & Crotts, J. C. (2005). Gastronomy Tourism: A Meaningful Travel Market Segment, *Journal of Culinary Science and Technology*, 4/2-3, p. 39-55.

Kivela, J. & Crotts, J. C. (2006). Tourism and Gastronomy: Gastronomy's Influence on How Tourists Experience a Destination, *Journal of Hospitality and Tourism Research*, 30/3, p. 354-377.

Kurnaz, A. & Arman, A. (2018). Sürdürülebilir Gastronomi, (Ed.: Akbaba A. & Çetinkaya N.). *Gastronomi ve Yiyecek Tarihi* içinde (p. 161-177). Ankara: Detay Yayıncılık.

Lang, J. H. (1988). *Larousse Gastronomique*. Londra: Crown.

Le´vi-Strauss, C. (1966). *The Savage Mind*. Chicago: Chicago University Press.

Lee, K.-H., Packer, J. & Scott, N. (2015). Travel Lifestyle Preferences and Destination Activity Choices of Slow Food Members and Non-Members. *Tourism Management*, 46, p. 1-10.

Londono, M. P. L. (2011). Gastronomy Tourism: An Opportunity for Local Development in Catalonia?, *A Stakeholder Analysis*, p. 1-24.

Long, M. L. (2005). *Culinary Tourism*. Kentucky: The University Press of Kentucky.

Mirosa, M. & Lawson, R. (2012). Revealing the Lifestyles of Local Food Consumers, *British Food Journal*, 114/6, p. 816–825.

Özgen, I. (2013). Uluslararası Gastronomiye Genel Bakış, (Ed.: Sarıışık, M.). *Uluslararası Gastronomi, Temel Özellikler-Örnek Menüler ve Reçeteler* İçerisinde (p. 1–26). Ankara: Detay Yayıncılık.

Pearson, D., Henryks, J., Trott, A., Jones, P., Parker, G., Dumaresq, D., & Dyball, R. (2011). Local Food: Understanding Consumer Motivations in Innovative Retail Formats, *British Food Journal*, 113/7, p. 886–899.

Pratt, J. (2008). Food Values: The Local and the Authentic, *Research in Economic Anthropology*, 28, p. 53–70.

Richards, G. (2002). Gastronomy: And Essential Ingredient in Tourism Production and Consumption?, (Ed.: Hjalager A. M. & Richards G.). *Tourism and Gastronomy* İçinde (p. 3–20). London: Routledge.

Santich, B. (2004). The Study of Gastronomy and Its Relevance to Hospitality Education and Training, *Hospitality Management*, 23, p. 15–24.

Sarıışık, M. & Özbay, G. (2015). Gastronomi Turizmi Üzerine Bir Literatür İncelemesi, *Anatolia: Turizm Araştırmaları Dergisi*, 26/2, p. 264–278.

Savarin, B. & Anthelme, J. (2016). *Lezzetin Fizyolojisi ya da Yüce Mutfak Üzerine Düşünceler*. İstanbul: Oğlak Yayıncılık.

Scarpato, R. (2000). *New Global Cuisine: The Perspective of Postmodern Gastronomy Studies*, Yayımlanmamış Yüksek Lisans Tezi, Melbourne: Royal Melbourne Institute of Technology Üniversitesi.

Scarpato, R. (2002). Gastronomy as a Tourist Product: The Perspectives of Gastronomy Studies, (Ed.: Hjalager A. M. & Richards G.). *Tourism and Gastronomy*, (p. 51–59). Londra: Routledge.

Schneider, S. (2008). Good, Clean, Fair: The Rhetoric of the Slow Food Movement, *College English*, 70/4, p. 384–402.

Sezgin, M. & Ünüvar, Ş. (2012). *Sürdürülebilirlik ve Şehir Planlaması Ekseninde Yavaş Şehir*. Konya: Çizgi Kitabevi.

Shenoy, S. S. (2005). *Food Tourism and the Culinary Tourist*, Güney Carolina: Clemson University, Yayımlanmamış Doktora Tezi.

Sims, R. (2009). Food, Place and Authenticity: Local Food and the Sustainable Tourism Experience, *Journal of Sustainable Tourism*, 17/3, p. 321–336.

Slow Food. https://www.slowfood.com/about-us/ (Accessed: 16 November 2019).

Slow Food. https://www.slowfood.com/about-us/our-history/ (Accessed: 16 November 2019).

Slow Food. https://www.slowfood.com/about-us/our-philosophy/ (Accessed: 16 November 2019).

Stolzenbach, S., Bredie, W. L., & Byrne, D. V. (2013). Consumer Concepts in New Product Development of Local Foods: Traditional Versus Novel Honeys, *Food Research International*, 52/1, p. 144-152.

Swarbrooke, J. (1999). *Sustainable Tourism Management*. New York: Cabi Publishing.

Şimşek, N. & Akdağ, G. (2017). Investigation of Green Genetic Restaurants within Content of Sustainable Gastronomy Tourısm, *International Journal of Social Science*, 60, p. 351-368.

United Nations Environment Programme ve World Tourism Organization. (2005). *Making Tourism More Sustainable: A Guide For Policy Makers*. World Tourism Organization Publications.

UNWTO (2005). *Sustainable Development of Tourism*.

UNWTO (2012). *Global Report on Food Tourism*. Madrid, Spain.

UNWTO (2016). Gastronomy Network Action Plan, http://cf.cdn.unwto.org/sites/all/files/docpdf/gastronomyactionplanenweb.pdf. (Accessed: 15 November 2019).

Uyar, H. & Zengin, B. (2015). Gastronomi Turizminin Alternatif Turizm Çeşidi Olarak Değerlendirilmesi Bağlamında Gastronomi Turizm İndeksinin Oluşturulması, *Akademik Sosyal Araştırmalar Dergisi*, 3/17, p. 355-376.

Üner, E. H. & Şahin, G. G. (2016). Türkiye Gastronomi Turizmi Potansiyelinin Her Şey Dâhil Satış Sistemi, *Journal of Tourism and Gastronomy Studies*, 4/3, p. 76-100.

Üner, E. H. (2014). *Her Şey Dâhil Sistemde Türkiye Gastronomi Turizmi Potansiyelinin Değerlendirilmesi*, Ankara: Atılım Üniversitesi Sosyal Bilimler Enstitüsü, Yayınlanmamış Yüksek Lisans Tezi.

Wolf, E. (2006). *Culinary Tourism the Hidden Harves*. Iowa: Kendall Hunt Publishing.

Yiakoumaki, V. (2006). Local, Ethnic and Rural Food: On the Emergence of Cultural Diversity in Post-EU-Accession Greece, *Journal of Modern Greek Studies*, 24, p. 415-445.

Yun, D., Hennessey, S. M. & MacDonald, R. (2011). Understanding Culinary Tourists: Segmentations Based on Past Culinary Experiences and Attitudes Toward Food-Related Behaviour. *International CHRIE Conference-Refereed Track*.

Yurtseven, R. (2011). Sustainable Gastronomic Tourism in Gökçeada (Imbros): Local and Authentic Perspectives, *International Journal of Humanities and Social Science*, 1/18, p. 17-26.

Yurtseven, H. R., & Karakaş, N. (2013). Creating a Sustainable Gastronomic Destination: The Case of CittaslowGokçeada-Turkey, American *International Journal of Contemporary Research*, 3/3, p. 91-100.

Yurtseven, R. & Kaya, O. (2010). Eko Gastronomi ve Sürdürülebilirlik, (Editör: Çolakoğlu, O. E.). *11. Ulusal Turizm Kongresi* (2-5 Aralık 2010 Kuşadası) içinde (p. 57-65). Ankara: Detay Yayıncılık.

Yüncü, H. R. (2010). Sürdürülebilir Turizm Açısından Gastronomi Turizmi ve Perşembe Yaylası, (Editör: Şengel, S.). *10. Aybastı-Kabataş Kurultayı* içinde (p. 27-34). Ankara: Detay Yayıncılık.

Mehmet Mert PASLI and Evren GÜÇER

10 To Determine the Recreational Potential of Trabzon[1]

Introduction

In a world that is developing and changing fast, as a result of the rapid devastation of big cities, the increase in construction, global warming and many other factors, the value people attribute to the natural environment is on the increase. With the increase in leisure time there is increasing interest in how people make use of it. What leisure means for people and families, how much time they allocate for travel and vacation, how much they spend, and the reasons for their satisfaction and dissatisfaction have been the subject of many studies.

All activities that people do in relation to their physical and mental lives during the time which they have out of their compulsory work life and which they use however they wish is defined as recreation (Karaküçük, 2005).

Sağcan (1986) describes recreation as any kind of activity and event in which people voluntarily participate in order to have fun in leisure time, to be mentally and emotionally satisfied and to eliminate their physical and mental fatigue and gain strength. Some similarities and differences can be observed from the internal and external motivational point of view of the behavior of the people who take part in tourism and recreation. Differences between attracting and repelling factors in the orientation of the preferred activities can be evaluated as a factor by keeping the leisure time as the focus point. (Özdemir, Güçer & Karaküçük, 2016).

People are looking for opportunities to escape the complexity and stress of daily life and work life and to improve the quality of their life. Individuals who want to relax, unwind and have the sense of oneness with nature visit some places or participate in tourist activities to achieve their personal goals. One of these activities is visiting national parks. National parks are the center of attraction

---

1 This study was produced from the doctorate dissertation of 'To Determine the Recreational Potential of Trabzon – Maçka Altındere Valley National Park and Measurement of Satisfaction Levels of People Who Participate in Tourism Activities According to Personalty Types'.

for tourists because of the rich natural beauty they contain and the values they possess. Owing to their cultural and natural values, forty-four national parks in Turkey are protected by law. Altındere Valley National Park is one of these protected national parks. The national parks in Turkey, including Altındere Valley National Park, are shown in Table 1.

Table 1: The National Parks in Turkey

National Park Name	Province
1. Yozgat Çamlığı	Yozgat
2. Karatepe - Aslantaş	Osmaniye
3. Soğuksu	Ankara
4. Kuşcenneti	Balıkesir
5. Uludağ	Bursa
6. Yedigöller	Bolu
7. Dilek Y. – B. Menderes D.	Aydın
8. Spil Dağı	Manisa
9. Kızıldağ	Isparta
10. Güllük Dağı - Termessos	Antalya
11. Kovada Gölü	Isparta
12. Munzur Vadisi	Tunceli
13. Beydağları Sahil	Antalya
14. Köprülü Kanyon	Antalya
15. Ilgaz Dağı	Kastamonu, Çankırı
16. Başkomutan TMP	Afyonkarahisar, Kütahya
17. Göreme TMP	Nevşehir
18. Altındere Vadisi	**Trabzon**
19. Boğazköy – Alacahöyük	Çorum
20. Nemrut Dağı	Adıyaman, Malatya
21. Beyşehir Gölü	Konya
22. Kazdağı	Balıkesir
23. Altınbeşik Mağarası	Antalya
24. Hatila Vadisi	Artvin
25. Karagöl – Sahara	Artvin
26. Kaçkar Dağları	Rize, Artvin
27. Aladağlar	Niğde, Adana

Table 1: (continued)

National Park Name	Province
28. Marmaris	Muğla
29. Saklıkent	Muğla, Antalya
30. Troya TMP	Çanakkale
31. Honaz Dağı	Denizli
32. Küre Dağları	Kastamonu, Bartın
33. Sarıkamış-Allahuekber Dağları	Kars, Erzurum
34. Ağrı Dağı	Ağrı, Iğdır
35. Gala Gölü	Edirne
36. Sultan Sazlığı	Kayseri
37. Tek Tek Dağları	Şanlıurfa
38. İğneada Longoz Ormanları	Kırklareli
39. Yumurtalık Lagünü	Adana
40. Nene Hatun TMP	Erzurum
41. Sakarya Meydan Muharebesi TMP	Ankara
42. Kop Dağı Müdafaası TMP	Bayburt, Erzurum
43. Malazgirt Meydan Muharebesi TMP	Muş
44. İstiklal Yolu Tarihi Milli Parkı	Kastamonu, Çankırı, Ankara

Source: Korunan Alan İstatistikler (2019)

National parks, having the status of protected area, are privileged with very rich biodiversity where sensitive ecosystems and rare species are protected, and they are used extensively by local and foreign visitors (Düzgüneş & Demirel, 2016).

In 2015, an analysis and planning for Altındere Valley National Park was carried out within the framework of Trabzon natural tourism master plan. Within this plan, infrastructure, accommodation, wildlife, accessibility analysis and facility analysis of the national park were performed. Altındere Valley National Park is considered to be unique and unequaled in terms of the rare values it offers as well as containing the site of Sumela Monastery. Sumela Monastery, which is on the UNESCO World Heritage tentative list, is an important element of attraction for the national park. Biodiversity and geological structure are among its other interesting features. The situation of the area is such that it could be included in many other tourist sightseeing tours. The area is in close proximity to areas such as Santa Ruins, Kustul and Vazelon Monasteries and Uzungol. The presence of the Trabzon Daisy as an endemic plant in the national park is another attraction element.

Altındere Valley National Park, which has natural, historical and cultural attractions, also has a potential for tourism and recreation. National parks throughout the world and also in Turkey are important destinations for tourism and recreational activities. The fact that a religious structure like Sumela Monastery with historical and cultural value is located in Altındere Valley National Park adds to the importance of the national park in terms of tourism. There are similar studies on the determination of recreational potential of Altındere Valley National Park. However, these studies were generally carried out by engineering and architecture departments, and the number of studies in terms of tourism and recreation is limited.

Demographic characteristics such as age, gender, educational level as well as personality characteristics are important factors in determining the type and frequency of recreation activities that people participate in to make use of their free time. Within the scope of the research the gender, age, educational background and personality types of the people who visit Altındere Valley National Park were examined, and the recreational activities in the national park and their frequency levels were analyzed.

Methods

For the purpose of collecting the data within the scope of the research, the survey method was preferred because the population was composed of a large mass, and it was necessary to collect a large number of data in a short time in a convenient and cheap way. It was important to ensure that all the participants were asked the same questions and that they were given a relative and convincing confidentiality guarantee (Karasar, 2012). In addition, in order to allow the data to be digitized and the results to be compared and generalized with other studies conducted on the subject, the survey method was preferred as the data collection tool (Nunnally & Bernstein, 1994). In order to determine the satisfaction and dissatisfaction factors in the questionnaire, the questionnaire which Uzun (2005) developed by using the studies of Dawson, Newman & Watson (1997) and Newman & Dawson (1998) was employed. In the questionnaire, users were primarily asked about the frequency of nineteen activities they did to determine their participation rates in recreational activities according to their demographic structure (gender, age, educational level, income). The first questionnaire was based on the twenty-four item Eysenck Personality Questionnaire Revised/Short Form. When the Eysenck personality theory was first developed, it consisted of neuroticism stability and extroversion-introversion dimensions, and later psychoticism dimension was added (Lewis et al. 2002). The study performed using the Turkish version of Eysenck Personality Questionnaire Revised Short Form not only proved the similarity of cross-cultural personality characteristics, but it demonstrated that EPQR-S was a reliable and valid scale that could be employed in the studies to be carried out on personality and in clinical applications in Turkey (Karancı, Dirik & Yorulmaz, 2007: 7). The Turkish version of the Eysenck Personality Questionnaire Revised Short Form was used to determine the personality types of the participants.

Table 2: Population, Sampling and Representative Number

Population	Target Number of Surveys	Valid Number of Polls
202400	384	497

A total of five hundred and eighty-seven questionnaires were distributed to tourists visiting the research area of Altındere Valley National Park, some of whom were interviewed face-to-face during the questionnaire administration. Ninety questionnaires were not included in the evaluation due to being incomplete and invalid. As a result, four hundred and ninety-seven questionnaires were used to collect the data of the study, which was then subjected to evaluation. The reliability coefficient varies between 0 and +1. If the reliability coefficient is close to 1, it means that the reliability and the internal consistency between the items are high, which is a desired condition. The reliability coefficients were found to be 0.895 for the activity criteria. It can be said that the calculated reliability coefficient is appropriate.

Table 3: Personality Types According to Gender

Gender		Average	Std Deviation	t	p
Extrovert	Male	4,05	1,81	1,123	,262
	Female	3,86	1,89		
Lie	Male	3,76	1,73	-1,603	,110
	Female	4,00	1,55		
Neuroticism	Male	3,09	1,74	-3,561	,000*
	Female	3,65	1,78		
Psychoticism	Male	1,87	1,24	2,741	,006*
	Female	1,55	1,31		

*p<0.05

For independent samples, t test was used to determine whether personality types differed according to gender, and the results are shown in Table 3. It was observed that extroversion and lying personality types did not vary according to gender (p>0.05). However, neuroticism and psychoticism personality types of individuals differed according to gender (p<0.05). The significance level was higher in neuroticism dimension for females compared to males, while it was higher in psychoticism dimension for males.

Table 4: Frequency of Activities in the National Park According to Gender Variable

	Gender (Average)		Mann-Whitney U	p
	Male	Female		
1- Hiking	2,7	2,9	27583,000	,039*
2- Jogging	2,2	2,0	27659,000	,041*
3- Doing Sports	2,4	2,3	29613,000	,457
4- Cycling	1,8	1,7	28325,000	,094
5- Fishing	1,7	1,2	23820,500	,000*
6- Resting	3,3	3,6	25865,500	,002*
7- Getting Fresh Air	3,4	3,8	24988,000	,000*
8- Unwinding	3,4	3,8	24997,500	,000*
9- Sightseeing	3,3	3,7	25626,000	,001*
10- Enjoying Scenery	3,5	3,7	26829,000	,011*
11- Picnicing	2,5	2,5	29931,500	,594
12- Reading	2,0	2,6	22287,000	,000*
13- Getting Away From Daily Routines	2,8	2,9	28369,500	,121
14- Exploring a Natural Environment	2,9	3,1	27449,500	,033*
15- Having the sense of oneness with Nature	2,9	3,0	29104,000	,288
16- Socialising	3,0	3,0	30209,500	,725
17- Passing Through	2,3	2,2	28448,000	,130
18- Playing	2,4	1,9	23638,000	,000*
19- Eating	3,1	2,9	28058,500	,084

*p<0.05

Mann Whitney U test was employed to find out whether the frequency of the activities performed by the participants in the national park varied in accordance with the gender variable, and the results are presented in Table 4. A statistically significant difference was found between the frequency of the participants' activities of hiking, jogging, fishing, relaxing, getting fresh air, unwinding, sightseeing, enjoying the scenery, reading books, exploring a natural environment and playing games in the national park in terms of their gender (p<0.05).

Females perform activities such as hiking, resting, getting fresh air, relaxing, sightseeing, enjoying the scenery, reading books and exploring the natural environment more frequently in comparison to males. Males, on the other hand, perform activities such as jogging, fishing, exploring the natural environment and playing games more often than females.

Table 5: Distribution of Personality Types According to Age Variable

		Average	Std Deviation	F	p
Extrovert	<18	4,08	1,83	2,005	,093
	19-25	3,75	1,89		
	26-40	4,28	1,72		
	41-60	4,03	1,89		
	61+	3,57	2,07		
Lie	<18	4,32	1,20	2,388	,050
	19-25	3,73	1,59		
	26-40	4,01	1,66		
	41-60	3,72	2,00		
	61+	5,00	1,41		
Neuroticism	<18	3,89	1,41	2,241	,064
	19-25	3,46	1,77		
	26-40	3,04	1,78		
	41-60	3,37	1,99		
	61+	3,00	1,15		
Psychoticism	<18	1,76	1,06	,405	,805
	19-25	1,76	1,21		
	26-40	1,64	1,18		
	41-60	1,67	1,72		
	61+	2,14	2,12		

*p>0.05

Extroversion, lying, neuroticism and psychoticism personality types of the participants were analyzed statistically in terms of their age. According to the results of this analysis, one-way analysis of variance according to the age of extroverted, lying, neuroticism and psychoticism personality types of the participants did not create a statistically significant difference.

Table 6: Frequency of Activities in the National Park According to Age Variable

	Age (Average)					p
	<18 Age	19-25	26-40	41-60	61+	
1- Hiking	2,6	2,8	2,7	2,9	2,0	,210
2- Jogging	2,4	2,1	2,0	1,8	1,4	,051
3- Doing Sports	2,5	2,5	2,2	2,2	1,6	,052
4- Cycling	2,0	2,0	1,5	1,1	1,4	,000*
5- Fishing	1,2	1,5	1,5	1,4	1,3	,520
6- Resting	3,2	3,5	3,3	3,5	2,9	,340
7- Getting Fresh Air	3,5	3,6	3,6	3,7	3,0	,684
8- Unwinding	3,4	3,6	3,5	3,7	3,1	,426
9- Sightseeing	3,2	3,5	3,5	3,6	3,1	,304
10- Enjoying Scenery	3,4	3,6	3,5	3,8	3,0	,345
11- Picnicing	2,5	2,5	2,6	2,5	1,7	,451
12- Reading	3,2	2,3	2,2	2,0	1,4	,001*
13- Getting Away From Daily Routines	2,4	2,7	3,0	3,1	2,6	,004*
14- Exploring a Natural Environment	2,8	3,0	3,1	3,2	2,9	,335
15- Having the sense of oneness with Nature	2,6	2,9	3,0	3,1	2,7	,233
16- Socialising	2,7	3,1	2,9	2,9	2,7	,343
17- Passing Through	2,1	2,4	2,2	1,8	1,9	,004*
18- Playing	2,4	2,4	1,9	1,5	1,9	,000*
19- Eating	2,9	3,1	3,1	2,7	2,4	,207

*p<0.05

A significant difference was detected between the frequency of the participants' activities of cycling, reading books, getting away from their daily routines, passing through and playing games in terms of their age (p<0.05). Visitors under the age of eighteen engage themselves more frequently in the activities of cycling, reading and playing games than other age groups. Visitors between the ages of forty-one to sixty mostly visit the national park in order to get away from their daily routines compared to other age groups.

Table 7: Distribution of Personality Types According to Education Variable

		Average	Std Deviation	F	p
Extrovert	Illiterate	5,00	1,41	,260	,904
	Primary Education	3,94	1,96		
	High School	3,90	1,81		
	University	4,00	1,85		
	Post Graduate	3,77	2,09		
Lie	Illiterate	5,00	1,41	1,406	,231
	Primary Education	4,02	1,82		
	High School	4,05	1,67		
	University	3,72	1,60		
	Post Graduate	4,00	1,63		
Neuroticism	Illiterate	5,50	0,71	4,515	,001*
	Primary Education	3,30	1,64		
	High School	3,30	1,86		
	University	3,46	1,75		
	Post Graduate	1,54	0,88		
Psychoticism	Illiterate	2,00	1,41	1,008	,403
	Primary Education	2,04	1,45		
	High School	1,67	1,37		
	University	1,67	1,18		
	Post Graduate	1,85	1,46		

*p<0.05

It was observed that extroversion, psychoticism and lying personality types of the participants did not differ according to educational level. However, neuroticism personality types of the individuals were found to differ in terms of educational level (p<0.05). One-way variance analysis showed a statistically significant difference between the participants with a postgraduate degree and those with other educational status.

Kruskal Wallis test was run in order to determine whether the frequency of the activities performed by the participants in the national park varied according to their educational level, and the results are shown in Table 8. A statistically significant difference was found between the frequency of the participants' activities of cycling, reading a book, passing through and playing games in the park in terms

Table 8: Frequency of Activities in the National Park According to Education Variable

	Educational Status Average					p
	Illiterate	Primary Education	High School	University	Post Graduate	
1- Hiking	2,5	2,7	2,8	2,8	2,8	,989
2- Jogging	2,0	2,0	2,1	2,1	2,2	,991
3- Doing Sports	2,5	2,3	2,3	2,4	2,1	,540
4- Cycling	1,0	1,8	1,4	1,9	2,0	,000*
5- Fishing	1,5	1,5	1,5	1,4	1,4	,906
6- Resting	3,0	3,3	3,4	3,5	3,6	,775
7- Getting Fresh Air	3,0	3,4	3,6	3,6	3,8	,623
8- Unwinding	3,0	3,5	3,5	3,6	3,8	,811
9- Sightseeing	3,0	3,4	3,4	3,5	3,9	,571
10- Enjoying Scenery	3,0	3,5	3,6	3,6	3,8	,885
11- Picnicing	2,0	2,2	2,6	2,5	3,5	,050
12- Reading	1,0	2,2	2,1	2,4	3,0	,039*
13- Getting Away From Daily Routines	2,5	2,8	2,8	2,8	3,6	,178
14- Exploring a Natural Environment	2,5	3,1	3,0	2,9	3,4	,620
15- Having the sense of oneness with Nature	2,5	2,7	3,0	2,9	3,6	,139
16- Socialising	2,5	2,7	2,9	3,0	3,7	,101
17- Passing Through	1,0	2,0	2,0	2,4	1,8	,001*
18- Playing	1,0	2,1	1,9	2,3	2,1	,003*
19- Eating	2,5	2,8	3,0	3,1	3,5	,574

*p<0.05

of their educational status (p<0.05). Participants with a university level of education mostly use the national park for passing through and playing games when compared to those with other educational status. On the other hand, participants with postgraduate education perform activities such as reading books and cycling in the park more than individuals with other educational levels.

Table 9: Correlation Analysis of the Relationship between the Personality Types of the Participants and the Activity Frequency in the National Park

	Extrovert	Lie	Neuroticism	Psychoticism
1- Hiking	,123**	-,172**	,139**	,022
2- Jogging	,122**	-,108*	,129**	,092*
3- Doing Sports	,137**	-,132**	,169**	,022
4- Cycling	,067	-,019	,018	,092*
5- Fishing	-,012	,025	-,114*	,129**
6- Resting	,037	-,109*	,125**	-,004
7- Getting Fresh Air	,085	-,055	,131**	,011
8- Unwinding	,065	-,064	,127**	,051
9- Sightseeing	,132**	-,093*	,106*	,039
10- Enjoying Scenery	,055	-,036	,111*	,007
11- Picnicing	,097*	,025	,027	-,059
12- Reading	,068	-,020	,135**	-,111*
13- Getting Away From Daily Routines	,124**	-,062	-,003	,015
14- Exploring a Natural Environment	,141**	-,005	,082	-,005
15- Having the sense of oneness with Nature	,036	-,004	,067	,039
16- Socialising	,184**	-,092*	-,047	-,005
17- Passing Through	-,060	-,030	,036	,036
18- Playing	,083	-,026	-,051	,129**
19- Eating	,115*	,048	-,036	,076

**p<0.001 *p<0.05

Some relationships were found between nineteen types of activities performed by the visitors in the park and four different personality types (psychoticism, extroversion, neuroticism and lying). These relationships show positive or negative variance depending on personality types. When we examined these relationships between activities and personality types, it was determined that there were positively significant but weak relationships between extroversion personality type and the activities of hiking, jogging, doing sports, sightseeing, eating, getting away from daily routines, exploring the natural environment and socializing with other people performed in the park. As extroversion personality type gets more dominant, the activities of hiking, jogging, doing sports, sightseeing, getting away from daily routines, exploring the natural environment, socializing with other people and eating in the park become more frequent. A negative, significant but weak relationship was found between lying personality type and the activities of hiking, jogging, doing sports, resting, sightseeing and socializing with other people in the park. Positively significant but weak relationships were found between neuroticism personality type and the activities of hiking, jogging, doing sports, resting, getting fresh air, relaxing, sightseeing, enjoying the scenery, and reading in the park. As neuroticism personality type gets more dominant, the activities of hiking, jogging, doing sports, relaxing, getting fresh air, unwinding, sightseeing, enjoying the scenery and reading books in the park become more frequent. It was found that there were positively significant but weak relationships between psychoticism personality type and the activities of hiking, jogging, cycling, fishing and playing games performed in the park. As psychoticism personality type becomes more predominant, hiking, jogging, cycling, fishing and playing activities in the park become more frequent. However, a significant negative correlation was found between psychoticism personality type and reading activity. In other words, as psychoticism personality type strengthens, the frequency of reading books in the park is reduced.

Results

The results of the research suggest that as extrovert personality type strengthens, the activities such as hiking, jogging, doing sports, sightseeing, getting away from daily routines, exploring the natural environment, socializing with other people and eating are performed more frequently. People with extrovert personality type represent sociality and impulsivity, and people with high scores in this dimension love communicating with people, are sociable and prefer to spend time with people rather than being alone. The results of the research coincide with the activities of individuals with extroversion personality type in the national park. The most concrete example of this is that individuals with extroversion personality type, who prefer to be with other people rather than being alone, come to the national park to socialize with other people. The neuroticism dimension indicates emotional consistency or over-reactivity and it is suggested that a person who scores high in this dimension could be anxious, depressed, tense, shy, excessively emotional and with low self-esteem. As neuroticism personality type gets more dominant, the activities of hiking, jogging, doing sports, relaxing, getting fresh air, unwinding, sightseeing, enjoying the scenery and reading books in the park become more frequent. In order to increase the satisfaction level of individuals with neuroticism personality type, it is necessary to improve or increase the hiking and running tracks as in other personality types. Furthermore, the view of the national park and its clean air can be advertised to ensure that people with this personality type visit the national park. On the other hand, the psychoticism dimension expresses more unusual personality traits such as being cold, distant, aggressive, insecure, insensitive, strange and unable to empathize, feeling guilt and insensitivity to other people. In the study, as psychoticism personality type becomes more predominant, hiking, jogging, cycling, fishing and playing activities in the park become more frequent. In order for visitors to spend more time in the national park and to increase their level of satisfaction, it is necessary to improve or increase the hiking and running tracks as in extraversion personality type. As lying personality type gets stronger, the activities of hiking, jogging, doing sports, resting, sightseeing and socializing with other people in the park decrease. It is thought that improving, encouraging and increasing the opportunities such as cycling and playing games, specifically for the psychoticism dimension will make a difference for people with the psychoticism personality type.

In addition to these results, the restoration work of the Sumela Monastery in 2015 in Altındere Valley National Park, and the fact that Orthodox ceremonies conducted under the leadership of the Fener Greek Patriarch Bartholomeos once a year since 2010 could not be performed due to restoration work, the terrorist attacks in Macka district, especially in the immediate vicinity of the park and the coup attempt on July 15 are all estimated to be the factors that contributed to the decrease in the number of visitors to the National Park. Cultural activities that will promote the national park should be supported and marketing activities should be carried out in a coordinated manner by considering the capacity and including public and private sector and all stakeholders.

In 2019, the reopening of Sumela Monastery is expected to increase the number of people visiting the national park. Altındere Valley National Park is a tourism destination with a high tourism potential due to its natural beauty and hosting such an important religious structure as Sumela (Meryemana) Monastery, which is accepted as one of the holy places for Orthodox Christians and the many opportunities it offers for recreational activities.

References

Dawson, C. P., Newman, P. & Watson, A., (1997). Cognitive dimensions of recreational user experiences in wilderness: An exploratory study in Adirondack Wilderness Areas. *Proceedings of the 1997 Northeastern Recreation Research Symposium*. New York. GTR-NE-241. 257–259.

Düzgüneş, E. & Demirel, Ö. (2016). Milli Parkların Koruma Yapısının Ekolojik Duyarlılık Analizi İle Ortaya Konması: Altındere Vadisi Milli Parkı (Trabzon/Türkiye)Örneği. *Kastamonu Üniversitesi Orman Fakültesi Dergisi*, 16(1).

Karaküçük, S. (2005). *Rekreasyon (Boş Zamanları Değerlendirme)*, Gazi Kitabevi, Ankara

Karanci, N., Dirik, D. & Yorulmaz, O. (2007). Eysenck Kişilik Anketi-Gözden Geçirilmiş Kısaltılmış Formunun (EKA-GGK) Türkiye'de Geçerlik ve Güvenilirlik Çalışması. *Türk Psikiyatri Dergisi*, 18(3), 1–8.

Karasar, N. (2012). *Bilimsel araştırma yöntemi*. Ankara: Nobel Akademik Yayıncılık

Korunan Alan İstatistikleri, (2019). Milli Parklar. Accessed: 21.12.2019. https://www.tarimorman.gov.tr/DKMP/Menu/18/Korunan-Alan-Istatistikleri.

Lewis C. A, Francis L. J, Shevlin M. & Forrest, S. (2002). Confirmatory factor analysis of the French translation of the abbreviated form of the Revised Eysenck Personality Questionnaire. Europe J Psychol Assess, 18: 79–85.

Newman, P. & Dawson, C. P., (1998). *The Human Dimensions of the Wilderness Experience in The High Peaks Wilderness Area*. Proceedings of the 1998 Northeastern Recreation Research Symposium. New York. GTR-NE-255.122–128.

Nunnally, J. C. & Bernstein, I. H. (1994). *Psychometric Theory*. New York: McGraw-Hill.

Özdemir, A. S., Güçer, E. and Karaküçük, S. (2016). Rekreasyon ve Turizm. Karaküçük, S. (Ed). *Rekreasyon Bilimi*. (313–393). Gazi Kitapevi: Ankara.

Sağcan, M. (1986). *Rekreasyon ve Turizm*. Cumhuriyet Basımevi: İzmir.

Uzun, S. (2005). *Kırsal Ve Kentsel Alanlardaki Parklarda Kullanıcı Memnuniyeti: Gölcük Orman İçi Dinlenme Alanı Ve İnönü Parkı Örneği*. Fen Bilimleri Enstitüsü, Peyzaj Mimarlığı Anabilim Dalı, Yayınlanmış Yüksek Lisans Tezi, Bolu.

Samet GÖKKAYA

11 Digital Transformation & Marketing in Tourism Industry

Introduction

In the period industry 4.0, where digital transformation has started to be implemented in all sectors, it is possible for enterprises to outperform their competitors and to evaluate the social communication environment effectively. In the 21st century, serial technological developments and computer networks, which are continuous in communication and information technology, especially under the leadership of the internet, eliminate geographical distances between the seller and the buyer, enlarge the market and bring an international quality to the field of consumption.

The virtual environment in which individuals frequently spend time and interact with each other has prepared the ground for many technological transformations and thus initiated digital transformation through instant information sharing and communication channels and user-friendly applications brought by web technology. In the traditional media, the technology that enables communication between people or between the masses has paved the way for digital media to convey information to a very large audience. With the digital media, the person has started to take an active role in producing and sharing information and significantly increased the communication superiority of the users (Engin, 2011: 35).

Together with the developing technology, the need of tourism enterprises for information systems has increased. Tourism enterprises, which want to be superior to their competitors by becoming a leader in the market and to maintain their sustainability, pay particular attention to the use of digital transformation elements created by the concept of tourism 4.0, which is the equivalent of tourism in industry 4.0. While the tourism enterprises want to increase the operational efficiency with the digital transformation elements used, to facilitate the preferability by the tourists who quickly adapt to the digitalization, to provide fast and reliable service to the tourists, the other side of the digital transformation in the tourism sector is the best, the most reliable and the fastest way. Want to experience tourist experiences.

In recent years, IT expenditures of enterprises have been used more strategically and as an important structural element within the organization rather than capital investments. (Bresnahan, 1998; Çetinkaya, 2007). Tourism enterprises to closely monitor and use the information systems movements; has contributed to management, production factors and organizational performance and has been an important factor in increasing productivity. Nowadays, information technology systems are seen as a strategic product that changes the sector structure and competition between companies instead of cost structure (Porter, 1980; Clemons & Row, 1991; Segars & Grover, 1995; Çetinkaya, 2007).

Tourism establishments have a structure that provides instant service and consumption at the same time as the sector in which it is located. Nowadays, the product and service produced, the tracking of the product and its benefit can be seen through the technological software brought by the digital transformation. In addition, technology-based It investments, which is one of the investments made by tourism enterprises, attract attention. Although these investments have high costs, they provide a look at the reporting and performance of the company in many areas such as revenue, expense, guest tracking, solution of technical problems.

In this section, the phases of digital transformation and the concept of digital marketing, applications for digital transformation in tourism and digital marketing applications are mentioned.

Digital Transformation and Digital Marketing

The emergence and development of the concept of Industry 4.0 (Digital Transformation), the last of the modern industrial development for the time being, has taken several hundred years and until today it has undergone three major industrial revolutions: 1.0, 2.0 and 3.0 (Atar, 2019: 102). The first of these stages, the first industrial revolution in England in the last quarter of the 18th century occurred. The invention of the steam engine and the replacement of manual labor triggered the transition to more capital-intensive production methods (Leighton, 1970: 3). The Industry 1.0 phase allowed production to increase and enter new markets with the creation of machinery and large-scale factories that were replacing manpower at the time. In particular, the systematic proliferation of production was one of the major objectives of industry 1.0.

The Second Industrial Revolution developed in the United States in the 1860s and 1890s (Leighton, 1970: 3). This time, business areas began to shift from manufacturing to services (Blinder, 2006: 116). The primary aim here is to ensure the transition to serial production Fordist. Technically Fordism; *"It is a form of production in which industrial production is realized in a significant mass production, the division of labor and job descriptions are made strictly, product standardization brings productivity increases and demand increases accelerate this standardization"* (Eraydın, 1992: 15).

As the third phase, the industry 3.0 phase which is closest to the concept of 4.0, which is accepted as the beginning of the digital transformation, has emerged from the fact that the production systems are made with the management systems. In other words, this phase enabled the advancements in communication technologies to participate in the production process, enabling the transition to flexible production systems that can adapt to changing conditions (Drath & Horch, 2014: 57; Sayer & Ülker, 2014: 67; Can & Kıymaz, 2016: 109; Qin & Liu & Grosvenor, 2016: 173; Lu, 2017: 3; Yüksel & Genç, 2018: 330; Ünlü & Atik, 2018: 436). The term Third Industrial Revolution (UED) was first used by economist Rifkin (2011). He stated that the Third Industrial Revolution was the integration of internet technology with renewable energy (Rifkin, 2011).

Recently, the concept of industry 4.0 has been frequently encountered in the literature. The concept actually points to the Fourth Industrial Revolution. It was used for the first time in 2011 at the Hannover Fair in Germany (Schwab, 2017: 16). The necessity of redesigning the production processes (Szegedi et. al., 2019) in order to respond quickly to consumer demands due to the constant change in consumer demands has brought this concept to the agenda. The Fourth Industrial Revolution (Industry 4.0) is a new industrial revolution (Xu et. al., 2018) that brings more intelligent action to the physical world to create a digital enterprise that incorporates intelligent digital technologies and advanced production and operation techniques, as well as communicating, analyzing and using data (Cotteler & Sniderman, 2017: 2). In the new systems aiming at production with automation systems, there are continuity in communication and cooperation and intelligent production systems formed by machines. It is aimed to create self-managed production processes both in order to reduce the need for human factor in production processes and to reduce human-induced errors. Thus, reduced costs, improved resource efficiency and switching to innovation-based growth is expected to increase its competitiveness. For this reason, they follow the process of transformation to Industry 4.0 in both developed and developing countries.

The concept of digital transformation is a harbinger not only of today but also of information and technology-based innovations and applications that are planned to be made in the future (Atar, 2019: 102). Because the said change and transformation occurs as a result of a certain accumulation and continues to occur with the rapid progress of science and technology. It is believed that the concept of industry 4.0, which is constantly changing and innovating, which takes communities to digitalization, will be different from its present state in terms of production, consumption structure, processes, and digitalization when we reach the 2050s (Bağcı, 2018: 124). Table 1 shows the emerging technologies and industrial revolutions.

Digital marketing can be defined as the digital marketing method that is realized by using digital channels and realizes all marketing applications in digital environment (Bulunmaz, 2016: 358). Digital marketing; It enables interaction with the target audience through the Internet, mobile platforms and social media channels. Digital marketing provides competitive advantages primarily due to differences such as being able to provide great financial benefit and being open to innovation, creativity and innovation and simultaneous interaction with the target audience.

Digital marketing is growing and developing around the World (Watson et. al., 2013). The basis of this is the widespread use of hand-held electronic communication devices such as mobile phones, digital music players and wireless internet access devices worldwide. With the increase in the usage of smartphones and tablets, a platform called mobile platforms has emerged and this area is used as a very important digital marketing channel. This situation necessitated two perspectives for institutions to make the content produced compatible with mobile platforms and the necessity of content-specific content production (Bulunmaz, 2016: 358). Nowadays, smartphones have become an active tool for social media. The use of smartphones, social networking applications such as Twitter, Facebook and LinkedIn; Video sharing sites such as YouTube; Photo sharing applications such as Instagram and information retrieval such as Wikipedia have become much more important than web pages (Safko, 2012: 464). This area, called social media, represents the social networks and social platforms used by individuals in their interactions and sharing with each other. In addition, this platform constitutes the digital marketing channel. Here, institutions communicate with their target audiences and carry out promotional activities through marketing practices.

Digital marketing consists of four steps: Acquire, Convert, Measure & Optimize, Retain & Grow. This marketing activity is based on gaining customers and creating awareness for their own content. In traditional marketing, however, the customer has to choose from an unlimited number of stimulants and variables. This feature separates the two. To explain activities called four steps. In the "Obtain" step, activities are carried out in order to attract the customer to the website or the page where the sales is performed; In the "Win" step, there are activities that will help to achieve the goals after the customer arrives at the website. While the "Measure-Optimize" step is very important for understanding what is wrong and what is done correctly and comparing the organization with the competitors, the last step is "Adopt, Enlarge" as the efforts made to satisfy the existing customers and make them permanent customers (Duncan & Everett, 1993).

With all these explanations, how is the integration of digital transformation and digital marketing with the tourism sector shaped? What is the reflection of this transformation, called Tourism 4.0, to the sector?

Digital Transformation in Tourism

The steam engine, invented in 1763, is also regarded as the beginning of the first industrial revolution, including tourist transport (Gierczak, 2011: 275). The technologies and inventions developed during this period also had a significant impact on the tourism industry. The application of steam engines to trains and ships can be seen as the most important invention for the tourism industry (Topsakal et. al., 2018). With the development of tourist transport in the 1820s San Sebastian destination SPA (Sanus Per Aquam) has been the destination of many tourists to use facilities (Larrinaga, 2005: 93). The seaside visit also started through SPA facilities. During these years, seaside travels started with demands for health, recovery and fashionable recognition (Beckerson & Walton, 2005: 55). During this period, with the change and ease of transportation of steam engines, Thomas Cook, a gardener in 1841, organized the first tour organization for Teetotalers Club members between Loughborough and Leicester. Thomas Cook after this tour events around the world in thirty years in 1845, has established a travel agency, which has become an agent (Enzensberger, 1996: 128).

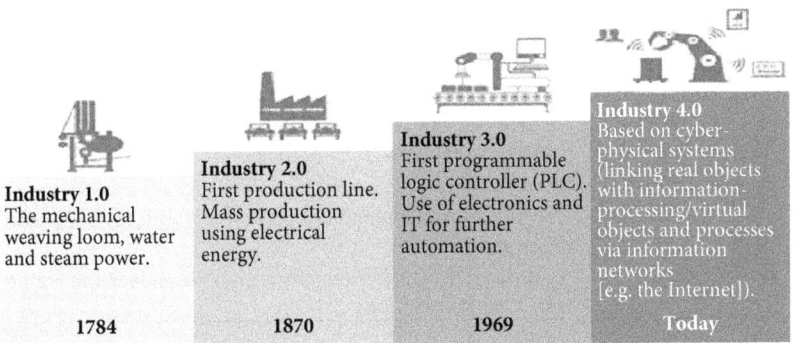

Figure 1: Historical Development of Industry 4.0. Source: (RSA Solutions. 2015)

Table 1: First, Second and Third İndustrial Transformation Technologies

Period	Date	Technology
First İndustrial Revolution	1631	David Ramsay has been patented by the British government for a steam-powered heated water pump.
	1712	Thomas Newcomen developed the first steam engine in the UK.
	1763	James Watt invented a steam engine.
	1764	James Hargreaves invented a new version of the traditional spinning machine.
	1765	Richard Reynolds introduced the first iron rails in the UK.
	1769	Richard Arkwright invented a water-spinning cotton machine.
	1802	Richard Trevithick developed the steam locomotive.
	1807	The American, Robert Fulton, used the steam engine on ships.
	1812	Steam engine was first used in locomotives.
	1825	George Stephenson has operated the Locomotion locomotive along a 32 km track in England. This is the first time a steam locomotive operated on a public railway track.
Second Industrial Revolution	1840	The first regular cross-ocean steamer flights began.
	1844	Samuel Morse opened the first commercial telegraph service in the USA.
	1876	Alexander Graham Bell - Telephone
	1877	Thomas Edison invented the phonograph.
	1879	Thomas Edison flooded the street half a mile in Menlo Park, New Jersey.
	1880's	Conveyor belts / assembly lines with electricity and light bulbs; Electric lamps = 24 hours working.
	1880's	Electric trams / cars 1889 German engineer Gottlieb Daimler has developed an engine similar to that of today's cars.
	1895	Guglielmo Marconi - Wireless Telegraph / Radio.
	1908	Henry Ford introduced the Model T.
	1910	Hydroelectric power.
Third Industrial Revolution	1960s	Computer.
	1980s	Personal computers.
	1990s	Internet
	2008	Google's Android operating system works with HTC Dream phone introduced.
	2010	Apple filled the gap between smart mobile phones and laptops with a tablet computer.

Source: (Topsakal et. all., 2018)

The second industrial revolution, which began at the end of the 19th century, led to significant technological progress in transport and production. Automobile, steamer, antenna and telegraph technologies have started to affect all kinds of developments from economy to social change (Roberts, 2015: 2). The main facilitating technologies of the second industrial revolution were the change in power supplies (electric power), transport (railways, automobiles); is the development of iron and steel production and the invention of the bulb. In the 19th century, the steam engine became widespread use by rail and steam vessels, followed by the invention of the gasoline engine and the development of hard surface roads, which allowed the development of motor tourism.

Cunard, a British company in 1840, made its first transatlantic flights. Passenger transport has been developed largely due to immigration and has been used for tourism purposes since World War I (Gierczak, 2011: 275). Transatlantic travel gained momentum in the 1860s and the concept of the big tour emerged. Translantic travels are motivated by improved transport, increased wealth, curiosity and the lure of new experience (Lickorish & Jenkins, 2006: 18). Developments in the sense of hotel management in 1870, Hotel Ezcurra in 1884 and Hotel Bermejo in 1884 for tourism purposes began to serve in Spain, in 1881, Hotel de Inglaterra and in 1884, Hotel Continental was put into service (Larrinaga, 2005: 97). In 1872, Thomas Cook organized his first world tour with a ship called 'Oceanic'. Traveling by airways started in 1918 and 1919 with regular lines on Paris-Brussels and London-Paris routes and the first regular flights on Berlin-Leipzig routes (Gierczak, 2011: 275).

After 1945, tourism has entered a new phase of development with mass travel. In 1953, the Vickers Viking, a twin-engine aircraft of 36 people, was designed to take tourists on a two-day tour of Lyon, Barcelona, Madrid, Tangiers, Casablanca and Agadir (Gierczak, 2011: 275). During this period, the development of computers brought technological innovations that led to major changes in the management of operations. Information technologies and especially the Internet have increased the demand and supply of tourism worldwide (Kiprutto et. al., 2011).

The third concept of industrial revolution was first used by economist Rifkin (2011). The first two industrial revolutions focused on goods (especially manufactured goods), while the third focused on services and goods (especially the integration of services and/or goods) (Tien, 2012: 262). The third period is both global and local, hence the term üy linguistic ortaya. TIR has begun to change the way we work, produce and have fun. It has radically changed the way we plan and manage cities and regions. It has also led to the globalization of production and the acceleration of business (Roberts, 2015: 2).

Although the technological developments in the 1990s did not change the organizational structures in the tourism sector, internet usage became widespread. The tourism sector, which is one of the most suitable sectors for the use of this internet, is benefiting from the internet as in the past in many areas related to tourism such as ticket sales for tourists, hotel reservations, car rental and guidance services (Kaya, 2009: 27).

Nowadays, Information and Communication Technology (ICT) is located within the fourth industrial revolution triggered by the development. Industry 4.0 is intelligent automation of technology-based and cyber-physical systems (Topsakal et. al., 2018). With Industry 4.0, the word olma being smart 'has entered the literature. Industry 4.0 has emerged with the introduction of new generation technologies such as robotic, analytical, artificial intelligence and cognitive technologies, nanotechnology, quantum informatics, wearable technologies and the internet of objects into the lives of people and businesses (Cotteler & Sniderman, 2017: 2). Since the tourism industry is rapidly adapting to technology, the use of Industry 4.0 technologies in the tourism industry has come to the fore and some academic studies have been started on this subject.

As the tourism industry is affected by new Technologies (Pamukçu & Tanrısever, 2019: 7), the tourism industry is rapidly adapting to new Technologies (Dominguez et. al., 2015). Especially the rapidly developing smartphone applications provide the convenience of tourists to get information about many issues such as making hotel or flight reservations, information about destinations and calculating the exchange rate. Smartphones, navigation, information search and social networks, etc. applications such as mobile applications support many different features (Wang et. al., 2012: 371). Furthermore, tourists can share their experiences not only while they are visiting, but also after the trip through smart phones (Topsakal et. al., 2018). In recent years, smartphone applications have emerged as a new tool that helps tourists create experiences. Therefore, when the smart phone and smart phone applications Given the potential impact is better understood how they could shape the travel experience for tourists of mobile applications (Wang et. al., 2011).

Smart tourism, which emerged with the new generation technologies developing with Industry 4.0, can be seen as the transition process from traditional tourism to e-tourism logic (Hwang et al., 2015: 164). Nowadays, with the integration of industry 4.0 into tourism, the information and communication oriented tourism oriented digital transformation elements; Destination Management System (DMS), Centralized Reservation System (CRS), Customer Relationship Management (CRM), Management Information Systems (Electronic Information Systems) used in airports, electronic material transfers, digital telephone networks, mobile communication equipment (Poon, 1993; Kaya, 2009: 29). In addition, in the last minute room reservation and entrance applications that can be made from mobile environments, as of 2012, it has started to serve the tourism sector in a way that will increase the quality of life of the tourists and facilitate their lives (Şanlıöz et. al., 2013: 251).

Virtual reality applications, smart remote video monitoring system, smart hotel management systems that affect the sustainability of businesses, smart ticket system, smart tour guide system, smart travel agency system, which enable tourists to see the destination center in advance (Atar, 2019: 102). With the emergence of the concept of tourism 4.0, it is seen as other information and technology based applications that have started to be used.

Some studies on this issue reveal that digital transformation is considered important for tourists. For instance; Kramer et. al. (2008) reveals that tourist preferences can be easily changed via smartphones. Studies on tourist behavior reveal that tourists often use their mobile phones for photographing and audio/video recording.

Digital Marketing in Tourism

There are many statements about digital marketing. . Some of those; internet marketing, online marketing, interactive marketing or e-marketing. As mentioned in the previous section, digital marketing basically consists of four steps. These steps are; Obtain, gain, measure and optimize, claim and grow (Duncan & Everett, 1993). These four methods are successive, one of which is the continuation of the other, and is of strategic importance for businesses aiming to reach their customers through digital marketing activities. The development of tourism, which started with mass tourism, shapes the present day of tourism marketing as a result of the cognitive activities of each individual.

The concept of digital change; It can be defined as the process of transitioning to business transformation forms with digital, social and mobile tools that add value to the customer by utilizing new technologies, aiming to develop the processes of putting the business to work and enabling the companies to increase their impact and competencies. Digital marketing practices vary according to customers, companies and sectors due to the characteristics of the digital world. This change is reflected in the products and services of companies or firms, their working methods, their working strategies, decision-making processes and their experiences (Koçak Alan et. al., 2018: 496). This area, which we call the digital world, provides access to product-services and information much faster than before. Mobile applications, social media and other applications used through smartphones and tablets are areas that allow consumers to access unlimited information while performing their purchasing and ordering activities (Smith, 2011: 492).

In the tourism sector, marketing activities are carried out through many different digital marketing channels. These channels can be listed as motor search engine marketing (SEM-SEO), e-mail marketing (E-mail marketing), mobile marketing (mobile marketing), digital content marketing (digital content marketing).

Search engine marketing (SEM-SEO) involves attracting visitors to websites and bringing the right visitor to the right product. The more accurate user behaviors are identified and improved by firms, the more accurately they attract visitors to their web pages. Search engines are the first resource in the internet environment that people often refer to in accessing information. Thanks to the Internet, consumers can obtain more information about a product or service before purchasing a product or service and compare it between substitute products and services. The most popular search engine providers in the world are Google, Yahoo, Bing, Baidu, Yandex, DuckDuckGo, and Google is the most popular (Sezgin & Parlak, 2019: 43).

E-mail marketing, when evaluated from the point of view of tourism enterprises, enables them to obtain multiple advantages by providing direct contact with their products and services via e-mail for existing and potential markets, these;

- Engaging new and existing customers of tourist products and services,
- To improve brand awareness and market position,
- Provide access to research data on all other business and marketing objectives,
- Unlike traditional marketing types, e-mail marketing offers much cheaper and more efficient reporting.

In line with these objectives, businesses carry out e-mail newsletter marketing independent of their physical offices in order to have full control and minimize the costs of their campaigns in order to maximize the results. (Cox & Koelzer, 2005).

The great habit of using mobile devices has become inevitable as a marketing tool with the presence of companies on digital marketing channels. Mobile promotions, short message services (SMS), in-app messaging, social media, e-mail notifications can reach consumers using many formats. Examining the psychological characteristics of consumers is of great importance in order to develop successful mobile marketing strategies in the tourism sector (Tan et. al., 2017). Consumers use the information they need during their travels to obtain prices or to make comparisons and to compare them during the purchase phase. The fact that smart phones are seen as a means of communication and sharing of mobile technologies has enabled it to be widely adopted in the tourism industry and used as a mobile guide for the discovery of location-based services (Tussyadiah, 2013). Especially for hotels and travel agencies, mobile devices provide the opportunity to interact with the consumer at any time during their digital marketing activities. Tourists with previous experience using smartphones will tend to use them again. Therefore, tourism marketing organizations should develop applications that can improve the quality of travel information, provide more effective services and increase the travel information of the users and contribute to the development of marketing objectives.

With the social media revolutionizing the communication and access of the media to the society, today, digital content marketing activities started to use creative snaps that encourage users to produce content by using the photos, videos, blogs and comments made by users on the internet rather than content production. One of the most striking benefits of such consumer-generated content is the ability to convey emotions and link them to a particular industry or brand experience. As an example of content marketing, we can cite Marriott's in-house content studio in 2014. In 2015, with their short film Kiss French Kiss ", they made a huge impact on the internet and experienced huge increases in reservations in less than two months. The "Two Bellmen" video series, which is also one of the company's video projects, has reached more than 9 million views.

Results

In the tourism sector, as in every sector to position itself in the global market and have fallen behind technologically leading position followed by tourism businesses who want to come and to adapt developments. Social media constitutes a rapidly expanding domain compared to traditional mass media. These areas, which we use as digital transformation, are important for bringing together people and communities and increasing the interaction and sharing between them. Digital transformation has become a part of our lives and even a focus. Individuals have started to use the internet more and more every day. For this reason, digital transformation is widely accepted as a necessity and importance element for enterprises. Because people often use digital channels to plan their travels. The absence of a tourism business in the digital world in such a period means that it lags behind all other competitors. Digital platforms can also be described as important elements for the recognition and branding of businesses.

With digital marketing channels in tourism today, it has become easier and faster to access information. Along with the increase in alternative tourism activities, tourism consumption patterns and habits have also changed. This transformation is of course related to technological developments and changes in the global economy. The most prominent purpose of businesses is to continue to exist. From this point of view, businesses should determine how to reach the target audience, the most appropriate method for the purpose and develop their advertising and promotion strategies. In this direction, necessary efforts should be made and digitalization should be utilized to the extent necessary. It is thought that the importance of digitalization in the stage of developing appropriate strategies to keep in mind as advertisement, promotion and marketing stages will benefit in the long term.

References

Atar, A. (2019). Dijital Dönüşüm ve Turizme Etkileri, (Ed.: Sezgin, M. & Özdemir Akgül, S. & Atar, A.) *Turizm 4.0 (Dijital Dönüşüm)* içinde (p. 99-114). Ankara: Detay Yayıncılık.

Bağcı, E. (2018). *Endüstri 4.0: Yeni Üretim Tarzını Anlamak.* Gümüşhane Üniversitesi Sosyal Bilimler Enstitüsü Elektronik Dergisi, 9(24), 122-146.

Beckerson, J. & Walton, J. K. (2005). Selling Air: Marketing the Intangible at British Resorts, (Editor: Walton, J. K.) In *Histories of Tourism Representation, Identity and Conflict* (p. 55–68). Clevedon: Channel View Publications.

Blinder, A. S. (2006). *Offshoring: The Next Industrial Revolution?*, Foreign Affairs, 85(2),113–128.

Bulunmaz, B. (2016). *Gelişen Teknolojiyle Birlikte Değişen Pazarlama Yöntemleri ve Dijital Pazarlama*, TRT Akademi, 1(2), 348-365.

Can, A. V. & Kıymaz, M. (2016). *Bilişim Teknolojilerinin Perakende Mağazacılık Sektörüne Yansımaları: Muhasebe Departmanlarında Endüstri 4.0 Etkisi.* Sosyal Bilimler Enstitüsü Dergisi, CİEP Özel Sayısı, 107–117.

Clemons, E. K. & Row, M. C. (1991). Information Technology at Rosenbluth Travel: Competitive Advantage in a Rapidly Growing Global Service Company. Journal of Management Information Systems, 8(2), 53–80.

Cotteler, M. & Sniderman, B. (2017). *Forces of Change: Industry 4.0.*, Deloitte Touche Tohmatsu Limited, New York.

Cox, B. & Koelzer, W. (2005). Internet Marketing Za Hotele, Restorane I Turizam. *M plus Zagreb*.

Çetinkaya, A. Ş. (2007). Bilişim Teknolojilerinin Konaklama İşletmeleri Performansına Etkileri: Beş Yıldızlı Otellere Yönelik Bir Araştırma. Selçuk Üniversitesi Sosyal Bilimler Enstitüsü. Konya.

Drath, R. & Horch, A. (2014). *Industrie 4.0: Hit or hype?* IEEE Industrial Electronics Magazine, 8(2), 56–58.

Dominguez, C. D., Hernandez, R. M., Talavera, A. S. & Lopez, E. P. (2015). *"Strategic Determinants in the Theoretical Framework of the "Smart Islands": The Case of the Island of El Hierro"*, t-FORUM Global Conference: Tourism Intelligence in Action, (s. 1–28). Naples, Italy.

Duncan, T. R. & Everett, S. E. (1993). *Client Perceptions of İntegrated Marketing Communications.* Journal of Advertising Research, 33(3), 30–39.

Engin, B. (2011). Yeni Medya ve Sosyal Hareketler. (Editörler: Binark, M. & Işık, B. F.) Cesur Yeni Medya "Wikileaks ve 2011 Arap İsyanları Üzerine Tartışmalar" içinde, (s. 33–37). Alternatif Bilişim: İstanbul.

Enzensberger, H. M. (1996). *A Theory of Tourism. New German Critique,* No. 68, Special Issue on Literature, 68(Spring – Summer), 117–135.

Eraydın, A. (1992). *Post-Fordizm ve Değişen Mekansal Öncelikler.* Orta Doğu Teknik Üniversitesi, Mimarlık Fakültesi.

Gierczak, B. (2011). *The History of Tourist Transport after the Modern Industrial Revolution,* Polish Journal of Sport Tourism, 18(4), 275–281.

Hwang, J., Park, H. Y. & Hunter, W. C. (2015). Constructivism in Smart Tourism Research: Seoul Destination İmage. Asia Pacific Journal of Information Systems, 25(1), 163–178.

Kagermann, H., Helbig, J., Hellinger, A. & Wahlster, W. (2013). Recommendations for İmplementing the Strategic İnitiative Industrie 4.0: Securing the Future of German Manufacturing İndustry; Final Report of the Industrie 4.0 Working Group. National Academy of Science and Engineering. 13–78.

Kaya, İ. (2009). *Otel İşletmelerinde Kullanılan Bilgi-İletişim Teknolojilerinin İşletmenin Farklı Boyutlarında Yarattığı Değişimler.* Cag University Journal of Social Sciences, 6(2).

Kiprutto, N., Kigio, F. W. & Riungu, G. K. (2011). *Evidence on the Adoption of e-Tourism Technologies in Nairobi,* Global Journal of Business Research, 5(3), 55–66.

Koçak Alan. A., Tümer K., E. & Erişke, T. (2018). *İletişimin Yeni Yüzü: Dijital Pazarlama ve Sosyal Medya Pazarlaması.* Electronic Journal of Social Sciences, 17(66).

Kramer, R., Modsching, M. & Ten Hagen, K. (2008). Development and Evaluation of A Context-Driven, Mobile Tourist Guide. International Journal of Pervasive Computing and Communications, 3(4), 378-399.

Larrinaga, C. (2005). A Century of Tourism in Northern Spain: The Development of High-quality Provision between 1815 and 1914, (Editor: Walton, J. K.) *In Histories of Tourism Representation, Identity and Conflict* (p. 88-103), Clevedon: Channel View Publications.

Leighton, D. S. R. (1970). *The Internationalization of American Business. The Third Industrial Revolution*, Journal of Marketing, 34(3), 3-6.

Lickorish, L. J. & Jenkins, C. L. (2006). *An Introduction to Tourism*, Butterworth-Heinemann, Jordan Hill, Oxford.

Lu, Y. (2017). *Industry 4.0: a Survey on Technologies, Applications and Open Research İssues.* Journal of Industrial Information Integration, 6, 1-10.

Pamukçu, H. & Tanrısever, C. (2019). Turizm Endüstrisinde Dijital Dönüşüm, Dijital Dönüşüm ve Turizme Etkileri, (Ed.: Sezgin, M. & Özdemir Akgül, S. & Atar, A.) Turizm 4.0 (Dijital Dönüşüm) içinde (p. 2-28). Ankara: Detay Yayıncılık.

Papp, I., Szegedi, Z. & Malouin, M. (2018). *The Effect of Industry 4.0 in Shaping the Strategy of Logıstıcs in Central-Eastern Europe.* International Journal of Management and Applied Science. 4(10), 10-15.

Poon, A. (1993). *Tourism, Technology and Competitive Strategies*, CABI Publishing, UK.87-96.

Porter, M. E. (1980). Competitive Strategy. New York: Free Press.

Qin, J., Liu, Y. & Grosvenor, R. (2016). *A Categorical Framework of Manufacturing for İndustry 4.0 and Beyond.* Procedia Cirp, 52, 173-178.

Rifkin, J. (2011). *The Third Industrial Revolution: How Lateral Power is Transforming Energy, the Economy, and the World*, Palgrave Macmillan, New York.

Roberts, B. H. (2015). *The Third Industrial Revolution: Implications for Planning Cities and Regions*, Urban Frontiers Working Paper 1.

Sezgin, M. & Parlak, O. (2019). Dijital Dönüşüm ve Turizme Etkileri, (Edit.: Sezgin, M. & Özdemir Akgül, S. & Atar, A.) Turizm 4.0 (Dijital Dönüşüm) içinde (p. 29-58). Ankara: Detay Yayıncılık.

Smith, K. T. (2011). Digital Marketing Strategies That Millennials Find Appealing, Motivating, Or Just Annoying. Journal of Strategic Marketing, 19(6), 489-499.

Sayer, S. & Ülker, A. (2014). *Ürün Yaşam Döngüsü Yönetimi*. Engineer & the Machinery Magazine, 55(657), 65-72.

Safko, L. (2012). *The Social Media Bible – Tactics, Tools & Strategies for Business Success*. Third Edition. John Wiley & Sons, Inc, New Jersey.

Schwab, K. (2017). *The Fourth İndustrial Revolution*. Currency. 10–17.

Segars, A. H. & Grover, V. (1995). The İndustry-Level İmpact of İnformation Technology: An Empirical Analysis of Three İndustries. Decision Sciences, 26(3), 337–368.

Szegedi, Z., Papp, I. & Nick, G. A. (2019). *The Appearance of Digitalization in the Strategies of SMEs in Central-Eastern Europe*.

Şanlıöz, K., Dilek, E. & Koçak, N. (2013). *Değişen Dünya, Dönüşen Pazarlama: Türkiye Turizm Sektöründen Öncü Bir Mobil Uygulama Örneği*. Anatolia: Turizm Araştırmaları Dergisi, 24(2), 250–260.

Tan, G. W. H., Lee, V. H., Lin, B. & Ooi, K. B. (2017). *Mobile Applications in Tourism: The Future of the Tourism İndustry?* Industrial Management & Data Systems, 117(3), 560–581.

Tien, J. M. (2012). *The Next Industrial Revolution: Integrated Services and Goods*, Journal of Systems Science and Systems Engineering, 21(3), 257–296.

Topsakal, Y. (2018). *"Akıllı Turizm Kapsamında Engelli Dostu Mobil Hizmetler: Türkiye 4.0 İçin Öneriler"*, Journal of Tourism Intelligence and Smartness, 1(1), 1–13.

Topsakal, Y. & Çelik, P. (2017). Turizmde Yeni Bir Strateji: Akıllı Destinasyonlar, (Editörler: Haşit, G., Çiftçi, H. ve Merter, M. E.) *Sosyo Ekonomik Stratejiler 1 İşletme* içinde (p. 96–106) IJOPEC Publication, Londra.

Topsakal, Y., Yüzbaşıoğlu, N. & Çuhadar, M. (2018). *Endüstri Devrimleri ve Turizm: Türkiye Turizm 4.0 Swot Analizi ve Geçiş Süreci Önerileri*. Süleyman Demirel Üniversitesi İktisadi ve İdari Bilimler Fakültesi Dergisi, 23, 1623–1638.

Tussyadiah, I. (2013). When Cell Phones Become Travel Buddies: Social Attribution to Mobile Phones in Travel. *Information and Communication Technologies in Tourism 2013* (p. 82–93). Springer, Berlin, Heidelberg.

Ünlü, F. & Atik, H. (2018). *Türkiye'deki İşletmelerin Endüstri 4.0'a Geçiş Performansı: Avrupa Birliği Ülkeleri ile Karşılaştırmalı Ampirik Analiz*. Ankara Avrupa Çalışmaları Dergisi, 17(2), 431–463.

Wang, D., Park, S. & Fesenmaier, D. R. (2011). *An Examination of Information Services and Smartphone Applications*, 16th Annual Graduate Student Research Conference in Hospitality and Tourism, 6-8 January 2011, Houston, Texas.

Wang, D., Park, S. & Fesenmaier, D. R. (2012). *The Role of Smartphones in Mediating the Touristic Experience*, Journal of Travel Research, 51(4),371-387.

Watson, C., McCarthy, J., & Rowley, J. (2013). *Consumer Attitudes towards Mobile Marketing in Smart Phone Era.* International Journal of Information Management, 33(5), 840-849.

Xu, L. D., Xu, E. L. & Li, L. (2018). Industry 4.0: State of The Art and Future Trends. International Journal of Production Research, 56(8), 2941-2962.

Yüksel, M. & Genç, K. Y. (30 Kasım 2018). *Endüstri 4.0 ve Liderlik.* Setscı-Conference Indexing System, 2nd Intenational Symposium on Innovative Approaches in Scientific Studies, 3, 338-341.

Hakan KENDİR

12 Cultural Heritage Tourism Inventory in Tokat Province

Introduction

Cultural heritage consists of concrete and abstract elements in terms of content. It contains important elements that should be taken from past generations and transferred to future generations. Cultural heritage also provides humanity with tangible information about cultures such as handicrafts, architectural-archaeological sites, clothing and food and drink as well as intangible elements such as life styles, traditions, belief rituals and traditional production methods (MacDonald & Jolliffe, 2003: 308) and creating touristic attractive (Dönmez & Türkmen, 2018). This has motivated the manifestation of the concept of cultural heritage tourism.

As a touristic activity, cultural heritage tourism is defined as the use of heritage assets as a tourism product to attract visitors to destinations that contain cultural and natural heritage resources. On the other hand, it is known that mass tourism causes the destruction of cultural and natural heritage sites. Therefore, it is essential to protect cultural heritage resources for future generations and to enable the continuation of their use as a tourism product (Fyall & Garrod, 1998: 213).

Supply sources such as archaeological sites, important architectural structures, ancient cities and museums are frequently visited by tourists participating in cultural heritage tourism. Furthermore, many activities such as participating in local festivals and observing the behavior of local people are also considered within this scope (Arslan et. al., 2019). The different cultural and natural heritage sites in the world are generally attracting tourists to a great extent. For example, the Colloseum in Rome and its environs were visited by more than 7 million tourists in 2017 (Statista, 2019a) while the Stonehenge ruins in England were visited by more than 1.58 million tourists (Statista, 2019b). The ancient city of Ephesus in Izmir Turkey was visited by 1.5 million domestic-foreign tourists while Konya Mevlana Museum hosted 2.8 million visitors, Nevsehir Goreme Open Air Museum 1.1 million (Dösim, 2018) and Safranbolu Houses were visited by approximately 315 thousand local and foreign tourists in 2018 (Safranbolu Tourist Information Office, 2018). These figures show the importance of cultural and natural heritage elements for tourism movements. In the light of this information, the aim of the study is to determine the concrete cultural heritage tourism inventory of Tokat province which attracts attention with its historical and natural resources. In addition, the study aims to contribute to the protection of cultural heritage resources of Tokat and to raise awareness for decision makers on this issue.

General Information about Tokat Province

Tokat province is located in the central part of the Black Sea Region bordered by Yozgat and Sivas provinces in the south and east, Samsun and Ordu provinces in the north, and Amasya province in the west. Tokat has 11 districts (Map 1). Tokat province covers an area of 10.072 km² and has a total population of 612.646 as of 2018 (TURKSTAT, 2019). The average altitude of Tokat center is 623 meters. However, altitude varies according to the districts.

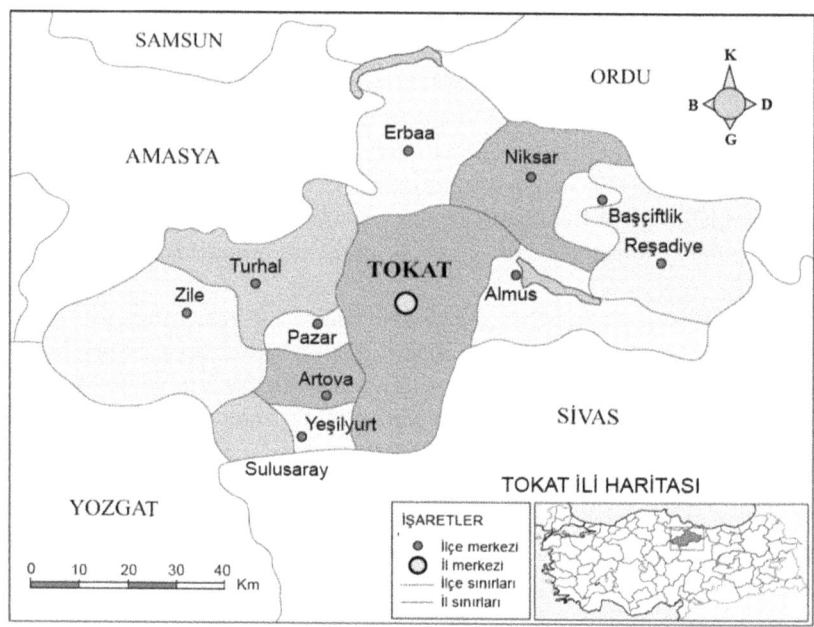

Map 1. Study Area Location (Coğrafya Harita, 2019)

All kinds of agricultural products suitable for the climate are grown in the fertile plains of Tokat. Forested areas covering 48.18 % of Tokat province have interesting natural beauties in terms of tourism. In the low altitude regions of Tokat (Erbaa-Niksar), maquis species such as Red Pine, Cedar and Lebanese Cedar which are suitable for the Mediterranean climate are observed while the higher parts include tree species such as larch, scotch pine, fir, hornbeam. The highways that connect the Black Sea to Central Anatolia and the Mediterranean and western Turkey with East Anatolia pass through Tokat. On the other hand, the railway connecting Samsun to the central parts passes through the districts of Artova, Zile and Turhal of Tokat. Furthermore, Tokat Airport, which is currently under construction with a capacity of 2 million passengers, is expected to open in mid-2020 (Tokat Provincial Directorate of Culture and Tourism, 2019a). Therefore, Tokat province has an important potential in terms of both trade and tourism as well as transportation.

Tokat province has hosted many different civilizations throughout history. Tokat has been dominated by Hattat, Hittite, Phrygian, Roman, Byzantine, Danishmendli, Ilkhanid, Seljuk and Ottoman rule throughout this term and referred to with different names. The most well-known names are Comano Pontika, Komana, Evdoxia, Dokia, Dokat, Kah-Cun, Sobaru, Darün-Nusret, Darün-Nasr and Tokat. One of the first settlements of the Turks who conquered Anatolia in the Battle of Malazgirt in 1071 was Tokat. The State of Danishmendli played a major role in the Turkification of the region and the development of its architectural structure. Subsequently Tokat was connected to the Seljuks and hosts many of the first works of Turkish architecture. Tokat Castle, which has played an important role throughout history and has an inaccessible quality, maintained its importance during the Ottoman period. Vlad Tepes, also known as Vlad the Impaler or Dracula, was imprisoned in Tokat Castle for a long time. On the other hand, architectural works were built and many important personalities were raised in Tokat during the Ottoman era. Tokat gained the status of province after the proclamation of the Republic in 1923 (Tokat Municipality, 2019). Atatürk showed the importance he attached to the region by visiting Tokat several times before and after the proclamation of the Republic (Tokat Governorship, 2019). Tokat developed rapidly in the Republican period to become one of the most prominent provinces of the region with its educational institutions, industrial facilities and touristic values.

Tourism in Tokat Province

Located in the Central Black Sea region, Tokat has many attractions in terms of tourism with its historical and natural beauties. The number of domestic and foreign visitors to Tokat was 246.634 as of 2018. The number of tourists coming to Tokat is increasing from year to year, the increase in recent years is especially remarkable (Tokat Provincial Directorate of Culture and Tourism, 2019b). The reason for this are the touristic investments made in Tokat (Channel Tokat, restoration works, Kaz Lake, new hotels, restaurants, facilities, etc.) and promotion activities (including Ballıca Cave and Niksar in Unesco's Tentative Heritage List, geographically marked products, social media, etc.).

Tokat, which has just started to reveal its touristic values, has initiated a breakthrough in terms of facilities. The province has increased its capacity with new hotel investments. As of 2018, there are 1 5-star, 2 4-star, 7 3-star, 2 2-star and 4 tourism certified apart-hotels in Tokat. In addition, 16 tourism certified hotels have a total bed capacity of 1513. There are 3702 beds in 64 hotels certified by the Municipality. On the other hand, there are 2 tourism certified restaurants and numerous independent food and beverage establishments in Tokat (KTB, 2019). In addition, Tokat Airport, which has been planned for an annual capacity of 2 million passengers and started to be built in 2016 is expected to open in 2020, replacing the former low-capacity airport. With this development, the number of tourists coming to Tokat will increase significantly. In terms of road transport, the Gerede-Gürbulak Motorway project will pass through Tokat (Hürriyet, 2014). This project, which is planned to be completed in 2023, will make a positive contribution to the tourism of Tokat.

Especially cultural tourism, faith tourism, nature and thermal tourism are prominent among the types of tourism in Tokat province. In addition to these tourism types, Tokat province is also suitable for cave, ornitho-tourism trekking, paragliding, rafting and angling activities (Tokat Provincial Directorate of Culture and Tourism, 2019a). Being included in the Unesco World Heritage Tentative List in 2019, Ballıca Cave attracts much attention both in terms of cave tourism and health tourism and constitutes one of the most important tourism values of Tokat. In 2018, 140 thousand people visited Ballıca Cave, which is thought to treat respiratory diseases such as asthma, bronchitis and COPD (Hürriyet, 2019). In terms of cultural tourism, Tokat is home to many civilizations and has a great number of touristic values. These places are also frequently visited by tourists coming to Tokat.

Tokat's Tangible Cultural Heritage Tourism Inventory

In Tokat, which has a history of thousands of years, it is possible to find works belonging to each period in the developing process. Tokat is like an open air museum with this feature. Table 1 shows the tangible (historical structures) cultural heritage inventory of Tokat province.

Cultural Heritage Tourism Inventory in Tokat Province

Table 1: Tokat's Tangible Cultural Heritage Inventory

Categories	Place/Building Name	Location	Time/Period
	Mevlevilodge	Tokat/Center	1638 (Ottoman)
	Garipler Mosque	Tokat/Center	1080–1090 (Danishmend)
	Tokat Grand Mosque	Tokat/Center	12nd cen. (Danishmend)
	Ali Pasha Mosque	Tokat/Center	1572 – 1573 (Ottoman)
	Mahmutpasha Mosque	Tokat/Center	17th cen. (Ottoman)
	Meydan Mosque	Tokat/Center	1485 (Ottoman)
	Behzad-I Veli Mosque	Tokat/Center	1536 (Ottoman)
	Takyeciler Mosque	Tokat/Center	15th cen. (Ottoman)
	Dodurga Village Mosque	Tokat/Center	15th cen. (Ottoman)
	Silahtar Ömer Pasha Mosque	Tokat/Erbaa	18th cen. (Ottoman)
	Niksar Grand Mosque	Tokat/Niksar	1145 (Danishmend)
	Çöreğibüyük Mosque	Tokat/Niksar	13rd cen. (Ilkhanid)
	Kesikbaş Mosque and Tomb	Tokat/Turhal	1759 (Ottoman)
	Zile Grand Mosque	Tokat/Zile	1591 (Ottoman)
Religious Buildings	Elbaşoğlu Mosque	Tokat/Zile	1801 (Ottoman)
	Beyazıt Bestami Mosque	Tokat/Zile	1206 (Seljuq)
	Boyacı Hasan Ağa Mosque	Tokat/Zile	1479 (Ottoman)
	Hubyar Dervish Lodge	Tokat/Almus	16th cen. (Ottoman)
	Esentimur Tomb	Tokat/Center	1314 (Ilkhanid)
	Sümbül Father Zawiya and Tomb	Tokat/Center	1292 (Ilkhanid)
	Ali Tusi Tomb	Tokat/Center	1230 (Seljuq)
	Burgaç Hatun Tomb	Tokat/Center	14th cen. (Ilkhanid)
	Ebu Şems House	Tokat/Center	1288 (Seljuq)
	Abdulmuttalip Zawiya	Tokat/Center	1318 (Ilkhanid)
	Şeyh Meknun Almshouse	Tokat/Center	13rd cen. (Seljuq)
	Horozoğlu Zawiyasi	Tokat/Center	15th cen. (Ottoman)
	Halef Sultan Zawiya	Tokat/Center	1291–1292 (Seljuq)
	Sefer Pasha Masjid and Tomb	Tokat/Center	1251 (Seljuq)
	Pir Ahmet Bey Tomb	Tokat/Center	15th cen. (Ottoman)
	Erenler Tomb	Tokat/Center	14th cen. (Ilkhanid)
	Sevdekar Murad Tomb	Tokat/Center	13rd cen. (Seljuq)
	Melikgazi Tomb	Tokat/Niksar	12nd cen. (Danishmend)

(continued on next page)

Table 1: (continued)

Categories	Place/Building Name	Location	Time/Period
	Sungurbey Zawiya and Tomb	Tokat/Niksar	12nd cen. (Danishmend)
	Kulak Dervish Lodge	Tokat/Niksar	12nd cen. (Danishmend)
	Şehit Komutan Mehmet Nurullah Tomb	Tokat/Turhal	1312 (Ilkhanid)
	Ahi Yusuf Tomb	Tokat/Turhal	1324 (Ilkhanid)
	Dazya Village - Ali Father Tomb	Tokat/Turhal	1370 (Eretna)
	Sheikh Nusreddin Tomb	Tokat/Zile	14th cen. (Eretna)
	Tokat Castle	Tokat/Center	(Roman)
	Niksar Castle	Tokat/Niksar	(Roman)
Castles	Turhal Castle	Tokat/Turhal	(Roman)
	Zile Castle	Tokat/Zile	(Roman)
	Boğazkesen Castle	Tokat/Erbaa	(Byzantine)
	Alipaşa Bathhouse	Tokat/Center	1572 (Ottoman)
	Pervane Bathhouse	Tokat/Center	13rd cen. (Ilkhanid)
	Sultan Bathhouse	Tokat/Center	13rd cen. (Ilkhanid)
Bathhouses	Paşa Bathhouse	Tokat/Center	1436 (Ottoman)
	Tekke Bathhouse	Tokat/Zile	17th cen. (Ottoman)
	Şehir Bathhouse	Tokat/Zile	1494 (Ottoman)
	Yeni Bathhouse	Tokat/Zile	16th cen. (Ottoman)
	Hatuniye (Meydan) Mosque Madrasah	Tokat/Center	1485 (Ottoman)
Madrasahs	GökMadrasah	Tokat/Center	1277 (Seljuq)
	Tokat Yağıbasan Madrasah	Tokat/Center	1148–1157 (Danishmend)
	Niksar Yağıbasan Madrasah	Tokat/Niksar	1158 (Danishmend)
	Comana Pontika	Tokat/Center	Early Bronze Age
	Maşat Tumulus	Tokat/Zile	B.C. 3000's
Ancient Cities	Horoztepe Tumulus	Tokat/Erbaa	B.C. 3000's
	Sebastapolis Ancient City and Roman Bath	Tokat/Sulusaray	B.C. 1st cen. (Roma)
	Artova Underground City	Tokat/Artova	Early Christianity
	Bolus-Aktepe Tumulus	Tokat/Center	Early Bronze Age

Cultural Heritage Tourism Inventory in Tokat Province 171

Table 1: (continued)

Categories	Place/Building Name	Location	Time/Period
Historical Mansions	Latifoglu Mansion	Tokat/Center	1746 (Ottoman)
	Tokat Culture House	Tokat/Center	18 – 19th cen. (Ottoman)
	Atatürk House and Etnografya Museum	Tokat/Center	19th cen. (Ottoman)
	Yüksek Kahve	Tokat/Center	20th cen. (Ottoman)
	Niksar Old Government House	Tokat/Niksar	1905 (Ottoman)
Inns and Caravansaries	Mahperi Hatun Caravansary	Tokat/Pazar	1238 (Seljuq)
	Taşhan – Voyvoda Inn	Tokat/Center	1626–1632 (Ottoman)
	Deveciler Inn	Tokat/Center	15th cen. (Ottoman)
	Sulu Inn	Tokat/Center	18th cen. (Ottoman)
	Pasha Inn	Tokat/Center	1752–1753 (Ottoman)
	Yazmacılar Inn	Tokat/Center	13rd cen. (Seljuq)
Bridges	Hıdırlık Bridge	Tokat/Center	1250 (Seljuq)
	Boğazkesen Bridge	Tokat/Erbaa	(Byzantine)
	Yer Bridge	Tokat/Erbaa	(Byzantine)
	Leylekli Bridge	Tokat/Niksar	(Roman)
	Talazan Bridge	Tokat/Niksar	13rd cen. (Seljuq)
	Hacı Boz Bridge	Tokat/Zile	17th cen. (Ottoman)
Other Cultural Buildins	Roman Arsenal	Tokat/Niksar	(Roman)
	Tokat Clock Tower	Tokat/Center	1902 (Ottoman)
	Sıkdişini Toilet	Tokat/Center	17th cen. (Ottoman)
	Arastalı Bedesten - Tokat Museum	Tokat/Center	15th cen. (Ottoman)
	Tombstone Museum	Tokat/Niksar	12nd cen. (Danishmend)
	Kırkkızlar Cupolas	Tokat/Niksar	1220 (Seljuq)
	Erzurumlu Emrah Tomb	Tokat/Niksar	1860 (Ottoman)

Source: Tokat Provincial Directorate of Culture and Tourism, 2019a; Tokat Governorship, 2019; Turkey's Culture Portal, 2019.

36 religious buildings, 5 castles, 7 bathhouses, 4 madrasahs, 6 ancient cities, 5 historical mansions, 6 inns and caravansaries, 6 bridges and finally 7 other cultural buildings are listed in Table 1 as cultural assets of Tokat province. All of these structures are still standing today and are being used at the same time. Apart from these, there are 118 registered buildings in the city center of Tokat (Akın & Özen, 2010) and 3600 historical houses in Zile district of Tokat (Tokat Provincial Directorate of Culture and Tourism, 2019a).

According to Table 1, the cultural assets mainly oriented to faith tourism in Tokat province stand out. Most of them are mosques and the rest are tombs and lodges. On the other hand, bathhouses are among the prominent cultural heritage assets of Tokat (Figure 1). The Roman Bath in the Sulusaray district is a historical site. Furthermore, the bathhouses, which were generally built after the 13th century, have been restored and are still in use.

Figure 1: Some Touristic Cultural Heritage Assets of Tokat (Turkey's Culture Portal, 2019)

The inns which have been restored and are used in Tokat attract many tourists. Especially Taşhan and Mahperi Hatun Caravanserai are among the places frequented by visitors to Tokat. The Yazmacılar Han, which was restored in 2019, is expected to attract attention. Furthermore, Latifoğlu Mansion and

Tokat Museum in Arastalı Bedesten are also attracting visitors. Finally, Yağıbasan Madrasahs, which were built in the Danishmends era (1071–1178) in the center of Niksar and Tokat qualify as the first Turkish educational institutions in Anatolia. The potential of these structures to attract domestic and foreign tourists is quite high.

Conclusion and Recommendations

Cultural heritage tourism does not only include the cultural assets that people want to visit during their touristic travels, it also covers intercultural interaction (Diker, 2016). Cultural heritage tourism, which has cultural aspects that promote and teach cultural differences, has been the subject of many academic studies in national and international literature (Poria et. al., 2001; McKercher & Du Cros, 2002; Bahçe, 2009; Wells et. al., 2015; Diker, 2016; De Jong & Wu, 2018; Alagöz et. al., 2018).

Cultural heritage tourism is one of the topics that UNESCO is interested in. Cultural sites and assets that are permanently included in the World Heritage List, which is updated every year, attract all humanity's attention and are of great importance in terms of tourism attractions (Drost, 1996: 481). As a result, the mentioned cultural areas and assets attract increasing attention in terms of cultural heritage tourism and are protected and transferred to future generations (Landorf, 2009). Turkey has important cultural heritage assets because it has hosted many civilizations. 18 of these assets are included in the UNESCO World Heritage List.

Tokat province, as well as the rest of Turkey hosts a number of cultural heritage assets. The tangible cultural heritage assets that are the subject of this study are spread to almost all parts of Tokat province. In particular, it has been determined that venues of faith come to the fore. However, Tokat province is also prominent with the historical bathhouses, castles, houses, bridges, etc. in its tangible cultural heritage inventory. Therefore, these assets must be protected to ensure their sustainability. Decision makers have important duties in this regard. In addition, effective promotion of these works is very important especially for the development of cultural heritage tourism in Tokat province. On the other hand, tourism infrastructure (transportation, touristic facilities, etc.) needs to be improved, especially the airport should start operating immediately. The increase in the number of academic studies within the scope of Tokat cultural heritage tourism will bring scientific approaches to this issue. Recent investments in tourism and the inclusion of Niksar in the UNESCO World Heritage Tentative List underline the importance of focusing on cultural heritage tourism.

References

Akın, E. S. & Özen, H. (2010). Tokat geleneksel evlerinin Beyhamam ve Bey Sokak örneğinde incelenmesi. *The Black Sea Journal of Social Sciences*, 2(2), 167–189.

Alagöz, G., Çalık, İ. & Güneş, E. (2018). Kültürel miras turizmi açısından Erzincan bakır işleme sanatının mevcut durumu ve sürdürülebilirliği. *Dumlupınar Üniversitesi Sosyal Bilimler Dergisi*, 55, 174–191.

Arslan, E., Kendir, H. & Özkoç, A. G. (2019). Niksar'ın Kültürel Miras Turizmi Potansiyelinin Değerlendirilmesi. VIII. National IV. International Eastern Mediterranean Tourism Symposium Book, Vol: 1, 19–20 April 2019.

Bahçe, A. S. (2009). Kırsal gelişimde kültür (mirası) turizmi modeli. *Dumlupınar Üniversitesi Sosyal Bilimler Dergisi*, 25, 1–12.

Coğrafya Harita, (2019). Tokat ili siyasi haritası. http://cografyaharita.com/turkiye_mulki_idare_haritalari5.html, D. A.: 13.11.2019.

De Jong, M. D. T. & Wu, Y. (2018). Functional complexity and web site design: Evaluating the online presence of UNESCO World Heritage Sites. *Journal of Business and Technical Communication*, 32(3), 347–372.

Diker, O. (2016). Kültürel miras ile kültürel miras turizmi kavramları üzerine kavramsal bir çalışma. *Akademik Sosyal Araştırmalar Dergisi*, 4(30), 365–374.

Dönmez, Y. & Türkmen, F. (2018). The relation between the landscape design and brand image in purchase preferences of tourists: The case of Safranbolu and Nevşehir, in Turkey. *Applied Ecology and Environmental Research*, 16(1), 629–643.

DÖSİM. (2018). *Müze ve örenyeri 2018 yılı toplam istatistikleri*. http://www.dosim.gov.tr/assets/documents/2018.pdf. D. A.: 11.11.2019.

Drost, A. (1996). Developing sustainable tourism for world heritage sites. *Annals of Tourism Research*. 23(2), 479–492.

Fyall, A. & Garrod, B. (1998). Heritage tourism: at what price?, *Managing Leisure*, 3, 213–228.

Hürriyet. (2014). *2023'te hedef 8 bin kilometre otoyol*. http://www.hurriyet.com.tr/ekonomi/2023te-hedef-8-bin-kilometre-otoyol-25547607. D. A.: 09.11.2019.

Hürriyet. (2019). *Ballıca Mağarası'na ziyaretçi akını*. http://www.hurriyet.com.tr/seyahat/ballica-magarasina-ziyaretci-akini-41201437. D. A.: 09.11.2019.

KTB. (2019). *Yatırım ve İşletmeler Genel Müdürlüğü Turizm İstatistikleri*, https://yigm.ktb.gov.tr/TR-9851/turizm-istatistikleri.html. D. A.: 05.11.2019.

Landorf, C. (2009). Managing for sustainable tourism: a review of six cultural World Heritage Sites. *Journal of Sustainable Tourism*. 17(1), 53–70.

Macdonald, R. & Jolliffe, L. (2003). Cultural rural tourism evidence from Canada. *Annals of Tourism Research*. 30(2), 308.

McKercher, B. & Du Cros, H. (2002). *Cultural tourism the partnership between tourism and cultural heritage management*. New York: Routledge.

Poria, Y., Butler, R. & Airey, D. (2001). Clarifying heritage tourism. *Annals of Tourism Research*, 28(4), 1047–1049.

Safranbolu Tourist Information Office, (2018). Safranbolu turist istatistik verileri. https://safranboluturizmdanismaburosu.ktb.gov.tr/TR-231191/turist-istatistik-verileri.html D. A.: 11.11.2019.

Statista. (2019a). *Number of visitors to the Colosseum and Roman Forum in Rome 2012-2017*.https://www.statista.com/statistics/515727/rome-colosseum-and-roman-forum-visitor-numbers-italy/ D. A.: 11.11.2019.

Statista. (2019b). *Number of visits made to Stonehenge in England from 2010 to 2017*. https://www.statista.com/statistics/586843/stonehenge-visitor-numbers-united-kingdom-uk/D. A.: 16.09.2019.

Tokat Governorship. (2019). *Atatürk ve Tokat*. http://tokat.gov.tr/ataturk-ve-tokatD. A.: 10.11.2019.

Tokat Municipality. (2019). *Tokat'ın tarihi*. http://tokat.bel.tr/sayfa/detay/44 D. A.: 10.11.2019.

Tokat Provincial Directorate of Culture and Tourism. (2019a). *Genel bilgiler*. https://tokat.ktb.gov.tr/TR-60574/genel-bilgiler.html D. A.: 09.11.2019.

Tokat Provincial Directorate of Culture and Tourism. (2019b). *Turist istatistikleri*. https://tokat.ktb.gov.tr/TR-231560/turist-istatistikleri.html D. A.: 09.11.2019.

Turkey's Culture Portal. (2019). *Tokat gezilecek yerler*. https://www.kulturportali.gov.tr/arama/tokat. D. A.: 10.11.2019.

TURKSTAT. (2019). *Adrese Dayalı Nüfus Kayıt Sistemi*.https://biruni.tuik.gov.tr/medas/?kn=95&locale=tr D. A.: 05.11.2019.

Wells, V. K., Manika, D., Gregory-Smith, D., Taheri, B. & McCowlen, C. (2015). Heritage tourism, CSR and the role of employee environmental behaviour. *Tourism Management*.48, 399–413.

Serdar SÜNNETÇİOĞLU

13 Effects of Digitalization in Tourism

Introduction

As in other sectors, the tourism sector is affected by technological developments and these developments change the way business partners work and the behavior of tourists. Dredge, Phi, Mahadevan, Meehan & Popescu, (2018) emphasized that digital technologies bring revolutionary products, experiences, business ecosystems and destinations in the tourism industry and digitalization transforms traditional roles of tourism businesses and tourists into new roles, relationships and business models. In this section, positive and negative results of digitalization and digitalization in tourism sector, which is considered as a labor intensive sector, are discussed. In this context; firstly, the concept of digitalization in tourism has been explained, digital application examples in tourism have been examined based on the related literature and case studies, and then the positive and negative results of these developments have been discussed.

Digitalization

Using digital technology leads to serious differences in business models. In order to make many innovations, production needs to be made more flexible and two factors are important for real-time evaluation of data: hardware and software solutions (Vuksanovic, Ugarak & Korçok, 2016: 293). Rojko (2017: 80) mentioned that today we are in the fourth industrial revolution triggered by the development of information and communication technologies (ICT). Soylu (2018) stated that Industry 4.0 is an environment in which information, communication and internet technologies have an intense impact on the production processes in a world that is experiencing a rapid digital transformation. He emphasized that, besides the effects of this situation at the firm level such as productivity, cost advantage, profitability in production, macro level effects such as growth, employment, human resources, education, investment and entrepreneurship are also in question. The term 'Industry 4.0' means a smart factory to which intelligent digital instruments are connected and communicate with raw materials, products, machines, implements, robots and people. Digital industry is characterized by flexibility, efficient use of resources and integration of customers and stakeholders into the business process (Vuksanovic et. al., 2016: 293). New generation technologies like robotization, Internet of Things, Artificial Intelligence, Sensors, Cognitive Technologies, Nanotechnology, Services of the Internet, Quantum Computing, Wearable Technologies, Augmented Reality, Intelligent Signaling, Intelligent Robots, Big Data, 3D and Intelligent Networks Technologies shaped the fourth industrial revolution which is called as Industry 4.0. Industry 4.0 technologies have started to change business environments and lifestyles by using them rapidly in the fields such as doing business, communication and education (Topsakal et. al., 2018: 3). Key features of Industry 4.0 are (1) horizontal integration through valuation networks to facilitate inter-agency collaboration, (2) vertical integration of on-site hierarchical subsystems to create a flexible and reconfigurable production system, and (3) engineering integration throughout the value chain to support product customization entirely (Wang et. al., 2016: 2). With the emerging Industry 4.0 concept, developments and applications such as Cyber-Physical Systems (CPS), Internet of Things (IoT), Service Internet (IoS), Robotics, Big Data, Cloud production and Augmented Reality have emerged (Pereira & Romero, 2017: 1207). It is thought that these developments affect the production sector as well as the tourism sector. These developments have brought about changes in both the concept of tourism and the concept of tourist.

Digitalization in Tourism

Information collection, storage and dissemination technologies are included in most tourist trips. These technologies affect the perceptions of place, distance, sociality, originality and other pre-conceptions about tourism (Jansson, 2007: 5). As people have access to the Internet and the use of mobile telephones and networks spread everywhere, social applications are undergoing a radical transformation in the field of travel, as in all other areas. Therefore, changes in the digital world revolutionize people's traditional relationships with time and space and develop new travel models (Dickinson et. al., 2014). The tourism industry is inevitably affected by technological developments. Both tourism destinations and enterprises are increasingly required to adopt innovative approaches and increase their competitiveness. On the demand side, the new, sophisticated, knowledgeable and demanding the tourist is becoming more and more familiar with increasing information technologies and requires flexible and interactive communication (Buhalis, 1998: 409). The role of information and communication technologies in the tourism industry cannot be underestimated and is a crucial driving force in the current information-driven society. Information and communication technologies provide new tools, create new distribution channels and thus create a new business environment. ICT tools facilitate business transactions in the industry by networking with trade partners, distributing product services, and providing information to consumers around the world. On the other hand, consumers use online information networks to obtain information and plan their travel (Shanker, 2008: 51).

Internet brings an unlimited business environment and a strongly competitive market. Especially in the hotel industry, managers try to be innovative and creative in order to differentiate their products among many competitors. Thanks to the interoperability and interconnectivity of all network partners, wisdom enables hospitality businesses to better understand their customers (Buhalis & Leung, 2018: 41). As technology evolves faster than ever, most travelers around the world are more interested in technology than in the past. The Internet has altered the tourism industry in the last 20–30 years. An increasing number of travelers search for information on the internet before making travel decisions. Therefore, it is important to adapt and improve the practices and skills of the workforce within the tourism industry to meet changing customer behavior (UNWTO, 2011: 9).

When new technologies about tourism are examined, it is emphasized that travelers usually search the Internet and also reach estimable travel information from different sources (Chung & Koo, 2015). Today, most travel information is provided on the internet and reservations and payments are made online. Even during travel, the internet provides travelers with rich, varied information due to their easy accessibility and connectivity with smartphones and other technologies (Huang et. al., 2017: 757). Huang et. al. (2017) state that some basic features of intelligent tourism technologies are more adopted and used in travel planning. Xiang et. al., (2015) reported that traditional internet use averages for travel planning are common across all customer groups, but high-level internet use (i.e., social media) is now more common, especially among Y generation travelers. Smart tourism, called Tourism 4.0, turned attention to the users of Industry 4.0 technologies, while the Community 5.0 philosophy emerged in Japan with Industry 4.0. At this point, it is stated that the emergence of the concept of 'Tourist 5.0 which will be named 'super smart tourist' is inevitable (Topsakal et. al., 2018: 3).

Technology is also a critical determinant of guest satisfaction. Tourism enterprises use technology as an added value opportunity to encourage differentiation and increase guest satisfaction (Çobanoğlu et. al., 2011). Douglas (2019) also stated that companies should spend money to develop practices that facilitate the experience of business travelers. Buhalis & Leung (2018: 43), in the study of smart hotel management from the perspective of connectivity and interoperability, said that technology supports dynamic supply chains and manages the efficiency of these systems by allowing hotels to cooperate with shareholders who can ensure suitable materials within time and price constraints. They also mentioned that being smart and interoperable means that hotels have access to many sub-ecosystems and that they can reach information effectively to ensure the best solution for their requirements. In addition, they emphasized that connectivity means minimizing the barriers to cooperation, effectively helping hotels to continually assess which subsystems serve their needs and strategies better.

In order to make tourism destinations smarter, it is necessary to establish a technological platform where information about tourism activities can be changed instantly and to ensure the dynamic interconnection of stakeholders (Buhalis & Amaranggana, 2015: 378). Smart systems are next-generation information systems that guarantee to ensure tourists and service providers with more detailed information, more mobility and finally more fun tourism experiences (Gretzel, 2011: 758). Alsetoohy, Ayoun, Arous, Megahed and Nabil (2019) mentioned that artificial intelligence applications such as logical models, robot brain architectures, sensor informatics, information engineering, neural networks and intelligent representation technology (IAT) changed the travel, tourism and accommodation industry. For example, it is emphasized that Marriott, Uber and Hilton use artificial intelligence solutions to understand the behavior of travelers, such as customer preferences, travel options, forms of travel and payment methods, and to identify qualifications to improve incomes and travelers' experience. Kabadayı et. al., (2019) discussed the process of smart service experience in tourism services and stated that an uninterrupted, accurate and entertaining smart service experience was created with technology and data based adaptive smart services based on personal privacy. As a result, they emphasized that smart service value, including customer value and service providers value, was achieved. It was also stated that the challenges of this process are individual/collective preferences; planned/unplanned preferences; visibility/invisibility; technology/human touch.

Table 1: Smart Application Examples from Different Tourism Businesses

Sector	Business Name	Service	Job Description
Hotel Management	Aloft, Cosmopolitan, M Social Singapore, Yotel	Service Robots	Reception – communication robots, robot bellboy, robot waiter, robot chef
Hotel Management	Hilton, Loews, Marriot, Peninsula	Smart Rooms	Personalized room features with tablets or infrared sensors
Tourism	Singapore, Malta	Smart City	Use of information data to increase customer experience and reach more information
Airport	Tokyo Haneda Airport, Miami International Airport	Smart Airport	Guidance-adapted applications and navigation and information transfer
Restaurant	Many Restaurant Chains	Smart Catering	Recommendations and personalized communication with virtual cashiers with the ability to recognize each customer

Resource: Kabadayi et. al., (2019: 327)

Similarly, Dredge et. al., (2018), explained how digitalization changed tourism industry. They evaluated transformation process with the scope of destination, management models and consumer behaviors (Table 2).

Alsetoohy et. al., (2019) stated that the rate of full use of smart technologies in hotels' food supply chain management is 5.7 %, but managers are looking forward to using this technology in the future. Buhalis & Leung (2018) stated that with the increase in the number of web activation applications, the exchange of data between the applications provides unprecedented convenience and mentioned that everything can be changed easily. Figure 1 shows the flow of data within the smart accommodation network and ecosystem.

TechTable, a platform connected with hospitality and technology, and Better Food Ventures, early-stage food technology and agricultural investment firm, and 2019 Restaurant Tech Eco System, evaluated the themes and trends that emerged after technological advances in the restaurant industry. Accordingly, the main technologies that will affect the restaurant are as follows:

Table 2: Transformation in Tourism as a Result of Digitalization

Disruption	New destination configurations	New business models, value chains, and ecosystems	Changing roles of consumers & producers	New roles for tourism organisations
Big data improves management Disruption to incumbent operators and pressure to reconceptualise traditional business models Rise of the platform economy, on-demand business New value creation opportunities Emergence of global value chains	Digitalisation allows greater customisation of visitor experiences, New customised destinations emerge	New actors such as online platforms act as information brokers and intermediaries (e.g. Expedia, TripAdvisor, etc) offer many services traditionally offered by tourism organisations. Digitalplatforms (e.g. Airbnb, Uber) are expanding beyond accommodation products to curate, coordinate, and facilitate visitor experiences in a destination.	Visitors have become prosumers actively producing and consuming their own experiences. They take on different roles, including booking,(self) guiding, reviewing, sharing and marketing the destination.	Destination marketing and product development, the traditional roles of Tourism organisations, are transformed, and these organisations find themselves increasingly infacilitation and capacity building roles with less and less direct influence over destination development, innovation, and marketing.

Resource: Dredge et al. (2018: 10)

- Voice Related Technologies: Phone order, search and discovery, brand marketing are used.
- New On-Site Ordering / Payments Technology: Kiosks, face recognition, mobile-guided ordering/payments, cashless payment.
- Managing and Minimizing Food Wastes: Various food recovery solutions to manage household kitchen waste management, 'bad food' supply, and deliver ready-to-eat foods to consumers and kitchens.
- Technologies Focusing on Increasing Customer Value: new CRM / loyalty technology, smart marketing usage, personalization to increase increased order value.

Effects of Digitalization in Tourism 183

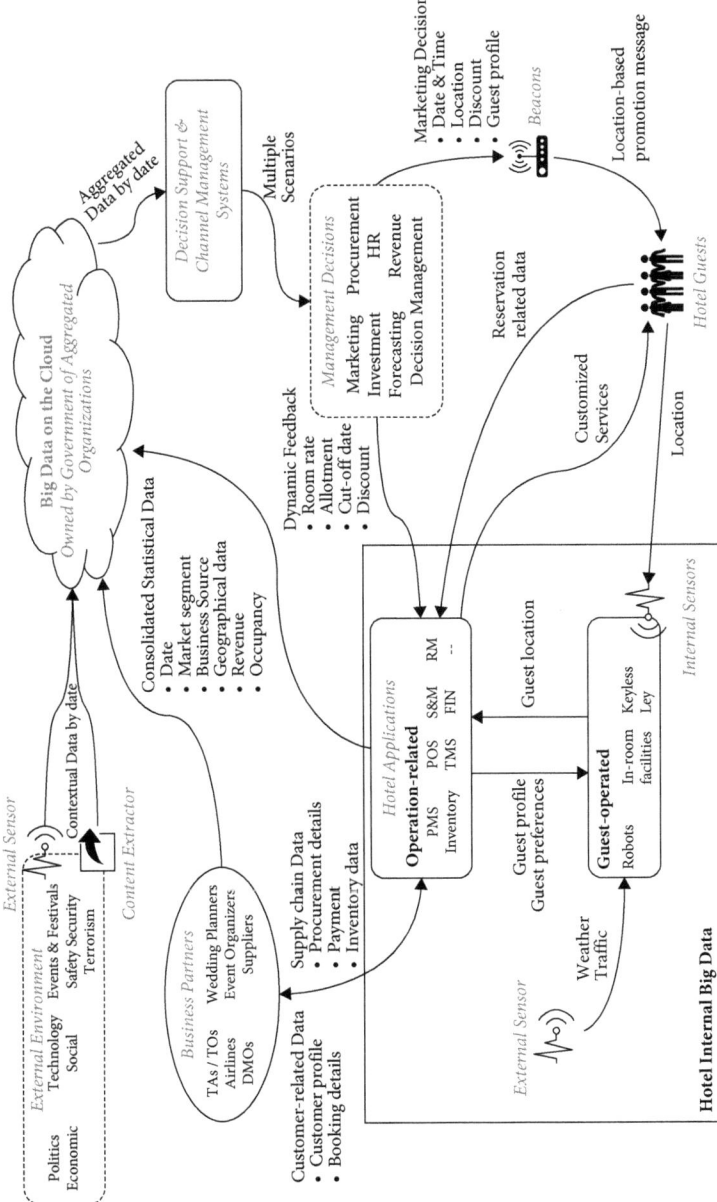

Figure 1: Information Flow in a Smart Accommodation Network. Source: Buhalis & Leung, 2018: 47

In addition, when the technological ecosystem of the restaurants is examined in Figure 2, it is seen that the blue colored section includes the operating systems such as smart kitchens, food waste management and the technologies used to carry out the back office works, and the red colored section includes systems for creating customer experiences such as research and discovery and marketing analysis.

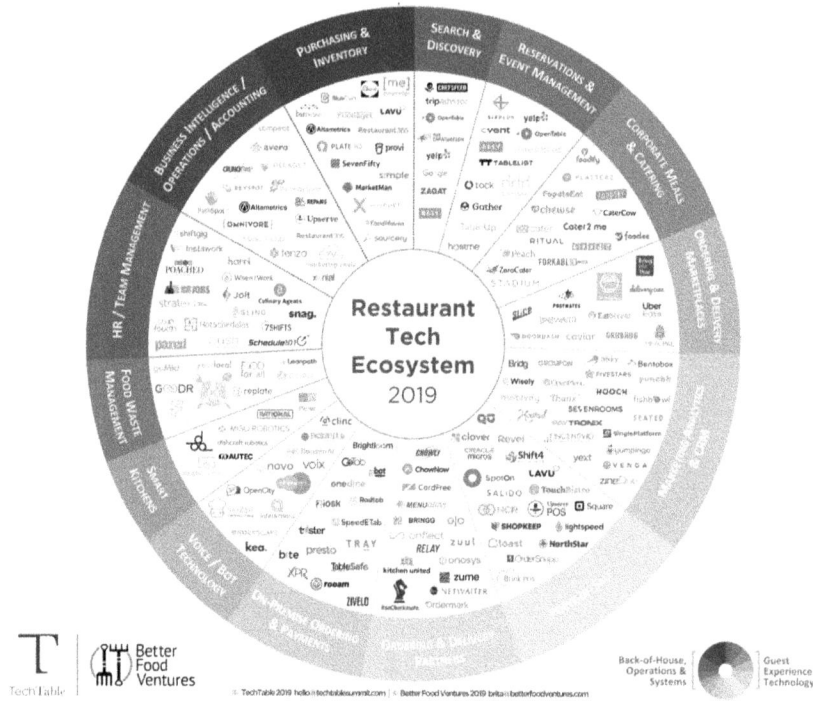

Figure 2: Technological Ecosystems of Restaurants. Source: Forbes (2019)

Effects of Digitalization in Tourism

As can be seen from the explanations given above, tourism sector is in a digital transformation and it is thought that there are positive and negative consequences of this inevitable transformation. As positive effects, speed, ease and quality of experience can be mentioned; as negative results and interpersonal interaction, personal information security. Similarly, Topsakal et. al., (2018: 255) stated that the application of new generation technologies in hotels can have both positive and negative effects. For example customer satisfaction, occupancy rate, time management can have positive effects on the hotel, as well as loss of personnel talent in the industry, job sites, may have negative effects on staff standards.

A survey of 4 and 5-star hotels found that communication technologies were considered as a way to reduce staff costs without affecting revenue or service quality. Front office managers and marketing managers who adapted communication technologies stated that they performed the same number of tasks (for example, changing information in check-in, check-in-out, distribution channels) with fewer personnel. Despite this accepted advantage, front desk managers stated that when unexpected events occurred, the number of staff was insufficient and affected customer satisfaction. Front office managers and hotel managers stated that the routine task automation created by communication technologies provides an opportunity to focus on customer interaction. They stated that employees in contact with the guests can spend more time understanding customers' needs, explaining hotel facilities and services and recognizing loyal customers (Melian-Gonzalez & Bulchand-Gidumal, 2016: 34). When the benefits of smart hotel applications are examined, it is stated that shortening and ease of processing times will prevent the stress formation caused by both tourists and staff. It is emphasized that personalized smart menu and similar applications are the factors that increase the service quality for the tourist. In addition, the development of the cloud system and mobile technologies allow tourists to create their own tourist products/packages. Instead of using a travel agency or tour operator, tourists can arrange their travel and accommodation to meet their needs. It can perform check-in, check-out on the computer or by phone, and it can also make various changes in travel at any time thanks to smart applications. Travel agencies provide the opportunity to visit the room and hotel where tourists will stay on the web with smart applications. Businesses that adapt to all these innovations and services make it easier for tourists and save time (Kızılırmak et. al., 2019: 760–765). One of the positive aspects brought by digital applications to tourism is the discovery of tourist profiles by recording customer characteristics, requests and needs with databases. For example, several restaurants have performed facial recognition technology experiments in their self-service kiosks and have achieved results. Caliburgeri UFood Grill of a KFC in Owings Mills, Maryland and Beijing stated that they provide technology that allows customers to easily look at their face recognition kiosks and re-order previous meals. Users who place an order with the kiosk for the first time are asked to receive a face scan to remember a future purchase order. Balancing a technologically innovative ordering experience with a rewards program for discounting more expensive dishes can help many customers smile and encourage them to use these biometrics to place future orders (www.pymnts.com). Similarly, it is stated that the time spent in the room, the monitored channels, the drinks used, additional services and comfort can be provided to increase the satisfaction of the guests in the following days. At this point, the personal actions of the guests are followed; as a precaution, it may be provided to authorize the guest to turn off this feature. (Topsakal et. al., 2018: 255)

In addition to many positive results brought to the tourism sector by digitalization has some negative consequences. Gretzel (2011: 770) interpreted this as the 'dark side of smart systems', and evaluations of smart systems in tourism later emphasized the need to focus more on not only their benefits but also their potential to harm users. In this regard, he stressed the need for critical perspectives and research on technology as to how tourists and tourism workers conceptualize and negotiate privacy. Mil & Dirican (2018) emphasized that with the development of robot technologies, most of the employees in the tourism sector may become unemployed and tourism education will need to be revised in the face of these developments. In another study, it was revealed that the chefs who are in the executive position expect that there will be a disadvantage for employment in the cookery profession due to the widespread use of smart kitchens due to technological developments (Yazıcıoğlu et. al., 2019: 836).

Fan et. al., (2019) emphasize that many tourism surveys conducted within the context of the fast improvement and dynamic structure of technology state that online social communication is a crucial part of the destination experience. However, it is emphasized that a small number of articles investigate the rich behavior patterns of social contact and mention that they associate them with the destination experience of tourists (Fan et al., 2019: 1). As shown in Figure 3, with the rapid development of the Internet, tourists experience only face-to-face contacts with different people at their destination. They also communicate digitally with other people while traveling through different social media platforms. Tourists can interact with their online and offline environments and also share suggestions and memories about their travels.

Effects of Digitalization in Tourism 187

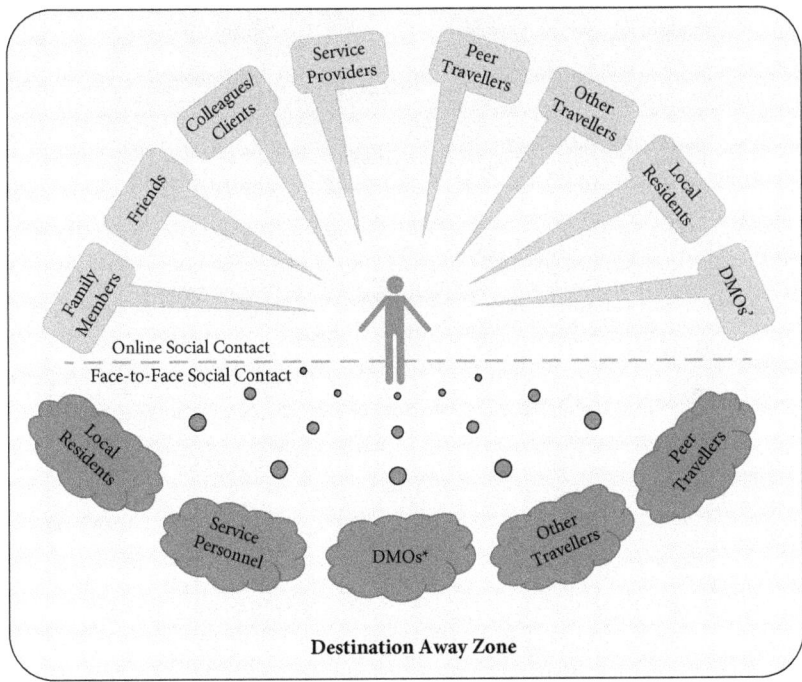

Figure 3: Tourists Face to Face and Online Interaction at Destination. Resource: Fan et. al. (2019)

Fan et. al., (2019) found a six-tiered typology of tourists in their study, which examined the online and face-to-face social relationships of tourists and their impact on travel experiences: Independent Adventurous Traveler, Digital Detox Traveler, Navigator Traveler, Dual Zone Traveler, Daily Life Controllers and Social Media Addicts. When these tourist typologies are evaluated, it is seen that digital detox and independent adventurous travelers have limited online communication and interact with the locals, employees and other tourists in the destination face to face. On the other hand, daily life controllers and social media addicts have the lowest level of face-to-face interaction and the highest level of online communication. Similarly, Sünnetçioğlu (2019) explained that one of the negative dimensions of digitalization in tourism is the decrease in communication between local and tourist relations as a result of the dehumanization process.

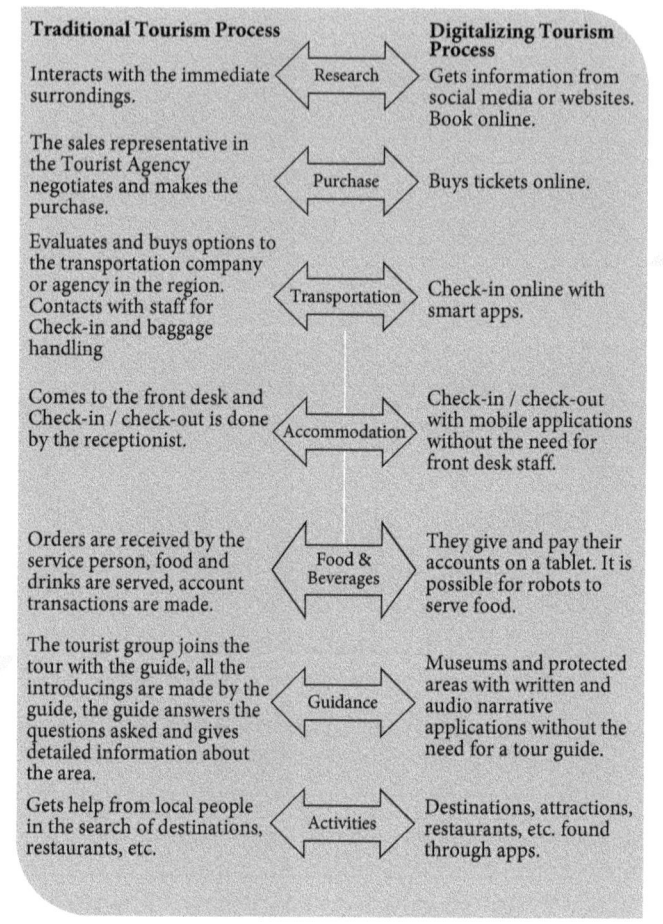

Figure 4: Social Interaction Model in Digital Tourism. Source: Sünnetçioğlu (2019: 614).

It is also stated that smart systems collect information about their environment and users and that this information may be personal information, including the physical location of a tourist. For this reason, smart systems are considered to be a potential threat to privacy and people are on a growing surveillance machine. (Gretzel, 2011: 770).

Conclusion

Since the tourism sector was a service-producing sector, it was accepted that it was a labor-intensive sector and that it could not be fully machined as production companies. However, with the digital revolution, internet and information technologies started to be used in the tourism sector and smart applications and artificial intelligence products started to be used frequently in tourism sector. Although it is not possible to say that all tourism enterprises have fully digitalized, adopted smart applications and used artificial intelligence infrastructure, it is obvious that tourism enterprises are affected by these developments. It is even likely that a tourist can make reservations for accommodation and travel from mobile devices, the smartphone application to make the check-in process with the QR codes, when the tourist goes to the hotel again s/he can check-in and check out with the QR codes, control the room temperature and lights from the mobile device, order via tablet in the restaurant and service by a robot. In this sense, all these developments have positive and negative consequences. While the convenience of collecting information about the destination and the place where people go before the holiday, booking for travel and accommodation and the speed of performing all the transactions with their mobile devices are considered to be positive;personal information confidentiality, decrease in employment in the sector and decrease in interpersonal interactions are among the negative consequences. In this context, tourism planners, destination managers, business managers and tourism educators should regulate their activities and plan in the light of these developments. It is considered that sector stakeholders should advance by improving the positive results more but providing solutions to the negative results.

References

Alsetoohy, O., Ayoun, B., Arous, S., Megahed, F. & Nabil, G. (2019). Intelligent agent technology: what affects its adoption in hotel food supply chain management?*Journal of Hospitality and Tourism Technology*, 10 (3), 286–310.

Buhalis, D. (1998). Strategic use of information technologies in the tourism industry. *Tourism Management*, 19 (5), 409–421.

Buhalis, D. & Amaranggana, A. (2015). Smart tourism destinations enhancing tourism experience through personalisation of services. In Information and communication technologies in tourism (pp. 377–389). Springer, Cham.

Buhalis, D. & Leung, R. (2018). Smart hospitality—Interconnectivity and interoperability towards an Ecosystem. *International Journal of Hospitality Management*, 71, 41–50.

Chung, N. & Koo, C: (2015). The use of social media in travel information search. *Telematics and Informatics*,32, 215–229.

Cobanoglu, C., Berezina, K., Kasavana, M. L. & Erdem, M. (2011). The impact of technology amenities on hotel guest overall satisfaction. *Journal of Quality Assurance in Hospitality & Tourism*, 12 (4), 272–288.

Douglas, A. (2019). Mobile business travel application usage Are South African men really from Mars and women from Venus. Journal of Hospitality and Tourism Technology, 10 (3), 269–285.

Dickinson, J. E., Ghali, K., Cherrett, T., Speed, C., Davies, N. & Norgate, S. (2014). Tourism and the smartphone app: capabilities, emerging practice and scope in the travel domain, *Current Issues in Tourism*, 17 (1), 84–10.

Dredge, D., Phi, G., Mahadevan, R., Meehan, E. & Popescu, E. S. (2018). Digitalisation in Tourism: In-depthanalysis of challengesandopportunities. Low Value procedure GRO-SME-17-C-091-A forExecutiveAgencyfor Small andMedium-sized Enterprises (EASME) Virtual Tourism Observatory. Aalborg University, Copenhagen.

Fan, D., Buhalis, D. & Lin, B. (2019). A tourist typology of online and face-to-face social contact: Destination immersion and tourism encapsulation/decapsulation. *Annals of Tourism Research*, 78, 1–16.

Gretzel, U. (2011). Intelligent systems in tourism asocial science perspective. *Annals of Tourism Research*, 38 (3),757–779.

Forbes (2019) Retrived on 15.11.2019 from https://www.forbes.com/sites/themixingbowl /2019/11/11/the-future-of-restaurant-tech-serving-the-next-course/#338a2a7816dd .

Pymnts (2019) Retrived on 15.11.2019 from https://www.pymnts.com/restaurant-innovation/2019/malibu-poke-facial-recognition-technology-self-service-kiosks/.

Huang, C. D., Goo, J., Nam, K. & Yoo, C. W. (2017). Smart tourism technologies in travel planning: The role of exploration and exploitation. *Information & Management*, 54, 757–770.

Jansson, A. (2007). A sense of tourism: New media and the dialectic of encapsulation/decapsulation. *Tourist Studies*, 7(1), 5–24.

Kabadayi, S., FaizanA., Choi, H., Joosten, H. & Lu, C. (2019). Smart service experience in hospitality and tourism services: A conceptualization and future research agenda. *Journal of Service Management*, 30 (3), 326–348.

Kızılırmak, İ., Güney, T., Çakmak, G., Kıran, E. & Ergan, K. (2019). Endüstri 4.0'ın Otelcilik Sektöründe Kullanılması: Yotel Örneği. The Third International Congress on Future of Tourism: Innovation, Entrepreneurship and Sustainability (Futourism 2019), 760–766.

Melián-González, S. & Bulchand-Gidumal, J. (2016). A model that connects information technology and hotel performance. *Tourism Management*, 53, 30-37.

Mil, B. & Dirican, C. (2018). Endüstri 4.0 Teknolojileri ve Turizme Etkileri. *Disiplinlerarası Akademik Turizm Dergisi*1 (3), 1-9.

Pereira, A. C. & Romero, F. (2017). A review of the meanings and the implications of the Industry 4.0 concept. *Procedia Manufacturing*13, 1206-1214.

Rojko, A. (2017). Industry 4.0 concept: Background and overview. *International Journal of Interactive Mobile Technologies*,11 (5), 77-90.

Shanker, D. (2008). ICT and tourism: Challenges and opportunities. Conference on Tourism in India - Challenges Ahead, 15-17 May 2008, IIMK.

Soylu, A. (2018). Endüstri 4.0 ve Girişimcilikte Yeni Yaklaşımlar. *Pamukkale Üniversitesi Sosyal Bilimler Enstitüsü Dergisi*, 32, 43-57.

Sünnetçioğlu, S. (2019). Turizmde Dijital Dönüşüm ve Turist Yerli Halk İlişkileri: Eleştirel Bir Yaklaşım. *The Third International Congress on Future of Tourism: Innovation, Entrepreneurship and Sustainability* (Futourism 2019), 610-616.

Topsakal, Y., Yüzbaşıoğlu, N., Çelik, P. & Bahar, M. (2018). Turizm 4.0 - Turist 5.0: İnsan Devriminin Neden Endüstri Devrimlerinden Bir Numara Önde Olduğuna İlişkin Bakış. *Journal Of Tourism Intelligence And Smartness*, 1 (2), 1-11.

UNWTO (2011). Affiliate Members Report Vol 1, Technology in Tourism.

Vuksanovic, D., Ugarak, J. & Korçok, D. (2016). Industry 4.0: the Future Concepts and New Visions of Factory of the Future Development. International Scientific Conference On Ict And E-Business Related Research, SINTEZA- Republic of Serbia, 293-298.

Wang, S., Wan, J., Li, D. & Zhang, C. (2016). Implementing smart factory of industrie 4.0 : An outlook. *InternationalJournal of Distributed Sensor Networks*, 6 (2), 1-10.

Xiang, V. P., Magnini, D. R. & Fesenmaier, D. R. (2015). Information technology and consumer behavior in travel and tourism: insights from travel planning using the internet, *Journal of Retailing and Consumer Services*. 22, 244-249.

Yazıcıoğlu, İ., Canbolat, C. & Doğan, S. (2019). Teknolojik Bakış Açısı İle Gastronominin Geleceği. *The Third International Congress on Future of Tourism: Innovation, Entrepreneurship and Sustainability* (Futourism 2019), 836-839.

Handan ÖZÇELİK BOZKURT

14 Climate Change and Tourism

Introduction

Climate scientists argue that climate change will occur at an unprecedented rate in the 21st century worldwide. Climate change is a significant statistical change in the average climate temperature or its variability over years. Climate change can be caused by natural internal processes, external coercion factors or continuous anthropogenic (human) changes in land use (Türkeş, 2008: 27). Experts identify deviations in the average annual temperature and the duration of these temperatures at many points of the globe. According to the Intergovernmental Panel on Climate Change (2001: 2), the global average surface temperature has increased by 0.6 ± 0.2 °C since 1861.

When the climate change in the world is mentioned, the changes such as decrease in snowfall, melting of glaciers, increasing water level and especially the flooding of small island countries come to mind. Besides these changes, climate change is expected to change the mountain flora and fauna. Climate change will indirectly affect mountain agriculture, hydroelectric power plants and mountain tourism (Bürki et. al., 2003: 1).

High altitude regions of Asia, including the Tibetan plateau and the Himalayan ranges, face ongoing climate change challenges such as glacial lake eruptions, glacial floods and glacial melting (Mool et. al., 2001: 227). Furthermore, the increase in minimum temperature in India due to climate change is higher than the increase in the last decade of the 20th century. Floods, droughts, forest fires, infectious diseases, etc. caused by climate change have the power to negatively affect the comfort and safety of tourists (Sonali & Kumar, 2013: 212).

It is predicted that per capita consumption will decrease by 20 % in the future due to climate change (Stern, 2006: 6). It is not possible to see climate change as a homogeneous process. Because changing climate conditions are likely to have biological, environmental, economic, psychological and sociological effects. Climate change, caused by global warming, has the power and potential to change the direction of demand in various sectors as well as psychosocially affecting the whole world. Tourism is one of the sectors affected by climate change and this level of impact is likely to increase over time. It is not possible to think of tourism activities separately and independently from the climatic conditions.

So much that, only the climatic conditions of some touristic destinations can be marketed as a touristic product.

In this study, climate change, closely related to all biological organisms, will be examined. In addition, the threats and opportunities that climate change is expected to create on the tourism sector will be discussed.

1. Climate Change

With a general point of view, climate change can be explained as inconsistencies in climate conditions. In other words, it is the regular changes that occur in a certain period (ten years or more) in climatic conditions such as flow, temperature and pressure in a region (Philander, 2008: 1130).

According to the Intergovernmental Panel on Climate Change (IPCC) and many climatologists, global warming and related climate change are the result of internal processes and human activities (IPCC, 2007: 30). The concept of climate change was first introduced by IPCC in 1988 and was examined as an economic and political problem (Kılıç, 2009: 20). According to the second report of the IPCC (1995: 39), the effects of human-induced climate change have been evident since last century. In the 21st century, it is predicted that more of these effects will be experienced. In the Second Assessment Report, it was emphasized that natural disasters due to human-induced climate change and their ecological, economic and social problems will be the biggest problems to be experienced in the 21st century. The third report focused on the possibility that most of the climate change over the last 50 years may be human-induced (IPCC, 2001: 5). In the fourth report of IPCC, for the first time, it was emphasized that 90 % of climate change is human-induced (Yönten, 2007: 26).

The main reason for the increase in greenhouse gas emissions as a result of human activities and these gases are as follows: the use of fossil fuels, the effects of aerosols, urbanization, destruction of forests and the reduction of green areas with each passing day, land use changes, etc., In other words, all these factors are listed among the causes of human-induced climate change (Garrett et. al., 2006: 490).

Melting of glaciers, rising sea water levels, severe hurricanes, floods, decrease of lake and ground water, mixing of clean water resources to sea and water problems, changes in rainfall, loss of soil in coastal areas, excessive evaporation and drought with high temperature, fires, changes in the ecosystem (extinction of some species in the habitat, rapid increase of some species), exceeding the capacity of local and universal transportation with the migration wave and consequent spread of problems, infectious diseases and health problems are expected to occur due to climate change (IPCC, 2007: 2).

2. Climate Change and Tourism

Current climatic conditions have the power to determine the direction of tourism. Similarly, changing climatic conditions are strong enough to change the direction of tourist destinations around the world (Rossello & Waqas, 2015: 4). For example, tour operators often use 'blue sky' in their brochures in summer or winter destination advertisements (Maddison, 2001: 3).

There are many stimulant that encourage people to tourism. Sea-sand-sun, congress, sports, adventure, synobism, etc. are the most important motivating factors. Tourists review push and pull factors in choosing a destination. At this point, climatic conditions can be a push or pull factor, but also have the power to play a decisive role in destination selection (Hamilton et. al., 2005: 246). Tourists can easily change their destination preferences depending on climate change. However, it is not so easy for suppliers to make changes in tourist products. Because tourist attractions such as hotels and resorts require high capital (Lise & Tol, 2002: 447).

The climate has a significant impact on the selection of tourist destinations and the duration of the stay (Hamilton & Lau, 2005: 252; Kozak et. al., 2008: 93). If the number of guests and overnight stays falls in a tourist destination, one of the reasons can be bad weather conditions (too cold, too rain, inadequate snow, etc.) (Lohmann & Kaim, 1999: 56). According to Giles & Perry (1998: 79), the main motivation factor for people living in England to travel to other countries is the bad weather conditions in England. According to Lohmann & Kaim (1999: 57), the importance of choosing tourist destinations for German citizens is landscape, price and weather conditions, respectively. Available resources determine where the recreation activity will take place, but implementation of the plans is entirely dependent on weather conditions (Amelung et. al., 2007: 286).

According to Smith (1993: 398), there are two different approaches to the selection of tourist destinations: climate-dependent and weather-sensitive. Climate-dependent approach has the power to attract tourists alone. The Mediterranean region is a good example of this approach. According to the weather-sensitive approach, climate alone does not have a decisive power; however, it has an indirect impact on destination selection. Because of changes in the climate, some current touristic areas become unsuitable due to storms, floods, sea level rise, etc. Furthermore, climate change affects and will negatively affect developing countries than developed countries (Rogerson, 2016: 5). Lack of capital, expert knowledge and insufficient policies reduce the adaptive capacity of developing countries (Hoogendoorn & Fitchett, 2016: 743).

According to the Intergovernmental Panel on Climate Change (IPCC) (2007: 2), water levels rose by an average of 1.8 mm per year worldwide between 1961 and 2003. This rise is estimated to be 17 cm in total during the 20th century. Some experts suggest that the average temperature around the world will increase between 1.1 and 6.4 °C by 2090–2099. The best estimate of 1.8 and 4.0 °C for the years 1980–1999 supports this prediction (Amelung & Nicholls, 2014: 229).

Australia is one of the most vulnerable developed countries to climate change. Daily temperatures have increased between 0.75 and 0.9 °C per year in Australia since 1910 (Australian Bureau of Meteorology, 2012). Australia heats between 0.4 and 0.7 °C and there has been an average 70 mm sea level rise since 1950 (Amelung & Nicholls, 2014: 229). According to another study, the surface temperature is estimated to increase in the range of 1.4 to 5.8 °C worldwide by 2100 (IPCC,2001: 83).

Climate change can make a tourist destination unattractive or more attractive. On the other hand, tourists who are not satisfied with the weather conditions in their own country and therefore make trips abroad can choose to have a holiday in their own country due to climate change. For example, climate conditions are changing due to climate change InEngland. This situation enables more tourists to come to the country while reducing the number of tourists going abroad (Berrittella et. al., 2006: 5). But according to Lise & Tol (2002: 447), it is difficult to determine the impact of climate change in destination selection. Because there is a possibility that the destination is not preferred for economic reasons, security problems or political reasons.

2.1 Climate Change and Coastal Tourism

According to Hall (2001: 602), coastal tourism covers all tourism activities carried out for recreation in coastal zone and offshore. These activities include accommodation, catering, food industry, etc,. Many touristic destinations owe their recreation activities and tourist intensity to the beaches they own. Hundreds of millions of people who participate in tourism activities prefer marine and coastal regions. At the same time, the intensity and diversity of coastal tourism is always increasing. For example, it is estimated that 60 % of European travels by European Union countries are spent at least four nights' stay at the seaside (Moreno & Becken, 2009: 474).

Effects of coastal erosion due to flooding, loss of coastal wetlands and mangroves, etc., reduce the touristic value of these destinations. Suitable weather conditions are not only necessary for sunbathing but also for windsurfing. Therefore, although climate change affects climate conditions, it will also affect many sources on which tourism resources are based (Moreno & Amelung, 2009: 1141).

It is accepted that tourism activity depends on weather conditions. The sea-sand-sun trio determines the tourist route, this can be understood from the current tourism destinations. Tourists generally prefer very hot (not too hot), sun-dominated, non-rain and light windy destinations (Lohmann & Kaim, 1999: 56). Skin cancer and radiation-related concerns in the Mediterranean region have so far caused behavior change (sunbathing time, protective cream usage, etc.) rather than changing destination choice (Perry, 2000: 3). However, it is estimated that the direction of tourism demand will change with the intolerable temperatures in the Mediterranean region (Amelung et. al., 2007: 285).

Tourism activities and tourist behaviors have changed coastal areas. Furthermore, changes in climate also complicate problems, particularly with regard to coastal areas (Gable, 1997: 50). Increased temperatures are a threat to biological organisms. Effects of sea level rise on beaches (Nicholls et. al., 2011: 170), degradation of coral reefs, jellyfish proliferation, etc., (Purcell, 2012: 209) should be measured in advance and measures should be taken in this regard (Nadal, 2014, 335).

It is predicted that temperature increases in Europe will vary according to countries. According to this; Spain, Italy and Greece are expected to warm up the most, while it is thought that the warming along Finland, northern Russia and the Atlantic coast will be weak (Nicholls, 2006: 153). Especially in southern Europe, a temperature increase of about 2 °C is foreseen by 2030. Spain and Greece are among the coastal countries affected by climate change. According to the forecasts of 2030, the maximum daytime temperatures in Greece will be above 29 °C and the number of tourists will decrease. The situation is similar for Spain, which has a low population density and a large number of beaches. However, Spain is expected to suffer fewer tourist losses than Greece due to low price practices and a relatively colder climate (Maddison, 2001: 19).

Increasing temperature melts the glaciers, changing the coastal boundaries and leaving the islands under water. Glacial melting caused an increase in sea level by 0.77 ± 0.27 mm between 1991-2004. The rise was 0.5 ± 1.8 mm in the 1961-2003 period (IPCC, 2007: 1182). This rise leads to border changes in the coastline and coastal areas, as seen in Cua Dai in central Vietnam. Significant coastline changes have occurred in Cua Dai during the last 50 years between 1964-1980 (Siddiqui & Imran, 2018: 72). It is predicted that a 1-meter rise at sea level will damage 49 %-60 % of the tourist centers in the region. The reconstruction of touristic centers, which were damaged by the rise in sea level in 2050, is estimated to cost 10 to 23.3 billion dollars (Nicholls, 2014: 6).

Coastal tourism is threatened by unbearably high temperatures, more and frequent rainfall, changes in wave dynamics and sea level rise (Hoogendoorn & Fitchett, 2016: 744). The Caribbean is the region with the highest concentration of small and developing countries. The number of these small countries around the Caribbean Sea is about 33. Although these countries are close to each other, they have different climatic characteristics (Gable, 1997: 50). Storms caused by rising sea levels and climate change may make tourism activities in the Caribbean impossible (Hendry, 1993: 115). Recent hurricanes and an increase in the number of tropical storms also indicate to these negativities (Lawrence & Gross, 1989: 2248).

Tourism is the main source of livelihood for small islands such as the Caribbean, the Pacific and the Indian Ocean. Changes caused by global warming such as climate change and sea level rise are factors that directly or indirectly affect tourism. Especially in small island countries, coasts will be inundated, freshwater aquifers will be salted, tropical structure will be damaged in coastal ecosystems, tropical storms will increase and loss of activation will be experienced in general (IPCC, 1997: 13). Small island countries are among the most affected and likely to be affected by climate change (Burns, 2000: 233). In addition, small island countries are the destinations most vulnerable to changing climatic conditions. Because with the rise of sea level, especially coastal cities are expected to be inundated. Furthermore drought, changes in the precipitation regime, especially in the northeastern part of the Australian continent and the changes in the monsoon rain regime of southeast Asian countries are among the possible consequences of climate change that threaten living things. These changes are expected to occur by 2050 or earlier. The number of people expected to be affected by the rising sea level is estimated at 26 million in Bangladesh, 12 million in Egypt, 73 million in China, 20 million in India and 31 million in other island countries. According to these numerical data, the total number of people expected to be affected primarily by climate change is around 162 million (Myers, 2001: 611; IPCC, 1997: 14). Even if measures are taken against climate change, the success rate will vary from region to region. Particularly on small islands, Africa and some part of Asia, coastal abandonment will occur (Nicholls et. al., 2011: 161).

2.2 Climate Change and Winter Tourism

According to Smith (1993: 398), there will be winners and losers in tourism sector as in other sectors due to climate change. According to the supply-demand balance, it is possible to change the products and services according to the demand. However, in the case of climate change, the elasticity of the supply decreases.

Especially in 1986 and 1987, the profits of the ropeway companies decreased by 20 % (Elasser & Bürki, 2002: 255).

It is not possible to get enough profit from winter tourism unless there is enough snow. The effects of climate change on winter tourism are being investigated by many countries such as Canada, USA, Australia, New Zealand, Austria, Switzerland, France and the U.K. In addition, many tourism suppliers have tried to adapt to climate change (Bürki et. al., 2003: 1). Despite the negativity, tour operators in Austria do not perceive climate change as a threat in low-altitude areas, they are developing alternative solutions such as making artificial snow (Wolfsegger et. al., 2008: 9). In Sweden, suppliers are worried that the cost of artificial snow increases and that the snow does not stay on the ground for a long time due to rising temperatures (Brouder & Lundmar, 2011: 921).

Even today, the power of global warming and climate change to determine the direction of existing markets has emerged. Because with the decrease in the amount of snow, artificial snow production facilities increase and ski centers build their facilities in higher regions (Elasser & Bürki, 2002: 255). But artificial snow production is a temporary solution. In addition, the use of fossil fuels in artificial snow production is likely to have harmful effects that accelerate climate change. However, skiers prefer real snow. Artificial snow production machines are costly and not suitable for small accommodation businesses. Nevertheless, thanks to artificial snow, it is estimated that only 4 out of 14 ski resorts in North America will be at risk. This number is expected to increase up to 10 between 2077–2099 (Nicholls, 2014: 9).

Changes in the snowfall routine, reduced glaciers and mild weather during the winter months have reduced the number of winter sport visitors in Europe and North America. Warming will reduce ski centers, especially in low-altitude areas, and will cause a short ski season (Nicholls, 2014: 7).

Switzerland economy is largely dependent on tourism. If the climate change forecast is correct, the amount of snow in the Alps will be considerably reduced and will negatively affect winter tourism activities in the region. According to the observations of a Swiss ski resort, it is necessary to snow for at least 100 days between December 1 and April 15 in 7 of the 10 winter seasons in order to be sure of the amount of snow. If climate change continues, winter tourism will only be possible in very high areas in the future, and ski resorts will sooner or later be withdrawn from the market. (Bürki et. al., 2003: 3).

2.3 Climate Change and City Tourism

British tourists prefer destinations with an average daytime temperature of 30.7 °C. They do not even prefer destinations that are a few degrees above this temperature. The amount of rain and the increase in rain play a decisive role in the choice of destination (Gössling & Hall, 2006: 164: Maddison, 2001: 17). Maddison (2001: 17) observed that the temperature of 29 °C is the ideal value for a tourist destination. During the Organization for Economic Cooperation and Development (OECD) researches, Lise & Tol (2002: 429) have determined that tourists prefer destination destinations with an average temperature of 21 °C during the hottest months of the year. For this evaluation, it does not matter which country the tourists come from.

It is accepted that the increasing temperature due to climate change decreases the quality of life especially in hot urban areas. Evaporation occurs in rural areas due to vegetation and this evaporation cools the environment. But heat absorption of buildings and differences of radiation, in particular, make the city centers more heated (Nicholls, 2006: 159). The air trapped between tall buildings and narrow streets can become more heated. Concrete surfaces, infrastructure features and industrial activities accelerate this process (Weng & Larson, 2005: 91). These changes in temperature may negatively affect activities such as festivals, concerts, sports events, etc. in the region (Nicholls, 2006: 159). Nikolopoulou (2001: 187) predicts that climate change will adversely affect urban outdoor use. But the warm-up can be advantageous for cool countries like the north of Europe. Because the decrease in rainfall and the increase in the amount of sunny days may turn into an opportunity for these cold countries in the near future (Nicholls, 2006: 159). Nicholls (2006: 21) claims that predicted climate change will provide more competitive advantage to northwest European countries in the summer compared to the Mediterranean.

The hurricane Gilbert occurred in Cancun, Mexico in 1998 and $ 87 million was lost in tourism revenues (Aguirre, 1991: 33). In addition, tourists have become reluctant to take a holiday in Cancun despite the passing of time (Davenport & Davenport, 2006: 281). Effective central warning systems, adequate communication and emergency response methods are extremely important in such disasters (Scott & Lemieux, 2010: 175). It is estimated that dengue will become widespread due to climate change and dense fog will occur due to temperature changes. Dengue fever is an infectious disease on a local and global scale and closely associated with climatic conditions. It is thought that such negativities will damage the tourist image especially in the metropolitan areas (WHO: 2012: 18; Siddiqui & Imran, 2018: 74).

Approximately 150 million people in the city centers are thought to suffer permanent water scarcity. It is estimated that this number will increase to 1 billion people by 2050. Despite all these negativities, city tourism is considered to be less affected by climate change compared to coastal and winter tourism. In this determination, pilgrimage, family visits or gambling tourism, etc., activities are considered within the scope of city tourism (Nicholls, 2014: 11–14).

Results

Global warming and climate change are a threat to all biological organisms. Therefore, all measures should be taken to prevent the destruction of vital resources and to reduce greenhouse gas emissions. All country representatives should come together to take environmentally friendly decisions and implement the decisions as soon as possible. Environmentally friendly practices are likely to damage many current economic resources. However, if future-oriented action is not taken, it will be inevitable that almost all sectors will be adversely affected by climate change.

Tourism is one of the sectors that will be most affected by climate change. Because a large part of the tourism product consists of outdoor activities. It is highly probable that there will be winners and losers in the tourism sector due to climate change. Because the change in climate will make these destinations warmer and more attractive. However, if climate change is not prevented, it is inevitable that these regions will face similar problems over time. Tourism has many positive effects, such as closing the current account deficit, providing employment and earning benefits to the sectors in which it is integrated. In this respect, many economic problems are likely to emerge in regions where tourist loss is experienced.

According to the data in this study, precipitation will decrease due to climate change and droughts will be experienced especially in warmer regions. Besides preventing climate change, it is very important to protect the available resources. Therefore, uncontrolled use of water should be prevented, especially in tourism activities. Because water and other energy use is maximized, especially due to all-inclusive concept applications. The use of energy cards has become common in hotels, but this is not enough. For example, water energy cards should be used in hotel rooms and this should be made widespread. In addition, necessary precautions should be taken to prevent the environment from being exposed to human wastes, especially in touristic areas. People should be made conscious in this sense.

Tourism is also one of the major sectors causing climate change. Because road and air transport are the two sub-sectors causing the highest emissions of greenhouse gases in the transport sector. Among the modes of transport, 70 % of global greenhouse gas emissions are road, 12 % are airway, 11 % is sea and 2 % is railway (Çalışkan et. al., 2017: 2). According to these ratios, it is very important to focus on railway transportation in the transportation sector.

The damages caused by climate change, especially to the agricultural and service industries, are extremely sad. If these predictions are realized, the emergence of water and basic food battles is not utopian. In addition, migration is likely to occur due to climate change. In this respect, all authorities and people should do their utmost to preserve existing resources, maintain and develop existing living standards and leave a livable world for future generations.

References

Aguirre, B. E. (1991). Evacuation in Cancun During Hurricane Gilbert. *International Journal of Mass Emergencies and Disasters*,9/1, p. 31–45.

Amelung, B. & Nicholls, S. (2014). Implications of Climate Change for Tourism in Australia. *Tour Manag*,41, p. 228–244.

Amelung, B. Nicholls, S. & Viner, D. (2007). Implications of Global Climate Change for Tourism Flows and Seasonality. *Journal of Travel Research*, 45, p. 285–296.

Australian Bureau of Meteorology, (2012). Annual Australian Climate Statement 2012. http://www.bom.gov.au/climate/current/annual/aus/2012/

Berrittella, M., Bigano, A., Roson, R., & Tol, R. S. J. (2006). A General Equilibrium Analysis of Climate Change Impacts on Tourism. *Tourism Management*, 27/5, p. 913–924.

Brouder, P. & Lundmark, L. (2011). Climate Change in Northern Sweden: Intra-Regional Perceptions of Vulnerability among Winter-Oriented Tourism Businesses. *Journal of Sustainable Tourism*,19/8, p. 919–933.

Burns, W. C. G. (2000). The Impact of Climate Change on Pacific Island Developing Countries in the 21st Century. (Edited by A. Gillespie & W. C. G. Burns) In *Climate Change in the South Pacific: Impacts and Responses in Australia*, New Zealand and Small Island States, (p. 233–250) Dordrecht, The Netherlands: Kluwer.

Bürki, R., Elsasser, H. & Abegg, B. (2003). Climate Change—Impacts on the Tourism Industry in Mountain Areas. *1st International Conference on Climate Change and Tourism*, Djerba, 9–11 April 2003.

Çalışkan, Z. D., Kurt, Ü. & Timur, M. C. (2017). İklim Değişikliği ve Ulaşım Sektörü İlişkisinin Ekonometrik Analizi: Türkiye Örneği. *International Congress of Energy, Economy and Policy*, 25–26 Mart, İstanbul.

Davenport, J. & Davenport, J. L. (2006). The Impact of Tourism and Personal Leisure Transport on Coastal Environments; A Review. Estuarine, *Coastal and Shelf Science*, 67, p. 280–292.

Elasser, H. & Bürki, R. (2002). Climate Change as a Threat to Tourism in the Alps. *Climate Research*, 20, p. 253–257.

Gable, F. J. (1997). Climate Change Impacts on Caribbean Coastal Areas and Tourism. *Journal of Coastal Research*, 24, p. 49–69.

Garrett, K. A., Dendy, S. P., Frank, E. E., Rouse, M. N. & Travers, S. E. (2006). Climate Change Effects on Plant Disease: Genomes and Ecosystems. *Annual Review of Phytopathology*,44, p. 489–509.

Giles, A. R & Perry, A. H (1998). The Use Of A Temporal Analogue To Investigate The Possible Impact Of Projected Global Warming On The UK Tourist Industry. *Tourism Management*, 19(1), p. 75–80.

Gössling, S. & Hall, M. (2006). Uncertainties in Predicting Tourist Flows Under Scenarios of Climate Change. *Climatic Change*, 79, p. 163–173.

Hall, M. C. (2001). Trends in Ocean and Coastal Tourism: The End of the Last Frontier?, *Ocean & Coastal Management*, 44/9–10, p. 601–618.

Hamilton, J. M. & Lau, M. A. (2005). The Role of Climate Information in Tourist Destination Choice Decision-Making. In: *Proceedings of the 17th International Congress of Biometeorology* (ICB 2005), Garmisch-Partenkirchen, Germany, 9–5 September 2005. Deutscher Wetterdienst, Offenbach am Main, p. 608–611.

Hamilton, J. M., Maddison, D. J. & Tol, R. S. J. (2005). Effects of Climate Change on International Travel. *Climate Research*, 29, p. 245–254.

Hendry, M. (1993). Sea-level Movements and Shoreline Changes. (Edited by: Maul, G. A.), In *Climatic Change in the Intra-Americas Sea*, (p. 115–161) Edward Arnold, London.

Hoogendoorn, G. & Fitchett, J. M. (2016). Tourism and Climate Change: A Review of Threats and Adaptation Strategies for Africa. *Current Issues in Tourism*, p. 1–19.

Hoogendoorn, G., Grant, B. & Fitchett, J. M. (2016). Disjunct Perceptions? Climate Change Threats in Two Low-Lying South African Coastal Towns. *Bulletin of Geography: SocioEconomic Series31*.

IPCC. (1995). Climate Change 1995: Summary for Policymakers: Scientific-Technical Analyses of Impacts, Adaptations and Mitigation of Climate Change. Cambridge University Press, Cambridge, UK.

IPCC. (1997). The Regional Impacts of Climate Change: An Assessment of Vulnerability. Cambridge University Press, Cambridge, UK.

IPCC. (2001). Climate Change 2001: The Physical Science Basis. Summary for Pilicimakers. Cambridge University Press, Cambridge, UK.

IPCC. (2007). Climate Change 2007: The Physical Science Basis. Contribution of Working Group I to the Fourth Assessment Report of the

Intergovernmental Panel on Climate Change. Cambridge University Press, Cambridge, UK.

Kılıç, C. (2009). Küresel İklim Değişikliği Çerçevesinde Sürdürülebilir Kalkınma Çabaları ve Türkiye, *C. Ü. İktisadi ve İdari Bilimler Dergisi*, 10/2, p. 19-41.

Kozak, N., Uysal, M. & Birkan, I. (2008). An Analysis of Cities Based on Tourism Supply and Climatic Conditions in Turkey. *Tourism Geographies*, 10/1, p. 81-97.

Lawrence, M. B. & Gross, J. M. (1989). Annual Summaries: Atlantic Hurricane Season of 1988. *Monthly Weather Review*,117/10, p. 2248-22.

Lise, W. & Tol, R. S. J. (2002). Impact of Climate on Tourism Demand. *Climate Change*, 55, p. 429-449.

Lohmann, M. & Kaim, E. (1999). Weather and Holiday Destination Preferences, Image, Attitude and Experience. *The Tourist Review*, 2, p. 54-64.

Maddisson, D. (2001). In Search of Warmer Climates? The Impact of Climate Change on Flows of British Tourists. *Climatic Change*,49, p. 193-208.

Mool, P. K., Bajracharrya, S. R. & Joshi, S. P. (2001). *Inventory of Glaciers, Glacial Lakes, Glacial Lake Outburst Floods: Monitoring and Early Warning Systems in the Hindu Kush-Himalayan Region*, Nepal. ICIMOD, Kathmandu, Nepal.

Moreno, A. & Amelung, B. (2009). Climate Change and Tourist Comfort on Europe's Beaches in Summer: A Reassessment. *Coastal Management*, 37, p. 550-568.

Moreno, A. & Becken, S. (2009). A Climate Change Vulnerability Assessment Methodology for Coastal Tourism. *Journal of Sustainable Tourism*, 17, p. 473-488.

Myers, N. (2002). Environmental Refugees: A Growing Phenomenon of the 21st Century, Philosophical Transactions of the Royal Society London. *Biological Sciences: Series B*,357(1420), p. 609-613.

Nadal, J. R. (2014). How to Evaluate The Effects of Climate Change on Tourism. *Tourism Management*, 42, p. 334-340.

Nicholls, S. (2006). Climate Change, Tourism and Outdoor Recreation In Europe. *Managing Leisure*, 11, p. 151-163.

Nicholls, S. & Amelung, B. (2008). Climate Change And Tourism In Northwestern Europe: Impacts and Adaptation. Tourism Analysis 13, p. 21-31.

Nicholls, M. (2014). IPCC Climate Science Business Briefings, Climate Change: Implications for Tourism, Key Findings from the Intergovernmental

Panel on Climate Change Fifth Assessment Report, University of Cambridge,(http://www.cisl.cam.ac.uk/business-action/low-carbon-transformation/ipcc-briefings/tourism, 19.10.2019).

Nicholls, R. J., Marinova, N., Lowe, J. A., Brown, S., Vellinga, P., de Gusmão, D., Hinkel, J. & Tol, R. S., (2011). Sea-Level Rise and Its Possible Impacts Given A 'Beyond 4 Degrees C World' in the Twenty-First Century. *Philos. Trans. R. Soc. London A*, 369, p. 161–181.

Nicholls, S. & Amelung, B. (2008). Climate Change and Tourism in Northwestern Europe: Impacts and Adaptation. *Tourism Analysis*13, p. 21–31.

Nikolopoulou, M. (2001). The Effect of Climate on the Use of Open Spaces in the Urban Environment: Relation to Tourism, (Edited by. A. Matzarakis & C. R. de Freitas), In *Proceedings of the First International Workshop on Climate, Tourism and Recreation*, Porto Carras, Neos Marmaras, Halkidiki, Greece, 5–10 October, p. 185–193.

Perry, A. (2000). Impacts of Climate Change on Tourism in the Mediterranean: Adaptive Responses, *Nota di Lavoro*35. 2000, Fondazione Eni Enrico Mattei, Milan, Italy.

Purcell, J. E. (2012). Jellyfish and Ctenophore Blooms Coincide with Human Proliferations and Environmental Perturbations, Annual Reviews, p. 209–235.

Philander, G. (2008). *Encyclopedia of Global Warming and Climate Change*, Vol. 1. Thousand Oaks, CA: Sage.

Rogerson, C. M. (2016). Climate Change, Tourism and Local Economic Development in South Africa. *Local Economy*,31, p. 322–331.

Rossello, J. & Waqas, A. (2015). The Use of Tourism Demand Models in the Estimation of the Impact of Climate Change on Tourism. *Revista Turismo em Analise*, 26/1, p. 4–20.

Scott, D. & Lemieux, C. (2010). Weather and Climate Information for Tourism. *Procedia Environmental Sciences*1, p. 146–183.

Siddiqui, S. & Imran, M. (2008). Impact of Climate Change on Tourism. *Environmental Impacts of Tourism in Developing Nations*, p. 68–83.

Smith, K. (1993). The Influence of Weather and Climate on Recreation and Tourism. *Weather*, 48/12, p. 398–403.

Sonali, P. & Nagesh Kumar, D. (2013). Review of Trend Detection Methods and Their Application to Detect Temperature Changes In India. *J. Hydrol.*476, p. 212–227.

Stern, N. (2006). *The Economics of Climate Change: The Stern Review*. Cambridge University Press, Cambridge.

Türkeş, M. (2008). Küresel İklim Değişikliği Nedir? Temel Kavramlar, Nedenleri, Gözlenen ve Öngörülen Değişiklikler, *İklim Değişikliği ve Çevre*,1: p. 45-64.

Weng, Q. & Larson, R. C. (2005). Satellite Remote Sensing of Urban Heat Islands: Current Practice and Prospects. *Geo-Spatial Technologies in Urban Environments*, p. 91-111.

WHO, World Health Organization, (2012). *Global Strategy for Dengue Prevention and Control World Health Organization*, Geneva. Retrieved October 27, 2017http://apps.who.int/iris/bitstre am/10665/75303/1/9789241504034_eng.pdf.

Wolfsegger, C., Gossling, S., & Scott, D. (2006). *Climate Change Risk Appraisal in the Austrian Ski Industry*. Tourism Rev Int.

Yönten, A. (2007). *Küresel Isınmanın Azaltılması Politikaları ve Stratejileri-Türkiye İçin Bir Yaklaşım*, Yayımlanmamış Yüksek Lisans Tezi, Dokuz Eylül Üniversitesi, Sosyal Bilimler Enstitüsü, İzmir.

İrem BOZKURT and Enes YILDIRIM

15 Overtourism

Introduction

Although the negative consequences of tourism on local people and the environment have been known for years, the concept of overtourism, which entered the literature in 2016 as a new concept, has started to attract the attention of researchers with the reactions of destinations to the tourism event (IPOL, 2018). The concept of overtourism, which has emerged to identify potential dangers for popular destinations, defends that tourism will have inevitable negative consequences if it is not managed in a sustainably (Sheivachman, 2019). Destinations are not only the places where the tourism event takes place, but also the centers where the local people live permanently. This situation enables social, cultural, economic and physical interactions between local resident and tourist. Due to the attractiveness of certain destinations, the tourist density is higher in those regions. Even if the intense flow of tourists provides economic positive effects, preserving the historical and cultural structures and traditions existing in the destinations becomes difficult, carrying capacity exceeding problems occur and it is on the edge of negative reaction of local resident (Çolak et. al., 2019: 986). With the increase of social media platforms, sharing of touristic areas leads to the density of tourists in destinations (Oklevik et. al., 2019: 3). In places where tourist density has started to increase, concepts such as overtourism, cities prone to overtourism and the fear of tourism appear to emerge. Overtourism is defined as a perceived negative impact on the quality of life and / or visitor quality of citizens in a destination or part of it (UNWTO, 2018). In places where the flow of tourists is intense and tourists dominate the city centers, 'crowding' situation occurs and under the circumstances the negative effects of tourism overshadow the positive effects (Oklevik et. al., 2019). In this section, the concept of overtourism and the factors that cause overtourism are reviewed andovertourism destinations and their measures are referred and ways of coping with overtourism are mentioned.

1. The Concept of Overtourism

Destinations provide visitors with multi-user, functional and complex environments, can accommodate a large number of local and foreign tourists, and can host business and friend-relative visits (VFR). The fact that cities have good infrastructural facilities and hosting a dynamic population shows that problems with the increasing number of tourists can cope with better than other cities. Tourism, which has been seen as an important driving force for economic recovery and growth especially after the 2008 economic crisis, has been seen as one of the sustainable economic growth strategies up to the present (Russo & Scarnat, 2018). However, the perception of urban tourism has changed conspicuously in recent years. Infrastructure, public transport, museums, roads, tourist attractions and entire other services that are primarily used for the host community are now at the service of increasing tourists. Increasingly popular online accommodation and transport services (AirBnB, HomeAway, Uber...), the desire to see and experience true, natural and authentic city life has made tourism activities more intertwined with local life (Pappalepore et. al., 2014). These disturbing developments related to the growth and expansions of tourism have attracted the attention of the local community and stakeholders of the region and have led to protests in destinations with high profile (especially hosting the most tourists). Although this situation is evident in European cities, tropical islands, backpacker ghettos and / or slums have exhibited similar thoughts and actions too. The concept of overtourism has also emerged to describe these deteriorations, thoughts and actions (Koens & Thomas, 2016).

Overtourism, which was used as a subject tag on Twitter for the first time in August 2012 in order to express people's concerns about tourism movements, has become increasingly common (Goodwin, 2017). In 2016, the concept was introduced into the literature by Skift, a media company providing news, marketing and research services for the travel industry, and subsequently became a trademark (Ali, 2016). Overtourism which is defined by UNWTO (2018) for evaluate as 'a concept that affects the impact of tourism on the destination or the impact of citizens on the quality of life and / or the negative perception of visitor experiences'. Participation in tourism, which is seen as an important consumption sector due to the developing standard of living of the community, has increased and caused overcrowding at the destinations. The concept of overtourism is used to indicate that there are too many visitors in the host destination and that the quality of life and quality of experience in the area is noticeably deteriorating. From this respect, it is completely contrary to the concept of responsible tourism, which uses tourism to make destinations more livable. Because the concept of responsible tourism states that the overtourism event unacceptably disrupts the quality of life and experience in the area where the host or guests, visitors or natives are too many (Goodwin, 2017). There is no clear definition for overtourism. However, in general, the phenomenon of overtourism is related to the increase in the number of tourists, the complexities in the relationship between visits and time, and the carrying capacity of destinations. It includes the views of various stakeholders such as local resident, visitors and businesses at the same time. According to a recent study, the challenges of overtourism can be line up corruption of society, deterioration in tourist behavior, exceeding carrying capacity, damages and threats to nature and cultural heritage (WTTC & McKinsey Company, 2017).

According to Colomb & Novy (2016); Milano (2018), overtourism means that the physical, psychological, economic, ecological social and/or political capacity at certain times and at certain places is exceeded by the impact of tourism. While Singh (2018) states that overtourism occurs when the number of visitors exceeds the number of local resident, Goodwin (2017) thinks that overtourism begins when local resident or visitors realize that excessive visits occur in the area and accordingly the writer sees the natural character of the region as the stage in which the thought that it loses its unique structure emerges and even the complaints expressed in the discomfort dimension occur. As for Milano, Novelli & Cheer (2018) defined overtourism as a permanent changes in the life style of the local resident in regions that cause overcrowding due to the seasonality of tourism outcome in general, affecting the general welfare and utilization of life opportunities and increasing the number of visitors, the negative consequences of overcrowding are listed as damage to nature, putting infrastructure under great pressure and exposing local resident to tourist market prices.

All concepts and components related to Overtourism are shown in Figure 1. The following conceptual model form (Figure 1) attempts to combine and summarize the main aspects of the phenomenon of tourism. In addition, this conceptual model provides a comprehensive perspective for the key elements of overtourism. If a definition is made for overtourism based on Figure 1, the carrying capacity is exceeded situation. Theconceptual model (Figure 1) contains all elements affecting destination and its physical environment, destination economy, local resident, cultural heritage and even visitors.

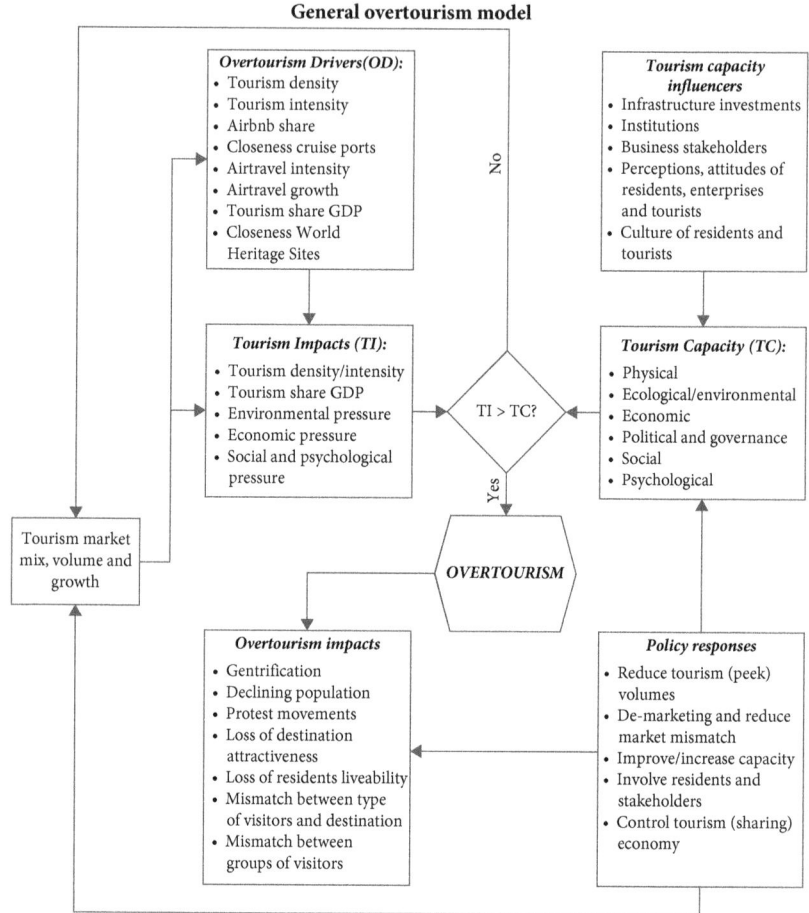

Figure 1: The Conceptual Model of Overtourism Source: (IPOL, 2018)

Figure 1 shows that each destination has an existing tourism marketing mix, tourism volume and growth. Various tourism developments, such as increase in tourist density or AirBnB accommodation, trigger overtourism. There are complex lower limits (bearing capacities) for each circle. These lower limits are not constant or equal for each circle. As it is mentioned earlier, these lower limits vary according to infrastructure investments and the developing perception and attitudes of local resident. In case of any of the lower limits is exceeded, the destination came up against overtourism event. Otherwise, the destination will continue its current growth process. If a destination is exposed to overtourism, it is possible to experience corruption as well as social and cultural conflicts between local resident with visitors. In addition to examining the effects of tourism marketing mix elements in terms of intensity and development, policy makers can also address the versatile effects of overtourism and address the overtourism dimension associated with the carrying capacity of destinations (IPOL, 2019).

2. Causative Factors of Overtourism

Tourism and travel is one of the fastest growing sectors in the world. In 2018, the number of tourists traveling around the world increased by 5 % compared to the previous year and reached 1.4 billion. (UNWTO, 2019). With the enrichment of the middle class in the world, travel becomes more accessible and tourism and travel (T & T) continues to grow. (ferdamalastofa.is). However, tourism and travel, which constitute the cornerstone of the global economy, have not spread evenly throughout the world. More popular and/or attractive destinations are exposed to potential environmental, social and aesthetic threats. Many factors are effective in the emergence of extreme tourism. These factors can be listed as follows (Duyar & Bayram, 2019; Goodwin, 2019):

- *Rise of the Middle Class:* Factors such as the ease of access to the internet worldwide and the increase in the number of people with a medium income can be shown as the reasons of the spread of accommodation and transportation opportunities. According to Webster and Ivanov, 'BRIC' (Brazil, Russia, India and China) and 'PINE' (Philippines, Indonesia, Nigeria and Ethiopia), which have a total population of 3.6 billion people both as tourist generating countries and as tourist destinations, countries are extremely attractive. Each of these countries is large enough to have a significant national tourism demand in the future, especially in the growing middle class market (Webster & Ivanov, 2015). Those living in the Asia-Pacific region mostly travel within the region, but this demand will increase with the increase in income and prosperity in these rapidly developing economies, and the number of travels from these countries to foreign countries is growing faster than the increase in (Usa Visa, 2018).
- *Sharing Economy:* The fact that people have a middle income level has revealed the concept of 'sharing economy'. The phenomenon of sharing

economics stems from a number of technological developments that simplify the sharing of physical and non-physical goods and services through various information systems on the Internet. (Hamari et. al., 2015). People use alternative accommodation methods such as Airbnb, Couchsurfing for accommodation and stay cheaper than hotels. Similarly, people are able to meet their transportation needs in a cheaper way than other road vehicles by using online internet applications that offer alternative sharing based transportation methods such as Uber and Zipcar for road transportation. However, with the increase in the number of tourists in apartment buildings where local people are located instead of certain points such as hotel businesses, the elements that make up the sharing economy also cause crowding in destinations (Duyar & Bayram, 2019).

- *The Effect of Media:* In the age of developing technology, it is very difficult to hide the attraction. Now people want to visit touristy and popular places, and are experiencing the craze of selfie collection / sharing. This event became one of the biggest motivation factors that led people to travel and made popular destinations face the problem of crowding (Goodwin, 2019).

- *The Low Cost of Travel:* The low cost of airlines and bus travel led people to travel. Where vacation time is limited, people take more city breaks. These breaks are mostly short-distance trips. Because they want to use the holiday break in the most appropriate way. Developing technology brought diversity in transportation network and reduced costs (Goodwin, 2019).

- *The Public Realm is Free:* One of the most important factors in crowding is the fact that the destinations visited are free for everyone to visit. Some destinations that notice this (such as Rome and Bhutan) both of charge city tax from tourists, thus paying the maintenance and repair costs of the destinations as well as changing the number and profile of visitors (Goodwin, 2019).

- *Seasonality:* It is one of the biggest problems experienced by some destinations. Because in destinations seasonality is caused to be crowded during certain periods and in specific regions. Intensive season to extend/ spread is desirable, but in this case it may not be unlikely for some destinations (IPOL, 2018).

Unless the destinations are visited within the framework of responsible tourism, it will be inevitable that they will experience overtourism. In the light of these factors, destinations and/or local residents experiencing overtourism show their discomfort in different ways. The study by Colomb & Novy (2016) provides an analysis of the experiences of this situation and the discontent, resistance and protests of the local people. According to Bock (2015), overtourism and tourismphobia have become the two biggest problems facing cities today. Cultural conflicts in social rules and norms due to overtourism are disturbing the local population seriously and directing them to move to less attractive cities. Examples of this situation, Barcelona, Prague, Berlin and Paris are seen in the city. Other overtourism effects expressed by the local people are that the grass is replaced by bare soils, the sandy beaches are filled with garbage, and the existing shops to meet the needs of the locals turn into shops selling expensive goods or selling flavor to tourists (Davies, 2017).

3. Overtourism Living Destinations

Nowadays, societies have turned to certain consumption patterns in order to gain status owner and to emphasize their individuality. People are competing with each other to get rid of monotonous city life, to keep pace with fashion or to catch some differences (Sevinç & Duran, 2018). Also one of the reasons for overtourism in destinations takes part as an 'imitation effect' in international literature. Since the effects and conceptual structure of Overtourism is multidimensional, it differs from destination to destination. Decision-makers and local governments may remain incapable of taking measure against these effects caused by overtourism. There is no common point of solutions. Because the wanted solutions must be required special to destination (Duyar & Bayram, 2019).

When the related literature is examined, the main effect of overtourism is that local resident no longer enjoys the place where they live. (Calzada, 2018). The effects of Overtourism on destinations and some of the discomforts can be listed as follows;

- Greece –In Santorini due to overtourism, fromalienation of local resident meet problems such asrealization of socio-cultural deterioration, increase in noncorporate accommodation typcs and imperilment of natural resources and infrastructure (Sarantakou & Terkenli, 2019).
- Spain – Barcelona, problems such as increase in the number of tourists, correspondingly increase in living costs, noise and air pollution, increases in waste and water use, loss of confidence feeling in local community occurred (Martin Martin et. al., 2018; Koens et. al., 2018).
- Germany – Berlincame up against the risk that local resident cannot use their own opportunities and negative perceptions may develop due to airbnb (Namberger et. al., 2019).

- Hungary–The increase in the number of tourists in Budapest due to overtourism, security problems in the region, price increase in housing, food and beverage businesses, noise pollution and decrease in quality of life caused the desire to leave Budapest in 18 % of the population (Pinke-Sziva et. al., 2019).
- Italy – Stated as negative impacts in Venice, increases in housing prices, decreases in public purchasing power and socio-cultural bonds, increase in crime rates, destruction of historical sites, emerge of pollution and traffic problems (Seraphin et. al., 2018).
- Slovenia–Ljubljana stated that as a result of the increase in the tourist, negative effects on the quality of life of the public occurred due to reasons such as air pollution, crowding and traffic density (Kuščer & Mihalic, 2019).
- Croatia – Dubrovnik explained that there are problems such as seasonality, traffic problems and increased living costs (Panatiopoulos & Pisano, 2019).

Tourism, which promotes social progress, is considered as an event causing social, cultural and environmental disturbances as well as an important source of income and employment creation economic activity. The development of tourism in destinations will pose various challenges, such as deterioration of the environment, cultural heritage, quality of life and social structure (Timur & Getz, 2009). According to the researches conducted, five main areas where tourism affects global environmental change are identified as energy use, vital change, biodiversity reduction, land cover and land use change, and changes in the perception and understanding of the environment through travel. Apart from all these, tourism can cause spread of diseases and some changes for attitudes of the local resident in the norms of ideas and values (increase in tobacco and alcohol use, the spread of night life, etc.) (Gössling, 2002). Overtourism living destinations has implemented a number of solution strategies and as a result of relevant literature review have been reached their solution strategies and tabularized below:

Table 1: Overtourism Living Destinations and Developed Solution Strategies

Destination	Solution Strategy
Italy – Venezia	In Venice, where more than 36 million tourists visited in 2017, while 150,000 people lived 50 years ago, the number has decreased to 53,000 with the effect of tourism today. Tourism causes overcrowding, environmental damages and expensiveness problems in the region. It was decided to ban certain size ships for the solution of the problems and police assistants to prohibit tourist behavior and charge entrance fee to the city from one-day tourists coming to the region.

(continued on next page)

Table 1: (continued)

Destination	Solution Strategy
Italy – Rome	With the new tourism arrangements in Rome, there is a ban on photographing with characters dressed as historical figures and selling and using drinks after 2 am, eating and sitting on the Spanish stairs. For tour buses, a ban on limiting the right of passage through the ancient city center was introduced.
Spain – Barcelona	It is seen that the city of Barcelona is beginning to lose its cultural heritage due to overtourism. However, due to the proliferation of pirated vendors, commodification problems have emerged in the culture. With the new regulation in Barcelona, the opening of new hotels in the city center is prohibited. In addition, it is on the agenda to stop the permissions given to online websites to convert apartments to accommodation. Because coming to the region only 8 million out of 32 million tourists put up at hotels. In Barcelona, an increase in house rents has seen due to overtourism. As a result of these, local community placed bill-posting in the region in order to react to the tourists in the city and a group of protesters attacked the tour bus and cut off the tires.
Holland – Amsterdam	While the number of visitors to Amsterdam was 18 million in 2018, this figure is expected to increase to 30 million in 2030, which is 50 times more than the population. The authorities to take measures for the density of tourists have started to take such measures lifting the 'Ben Amsterdam' sign in urban square to uptown, preparing guideposts on tulip gardens to avoid damaging tourists, suspension of promotions for the region and termination for tours to Red Light region. Local residents in Amsterdam have begun to leave the city due to noise in the city, parties organized by tourists and environmental damage. It also prohibited the opening of new hotels and souvenir shops in Amsterdam.
Peru – Machu Picchu	More than 5000 tourists visit the Machu region of Peru per day. There are many problems in the region due to tourist behavior. In order to solve the problems, the entrance times of the region, the types of visits and the inward periods are limited.
India – Taj Mahal	In Taj Mahal, the effects of overtourism are felt at a high level, with an average of 35,000–40,000 per day and 70,000 people on weekends. In order to solve the problem, the authorities limited the time spent by tourists in the region and placed tourniquets at the entry points to protect the region.
America – California	Residents of Big Sur in California, especially after a TV show, increased the popularity of the Bixby bridge, and as a result of this, they put up a banner saying 'Overtourism kills Big Sur' to show their frustration with tourism. Daffodil Hill, a field of beautiful yellow and white flowers outside Sacramento, has been announced to close indefinitely due to unexpected popularity in social media and overtourism.

Source:(Çolak et. al., 2019).

Overtourism is not the same in all destinations. The problems identified in a survey of the effects of over-tourism in European cities on tourism stakeholders and local communities are as follows (Koens et. al., 2018):

- Local resident's life way is getting difficult due to tourist activities,
- Overcrowding in streets, public transport and traffic,
- Difficulty and expensive parking space for vehicles in city centers,
- Less housing possibility for local resident due to online platforms like Airbnb,
- Price increases in house rents and the community leaving the city,
- Loss of community identity and sense of security,
- Air pollution, noise and water pollution caused by waste, polluted water and cruise ships, and climate change emergence in the long term.

4. Dealing Ways with Overtourism

There is no single solution way for Overtourism. Because each destination has its own characteristics and management style. The situation as a precaution qualification for one destination may result in worse consequences for another destination. Destinations should initially decide what they want to deal with overtourism. Do they want to protect quality of life? Do they improve tourist behavior and experiences? Do they protect cultural heritage sites? Otherwise all of them? Subsequent to the desired is determined; another important question to be answered is who will do for the destinations to these requests. Within the scope of study, the ways of coping with overtourism will be examined under five main headings (WTTC & McKinsey & Company, 2017):

4.1 Smooth Visitors Over Time

Many destinations suffer from demand fluctuations. Correction of these fluctuations has great importance for destinations that are subject to an overloaded infrastructure, cultural heritage and threats for nature and disrupted tourist experience. In some cases it makes sense to limit the number of visitors within certain days. However, there are also destinations that are increasingly composing reservation and ticketing systems, using a number of inputs to guide visitor behavior and changing promotion strategies (WTTC & McKinsey & Company, 2017).

- *__Limit the Number of Tourists:__* Restricting tourism event is a difficult task. In this case, the provocations of the oppositions who lose tourism revenues or fail to realize growth can be observed or some private sector representatives may admit that the destination really exceeds its limits. Some destinations have begun to put daily quotas on the number of tourists visiting the sites to protect their natural and cultural assets. For example, the Galapagos Islands in Ecuador, which is at UNESCO World Heritage Site an important location in terms of plant and animal species, have been limited of 15 days and 14 nights for each vessel due to exposure for excessive visitor and these vessels cannot visit the same area twice during the visit (igtoa.org).

- *__Distribute Reservations and Ticketing Systems In Intensive Periods:__* Theme parks and some tourist attractions have actually been using this system for a long time. The main point of the system is to limit the number of visitors on a daily basis during peak periods of season and to not bring any no limitation the number of visitors during the more stable periods of the season. For example, while the walking track in the rainforest is limited to 90 hikers per day in high season, but there is no such limitation in low seasoning New Zealand (doc.govt.nz).
- *__Benefit From Technology To Manage Visitor Density:__* Destinations benefit from technology to manage the intensity at times of occurring tourism event. Amsterdam, for example, gives place to the waiting times of the most popular touristic attractions on its website. In addition, it has created an application that directs to alternative areas where there are long queues (forbes.com).
- *__Prolong Seasonand Change Promotion Focus:__* It is a very important situation for tourism to spread the demand all season. Because densities are not trapped within at a given period of time. The island of Santorini in Greece, which do the honors both night visits and daily life period, hosts approximately 2 million visitors annually. This intensity has adversely affected island infrastructure, land and water use. With this the island of Santorini in Greece, which has been closed for visits since November, continues to host visitors in November and December. Thus, demand is also constituted during off-season periods (greece.greekreporter.com).

4.2 Spread Visitors Across Sites

Distribution efforts for visitors geographically are reducing situation excessive intensity. Because the waiting of tourists for a long time by creating queues to visit a place is a compelling factor for infrastructure (WTTC & McKinsey & Company, 2017).

- *__Introduce Less Visited Touristic Attractions:__* Some destinations, remove from the most visited places 'promotion focus'. For example, Iceland is trying to distribute the intensity of the visit by emphasizing the promotion of the town of Akureyri, which includes waterfalls and hot springs. Touristic destination authorities can make agreements with tour operators, agency representatives and journalists to promote less visited places (Moore, 2017).

- *Develop New Routes and Attractions:* Some destinations determine new tourist attractions and routes to distribute the concentration of tourists focusing on specific areas. Showing success in this method pass through the cooperation and creativity of the public and private sectors. Iceland has established the Fund for the Protection of Touristic Places. Another purpose of this fund is to encourage visitors to see unknown places in the country. This fund identifies regional flight routes at airports for the towns of Akureyri and Egilsstaðir and encourages people to visit these places (isavia.is).

4.3 Adjust Pricing to Balance Supply and Demand

Pricing is an important criterion for supply and demand balance. Increasing the prices of touristic attractions will not only decrease the number of visitors, but will also appeal to high-income tourists. However, this situation may cause the reaction of the local resident who have low income group. Therefore, pricing should be done in line with the interests of all stakeholders (including local resident) (WTTC & McKinsey & Company, 2017).

- *Application of Special Taxes and Fees:* Making a destination more expensive but not for everyone is seen as a high value / low volume strategy. One of the best-known examples of this is Bhutan fort this situation. Bhutan, a country with limited tourism policy in South Asia, demands $ 200 in low season and $ 250 in high season for each day of their stay in Bhutan from foreigners who will visit this country for tourism purposes to tourism agencies recognized by the Government of Bhutan (tourism.gov.bt).
- *Collect the Actual Fee:* Some destinations may determine some prices to ensure long-term sustainability of supply, rather than balance supply and demand. These prices are usually reflected in ticket/entrance fees. For example, by the end of 2017, Eiffel Tower announced a 50 % increase in ticket prices and financed some of the modifications of tower in this way. The Pantheon Temple in Rome hosts approximately 7 million visitors annually. The Roman Government has announced that they will also implement entrance fees for this temple (Nozari, 2017).
- *Price Diversification or Stratified Price Tariff Application:* It is a common method to apply different pricing procedures for different visitors. Many museums and/or archaeological sites entrances have discounted applications for students, the elderly or the disabled. For example, while admission is free for locals residing in Barcelona Park Güell and MontjuïcCastle region, admission fees are determined for tourists and other visitors. Also New Zealand including the Milford Track raised admission prices for foreign visitors in

seven of the Great Walks went on a fixed price application for its own community (Davison, 2017).

4.4 Regulate Accommodation Supply

Increasing the capacity of accommodation creates the opportunity to attract more tourists to destinations. However, destinations that complain about the density of tourists go to various restrictions and/or regulations instead of consulting to this method. Recently Airbnb application brought by the sharing economy which is in the foreground is seen as one of the factors affecting overtourism. Explosive growth in housing sharing with the Airbnb application accelerated the increase in the number of visitors. Rental housing prices have risen considerably in addition to the increase in the number of visitors and local community has difficulty in maintaining their vital activities in their own country. These economic problems also result in people leaving their homes. Spain, Barcelona, Ibiza, Majorca and Balearic Islands, which are among the most damaged destinations, have passed restrictions on leasable housing. These restrictions cover the number of stay overnight, accommodation price, number of person and housing rent with related to regulations (Shankman, 2017).

4.5 Limit Access and Activities

It is inevitable that there will be deterioration in the quality of life and destruction of nature and cultural areas due to Overtourism. In order to avoid the negative consequences, the destinations are faced towards different applications. One of these applications is the prohibition or limitation of tourism-oriented activities. The examples of these applications implemented by the destinations can be listed as follows (WTTC & McKinsey & Company, 2017):

- It is forbidden to open tourist-oriented shops in Amsterdam city center, including souvenir shops, bicycle rental companies and fast food restaurants.
- In Rome, alcohol use is prohibited, especially after 2 am on the side streets.
- Barcelona has banned the use of Segways and electric scooters in the city center and at the seaside to eliminate density and congestion in the transport network.
- Amsterdam has banned famous beer bikes to find solutions to traffic problems.
- Rome imposed a ban on tour buses entering to the city center.

Prohibitions and/or restrictions have always been a difficult process for people to accept. Because people prefer to break the bans instead of accepting this situation at once. In order to get over this situation more easily, the Roman government supports it with harsh fines. In addition, retired police officers voluntarily joined and volunteered to patrol famous sights in high season (WTTC & McKinsey & Company, 2017).

Conclusion

The definition and effects of extreme tourism are multidimensional and vary from destination to destination. Some articles and books deal with extreme tourism for the negative effects on local people, visitors and destinations, while some articles and books address natural and cultural heritage and destination carrying capacity. Seven different myths have been introduced to make the overtourism event and/or its concept more comprehensible. These myths are, (1) although the concept of overtourism is new, the effects are a long-standing condition, (2) that extreme tourism does not have the same meaning as mass tourism, (3) The overtourism incident is experienced in popular areas of the city, (4) overtourism is also a problem arising from the excessive use of resources, infrastructure and facilities in the destination, (5) technology alone will not be enough to solve the overtourism, (6) there is no single solution for all destinations and (7) is the fact that overtourism can occur not only in cities but also in rural areas (Koens et. al., 2018).

The reasons for overtourism in destinations are the deficiencies in planning, lack of communication between stakeholders and the fact that some destinations are in the forefront. Although there is no clear solution for the situation of overtourism, these solutions may be separate for each destination (Duyar & Bayram, 2019). Some of the solutions offered to different destinations can be listed as follows, (a) adopt responsible tourism understanding, (b) to use sustainability accounting applications, (c) to calculate the carrying capacity of destinations, (d) de-marketing practices to implement, (e) reduce demand by using pricing during peak periods, (f) tourist tax, (g) determine limits of acceptable change (LAC Model) in tourism (responsible travel).

As a result, the resources that make up the supply of the tourism sector can be protected if their sustainability is achieved. Although extreme tourism impedes the sustainability of resources, any destination that wishes to cope with this situation should choose the appropriate solution or coping method and apply it to them.

References

Ali, R. (2016). 'Exploring the Coming Perils of Over tourism', www.skift.com (02.11.2019).

Bock, K. (2015). The changing nature of city tourism and its possible implications for the future of cities. *European Journal of Futures Research*, 3/1, p. 1–8.

Calzada, I. (2018). Local entrepreneurship through a multistakeholders' tourism living lab in the post-violence/peripheral era in the Basque Country. *Regional Science Association International*,11/3.

Colomb, C. & Novy, J. (2016). Urban tourism and its discontents: an introduction. In C. Colomb & J. Novy (Eds.), *Protest and Resistance in the Tourist City* (p. 15–44). London: Routledge.

Çolak, O., Kiper, O. V. & Batman, O. (2019). Kent Destinasyonlarında Over Tourism'e (Ölçüsüz Turizm) Dair Kavramsal Bir Yaklaşım.*20. Ulusal Turizm Kongresi Bildiriler Kitabı*.16–19 Ekim, Eskişehir.

Davies, P. (2017). Intrepid offers 'lesser-known' destinations to tackle over-tourism. http://www.travelweekly.co.uk/articles/289988/intrepid-offers-lesser-known-destinations-to-tackle-over-tourism (21.11.2019).

Davison, I. (2017). Foreigners to pay double to tramp New Zealand's Great Walks. *New Zealand Herald*.nzherald.co.nz. (12.11.2019).

Doc.govt.nz, (2019). Milford Track, Department of Conservation-Te Papa Atawhai. https://www.doc.govt.nz/globalassets/documents/about-doc/annual-reports/annual-report-2018/annual-report-2018.pdf (11.11.2019)

Duyar, M. & Bayram, M. (2019). Overtourism and tourismphobia: evolution of host and tourism relationship. *International Journal of Geography and Geography Education (IGGE)*, 40, 347–362.

Ferdamalastofa.is. (2019). Tourism in Iceland in figures, Icelandic Tourist Board, June 2017. https://www.ferdamalastofa.is/en/recearch-and-statistics/tourism-in-iceland-in-figures. (20.11.2019).

Forbes.com. (2019). Blacklisting Venice to save it from too many tourists and too few Venetians. https://www.forbes.com/sites/ceciliarodriguez. (13.11.2019).

Goodwin, H. (2017). The challenge of overtourism. *Responsible Tourism Partnership Working Paper*.

Goodwin, H. (2019). Overtourism: causes, symptoms and treatment. *TourismusWissen*, p. 110–114.

Gössling, S. (2002). Global environmental consequences of tourism. Global Environmental Change, 12/4, p. 283–302.

Greek Reporter. (2019). https://greece.greekreporter.com/santorini-emerges-as-one-of-the-worlds-top-tourism-destinations/. (11.11.2019).

Hamari, J., Sjöklint, M. & Ukkonen, A. (2015). The Sharing Economy: Why people participate in collaborative consumption?. *Journal of the Association for Information Science and Technology*, 67(9), p. 2047–2059.

Igtoa. (2019). Challenges facing the Galápagos Islands, International Galapagos Tour Operators, https://www.igtoa.org/travel_guide/challenges. (11.11.2019).

IPOL. (2018). Policy Department for Structural and Cohesion Policies: Overtourism: impact and possible policy responses. http://www.europarl.europa.eu/. (06.11.2019).

Isavia.is. (2019) The Icelandic Route Development Fund. isavia.is. (13.11.2019).

Koens, K., Postma, A., & Papp, B. (2018). Is overtourism overused? Understanding the impact of tourism in a city context. *Sustainability*, 10/12, p. 43–48.

Koens, K. & Thomas, R. (2016). You know that's a rip-off: Policies and practices surrounding micro-enterprises and poverty alleviation in South African township tourism. *Journal of Sustainable Tourism*, 24/12, p. 1641–1654.

Kuščer, K., & Mihalič, T. (2019). Residents attitudes towards overtourism from the perspective of tourism impacts and cooperation–The case of Ljubljana. *Sustainability*, 11/6, p. 18–23.

Martín Martín, J., GuaitaMartíncz, J., & Salinas Fernández, J. (2018). An analysis of the factors behind the citizen's attitude of rejection towards tourism in a context of overtourism and economic dependence on this activity. *Sustainability*, 10/8, p. 28–51.

Milano, C. (2018). Overtourism, malestar social y turismofobia. Un debate controvertido. PASOS. *Revista de Turismo y Patrimonio Cultural*, 6/3, p. 551–564.

Milano, C., Novelli, M. & Cheer, J. M. (2018). Overtourism: a growing global problem. https://www.researchgate.net/publication/326573468 (06.11.2019).

Moore, T. (2017). Iceland's tourism boom—and backlash, *Financial Times*. ft.com. (13.11.2019).

Namberger, P., Jackisch, S., Schmude, J., & Karl, M. (2019). Overcrowding, overtourism and local level disturbance: how much can Munich handle?. *Tourism Planning & Development*, 16/4, p. 452–472.

Nozari, E. (2017). Rome's Pantheon will soon charge an entrance fee. *Condé Nast Traveler*, cntraveler.com. (12.11.2019).

Oklevik, O., Gössling, S. C., Michael, H., JensKristian, S. J., IvarPetter, G. & McCabe, S. (2019). Overtourism, optimisation, and destination performance indicators: A case study of activities in Fjord Norway. *Journal of Sustainable Tourism*, p. 1–21.

Panayiotopoulos, A. & Pisano, C. (2019). Overtourism dystopias and socialist utopias: Towards an urban armature for Dubrovnik. *Tourism Planning & Development*, p. 1–18.

Pappalepore, I., Maitland, R. & Smith, A. (2014). Prosuming creative urban areas. Evidence from East London. *Annual Tourism Researchs*, 44, p. 227–240.

Pinke-Sziva, I., Smith, M., Olt, G. & Berezvai, Z. (2019). Overtourism and the night-time economy: a case study of Budapest. *International Journal of Tourism Cities*, 5/1, p. 1–16.

Responsible Travel. (2019,). *Overtourism Solutions. https://www. responsibletravel.com/copy/overtourism-solutions. (20.11.2019).*

Russo, A.P. & Scarnato, A. (2018). Barcelona in common: a new urban regime for the 21st-century tourist city?, *Journal of Urban Affairs.*, 40/4, p. 455–474.

Sarantakou, E. & Terkenli, T. (2019). Non-institutionalized forms of tourism accommodation and overtourism impacts on the landscape: The case of Santorini, Greece. *Tourism Planning & Development*, 16/ 4, p. 411–433.

Seraphin, H., Sheeran, P. & Pilato, M. (2018). Over-tourism and the fall of Venice as a destination, *Journal of Destination Marketing & Management*,9/3, p. 374–376.

Sevinç, F. & Duran, E. (2018). SürdürülebilirDenizTurizmiveTüketimParad oksu: TüketirenTükenmek. *TüketicliveTüketimAraştırmalarıDergisi*, 10/2, p. 173–196.

Shankman, S. (2017). *Barcelona overtourism: Airbnb and short-term rentals,* www.skift.com. (12.11.2019).

Sheivachman, A. (2019). *Iceland and The Trials of 21st Century Tourism. https:// skift.com/iceland-tourism/. (06.11.2019).*

Singh, T. (2018). Is over-tourism downside of mass tourism?*Tourism Recreation Research*,43/4, p. 415–416.

Timur, S. & Getz, D. (2009). Sustainable tourism development: how do destination stakeholders perceive sustainable urban tourism?, *Sustainable Development*, 17/4, p. 220–232.

Tourism.gov.tr. (2019). *Minimum daily package, Tourism Council of Bhutan. tourism.gov.bt. (13.11.2019).*

UNWTO. (2018). Overtourism? – Understanding and managing urban tourism growth beyond perceptions: executive summary, https://www.e-unwto.org/doi/pdf/10.18111/9789284420070. (09.11.2019).

UNWTO. (2019). International tourism highlights, https://www.e-unwto.org/doi/pdf/10.18111/9789284421152 (20.11.2019).

Usa Visa. (2018). Asian travelers could alter established destination patterns as Asya-Pasifik traveling class rapidly develops. https://usa.visa.com/content/dam/VCOM/global/partner-with-us/documents/travel-insights-newsletter-q1-2018.pdf. (20.11.2019).

Webster, C. & Ivanov, S. (2015). Geopolitical drivers of future tourist flows. *Journal of Tourism Futures*, 1(1), p. 58–68.

WTTC & McKinsey & Company, (2017). Coping with success: managing overcrowding in tourism destinations. https://www.wttc.org/-/media/files/reports/policy-research/coping-with-success---managing-overcrowding-in-tourism-destinations-2017.pdf/ (03.11.2019).

Kansu GENÇER

16 Qualitative Approaches for Tourism Research

Introduction

Researches on tourism as a social phenomenon can be carried out using both quantitative and qualitative methods. Although qualitative and quantitative researches have advantageous or disadvantageous features when compared, it is observed that there is an increase in the number of researchers who prefer qualitative methods. Semiotic approach, hermeneutic approach and phenomenology are the methods which have different perspectives than positivism in social sciences and are frequently preferred by researchers. A new perspective is provided to understand people through qualitative approaches in tourism research. The fact that tourism is a human-centered social phenomenon makes these approaches one of the most important methods for tourism research.

Semiotic, hermeneutical and phenomenological approaches to understand social phenomenon provide researchers with the freedom to interpret and subjectivity. Nevertheless, this area of freedom takes place within the systematic of qualitative research. In this section, semiotic, hermeneutical and phenomenological approaches, which are becoming widespread today and which draw attention to the importance of subjectivity, are examined.

Semiotic Approach

According to Chandler (2007), human beings are driven by the desire to create meaning. In his own words, 'homo significans' refers to the person he calls 'meaning-maker, meaning giver'. Meaning is made by human creativity and interpretation of the signs. (Chandler, 2007). According to Pierce (1955), one thinks only in signs and nothing is a sign unless it is interpreted as a sign. Signs are words, pictures, sounds, smells, movements, and objects that a person gives meaning to, and can only be signs when a person attaches meaning to them. Anything that points to something, shows it, expresses something other than its own essence can be a sign (Chandler, 2007). Semiotics is the science that examines signs and sign systems in its most general and well-known definition. However, this definition is made according to the subject covered by semiotics. Semiotics can also be defined according to the method used. According to this, semiotics is science which applies linguistic methods to objects, describes every linguistic thing (games, gestures, facial expressions, religious rituals, literary works, pieces of music) and all non-linguistic phenomena that tries to explain the language metaphor (Dervişcemaloğlu, 2008). In Semiotics, the relationship between things rather than the essence of things is examined (Johansen & Larsen, 2002). This is due to the structure of the signs showing something. Because the signs have the meaning interpreted, and the interpretation of what they tell, and the interpretation of their essence are different.

According to Saussure (1916), the sign 'combines a concept with a communication symbol. The image of the hearing is the sound structure of the sign and the concept is its semantic content'. This definition is based on language, but the concept of the sign can also spread to non-linguistic communication strings, which can load the communication task and form strings for it. According to Saussure (1916), the signer and the signed are combined to form the sign. With these concepts, it is suggested that one thing can replace another, and at the same time, it refers to understanding. The sign consisting of the combination of a concept and hearing image is an imaginary concept that occurs as a result of the image formed in the mind and the voices produced by the word (Bayav, 2006).

Pierce (1955) laid the foundation for a semiotic of logical origin. According to him, the representamen is what replaces anything in any respect or capacity for a person. This sign for someone creates a sign that is equivalent to the thought of this person or a more developed sign. He stated that a thought without a sign cannot be perceived and that a thought that cannot be perceived cannot exist. Therefore, all thoughts should be explained in the sign. According to this, the sign and object are interpreted by a third party. According to Pierce (1955), semiotic, which is the 'formal teaching of signs', is another name for logic. Pierce (1955), who prepared a theory about linguistic and non-linguistic signs, is an important triad distinction (Bayav, 2006).

Charles Morris is a researcher who adapted Pierce's work to a form of behaviorism. According to Morris, there are four elements of the period which he calls 'semiosis'. These are: an object that directs the person to the target, an interpreter performing semiotic activity, the designatium of the symbol object, and the interpretant in the mind of the interpreter. According to Morris (1971), it consists of 'semantics' that examine the relationships between signs and the world of things, 'syntactics' that examine the relationships between signs, and 'pragmatics' that examine how indicators affect human behavior (Ryder, 2004).

Examples from Semiotic Research

In the field of sociology, Gottdiener (1985), in his study 'Hegemony and Mass Culture: A Semiotic Approach', examined the issue of mass culture, relations between cultural objects that make up mass culture and hegemony in a semiotic analysis. In the field of mathematics, Radford (2000), in his study 'Signs and Meanings in Students' Emergent Algebraic Thinking: A Semiotic Analysis' used the semiotic method to find out how students use signs, how they load meaning and provide algebraic thinking to students in their first encounter with algebraic generalizations and patterns. In the study of music, Spychiger (2001) followed a semiotic approach to understand musical activity and musical learning as indicative processes in his study 'Tow Understanding Musical Activity and Musical Learning as Sign Processes: Toward a Semiotic'. In the field of tourism, Petr (2002), in his study titled 'Tourist Apprehension of Heritage: A Semiotic Approach to Behavior Patterns', used semiotic analysis to determine the typology and behavioral patterns of tourist visitor routes and to identify the distinctive characteristics of each segment. In the field of tourism, Nelson (2005) used the semiotic approach to examine the pictures used in the promotional materials of the Grenada region in his study titled 'Representation and Images of People, Place and Nature in Grenada's Tourism'. Semiotic was a suitable choice for him because the pictures to be interpreted were signs and he thought that the semiotic would be explanatory in this regard. Yet another study in the field of tourism belongs to Ryan (2002). In his study 'Tourism and Cultural Proximity: Examples from New Zeland', Ryan examined the concepts of tourism and cultural proximity in New Zealand and again benefited from the semiotic approach.

Hermeneutic Approach

Hermeneutics is a concept whose origins are based on ancient Greek civilization. The Greek word 'hermeneuein' means expressing, interpreting and translating (Schmidt, 2006), while hermeneutic is the adaptation of the word to the present alphabet and colloquial language.

The word Hermeneuein is thought to derive from Hermes, the messenger and wind god of ancient Greek civilization. Hermes is the son of Zeus and conveys messages from the gods to people. From this perspective, it can be thought that the word 'hermeneuein' has a great effect on its formation. In Late Greek, hermeneuia is referred to as 'wise explanation' and hermeneios as 'explaining', 'translating'. Thus, the art of hermeneuia, that is, the hermeneutic is, to explain the sacred expedition, especially the sacred and authoritative will, to the mortal, that is the listener (Felsefe.gen.tr, 2019). In Hermeneutics, the integrity of meaning from another world and another language is translated into the language and culture of the individual (Topakkaya, 2007). 'Interpretatio', which is the Latin translation of the word 'Hermeneuein', forms the origins of the word 'interpretation' in English (Schmidt, 2006).

Basic Principles of Hermeneutic Approach

Hermeneutic analysis is the name of various analysis methods based on interpretation. However, according to Arnold & Fisher (1994), the basic principles of Hermeneutic approach can be summarized as prejudices, social dialogue, commitment of understanding to the language and non-objective interpretation.

Prejudices

One of the important issues emphasized in Hermeneutics is prejudice. People are part of a cultural world, and therefore interpretors and interpretations are closely linked to traditions, theories, events, ideologies and beliefs. Therefore, it is impossible to put forward an interpretation independent of the things that are attached. According to the Hermeneutical approach, prejudices formed on these grounds are not restrictive for the interpreter, but rather supportive. Prejudices and preconceptions are the windows of people to the world. Without prejudices, it cannot be possible to give meaning to any word, event or object observed. In hermeneutics, prejudices are not seen as a situation to be avoided because preconceptions take place with prejudices and therefore preconceptions are not prevented, and the prejudices are accepted as an impressive role.

In the researches about consumption experience, there are two experiences; as consumer experience and experience as researcher. Prejudices occur as a result of past research or past information. Concepts related to consumer behavior such as theoretical information, attitudes and socialization process provide information about consumer behavior to the researcher. At the same time, this information is shaped by the social and cultural accumulation of the researcher and gains meaning. Therefore, the consumer's experience will be perceived in the researcher by being shaped by the researcher's experiences. This approach of hermeneutics about prejudices is parallel to Mick's (1986) semiotic approach, some forms of structuralism and more integrative forms of literary criticism (Arnold & Fisher, 1994).

Social Dialogue

According to Arnold & Fischer (1994), another principle of hermeneutics is social dialogue. The strengthening of social dialogue seems to be critical to philosophers interested in hermeneutics. As mentioned in the section on prejudices, people can begin to interpret through past experiences. Because different people have different experiences, this leads to different interpretations and understandings. For this reason, the strengthening of social dialogue will lead to the communication between people and thus the sharing of experience, and these experiences will play an important role in understanding each other. Conducting a research in a group dialogue appears to be hermeneutically advantageous. Because there is a possibility that another person will notice details about an event or object that one person cannot make sense of and therefore cannot realize. This improves interpretation and enriches research.

Commitment of Understanding to Language

In Hermeneutical philosophy, language has a special importance. Experiences pass through the language filter, are coded and communicated through dialogue. Thus, language bridges between past and present, between texts and interpreter. According to Heiddeger (1949), no text can be interpreted impartially. The interpreter belongs to a world and has a heritage of understanding, formed through language (Arnold & Fischer, 1994).

Non-Objective Comments

People may share some traditions, but not all parts of that tradition. The world we know is the language we use. As a result, insights are also sensitive to certain issues and may not be relevant to other issues. Because of this devotion to language, a text has no objective meaning. It has a multilateral meaning. Therefore, a text cannot be understood in all its parts. But this does not mean that all interpretations should be equally credible. Consensus in dialogue of societies allows some comments to dominate others (Arnold & Fischer, 1994).

Understanding

Hermeneutic understanding is self-understanding, self-reflection and self-improvement. But the 'self' here is not a researcher. There is a high-level 'self' state that emerges as a result of the combination of many assumptions in Hermeneutical philosophy. In the process of hermeneutic understanding, which is connected with social dialogue, the boundaries between the object and the

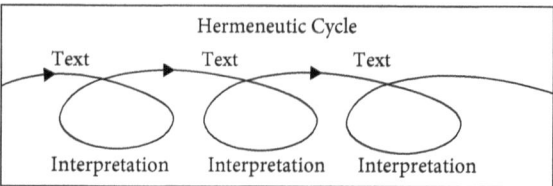

Figure 1: Hermeneutic Cycle. Source: Sfu.ca (2019)

subject are lifted. While researchers try to understand the others, the others also try to understand them. In this process, the structure of comprehension is formed and the foreigner becomes familiar. Another important point is that the same events can be perceived differently by different people as witnesses observing the same event can give different expressions even when they tell only what they see (Klein & Myers, 1999) and as a result multifaceted interpretations and meanings are formed.

Hermeneutic Cycle

According to Palmer (1969), comprehension is simply a reference process, we understand things by comparing them to something we know before. What we understand is shaped as a systemic integrity. The whole is a union of units as parts and a combination of these units. For example, if a sentence is a whole, it is the words in the parts that make up the whole. We understand and interpret the meaning of a word with reference to the holistic meaning of the sentence. However, we deduce the meaning of the sentence from the meanings of words as units. Therefore, the holistic meaning of the sentence gives meaning to the word, and the unitary meaning of the word gives meaning to the sentence, and a dependent cycle emerges. This cycle is called hermeneutic cycle in hermeneutics.

The importance of the hermeneutic cycle stems from: in order to give meaning to the whole, one has to understand the part and the meaning of the part depends on the whole, in this case a situation occurs as if it is impossible to understand (Palmer, 1969). For this reason, the hermeneutic cycle must be formed to ensure understanding. The meaning is not realized until the parts are examined with a whole and all parts with a repetitive cycle. According to Gadamer (2004), understanding occurs when all parts are blended with the whole, and in the event of failure of this blending, understanding also fails. In the Hermeneutic cycle, the meaning of the text as a whole comes from the parts in the text (Bernstein, 1983) and the meaning of the parts comes from the whole. For this reason, the contradictions between the parts and the whole are analyzed, the parts are examined repeatedly, interpreted and the relationship with the whole is established.

Some Researches Conducted with Hermeneutic Approach

There is a wide range of research from the Hermeneutical perspective, and texts, stories, photographs, audio recordings and diaries can be used for interpretation. In order to provide healthy critical thinking in a research, social and historical background of research conditions should be given, thus providing more explanatory information about how the situation under research develops (Klein & Myers, 1999). For example, Ciborra et al. (1996) discussed the historical driving forces that pushed the Fiat brand to build a new assembly plant to demonstrate how effective the old Fordist understanding is, despite radical changes in business organization and operations. Bearing in mind that understanding can occur through the change of person to person and by blending these meanings in the whole sense, Levine & Rossmore (1993) found that Bremerton employees had very inconsistent expectations in their research on the Threshold system, which was the subject of interpretative research. Therefore, the reasons for each expectation will be interpreted separately and form the meaning of the whole, including in the hermeneutic cycle.

Hermeneutical research can be used in various disciplines. Thompson (1997) tried to derive a marketing understanding by interpreting consumer texts about consumption in a hermeneutical framework. In this marketing study, Thompson used a hermeneutical framework to interpret consumers' written stories about goods, services, brand images and shopping. Anthropologist researchers Young, Pompana, Spitzer, and Candler (1997) attempted to re-frame ideas to understand the concept of grandfathers, a key component in Indian religious groups. Thus, it would be ensured that the concept of local integrity was not compromised when comparing with similar ideas in different cultures, the concept would be better understood. The theologian researcher Payne (1999), who worked on Buddhist studies, tried to interpret the Japanese story 'A long tale for an autumn night'. The need for interpretation, especially in the social sciences, makes hermeneutic philosophy remarkable for researchers.

Obenour (2004) in the field of tourism, low-budget tourists on how to understand the concept of travel and has been determined by determining the characteristics. Hermeneutic understanding was also used in this research and the researcher interpreted the stories he obtained through interviews with low-budget tourists. In another study in the field of tourism, Obenour, Patterson, Pedersen & Pearson (2006) used the hermeneutical approach to explain the quality of service in consumers' studies in their study titled 'Conceptualization of Meaning Based Research Approach for Tourism Service Experiences'.

Phenomenological Approach

According to Husserl, who is considered the founder of the phenomenological approach, positivism has closed itself to other data except for certain data, especially sensational data (Sinha, 1963). In contrast, phenomenology attaches importance to sensory data. Phenomenology, as an alternative to positivism, concerns individuals' perceptions, attitudes and beliefs, intuitions and emotions (Denscombe, 2010). In contrast to positivism, it emphasizes (Denscombe, 2010):

- Subjectivity (rather than objectivity)
- Description (rather than analysis)
- Interpretation (rather than measurement)
- Agency (rather than structure)

Phenomenology in the social sciences tries to understand how social phenomena are experienced by humans. It seeks to capture a clear and free gaze that can understand the essence of an experience. In order to be able to speak of the existence of the object, the subject that speaks of the object is required. The object must be detected and transmitted by a subject. The subject is needed to know the existence of the object.

Consciousness is an absolute entity that cannot emerge into the field of view in terms of its existence and cannot be given as an image. As such, nature is not an absolute entity in itself; it has no absolute existence. It is the presence of a 'thing'. Thing, by nature, is an entity that is dependent on orientation, that is, only visible to consciousness. Therefore, according to Husserl, consciousness is not a part of nature (Öktem, 2005). In order for words that express consciousness acts such as hear, love, think, to gain a meaning, these experiences must be connected to something and to have a relationship with something. 'I love something', or 'I think of something' and so on. What is important here is the relationship between the act of consciousness and the subject of consciousness. There is no physical orientation; a rock continues to exist in itself without making contact with anything. Orientational experiences are spiritual and conscious events (Topakkaya, 2009).

According to Husserl, the body is a 'center of orientation'. According to him, no matter what I perceive, no matter where I go, I always try everything in relation to my body. Such as 'up', 'down', 'right', 'left' and so on like all dimensions always gain meaning according to my body. The purpose of body orientation is

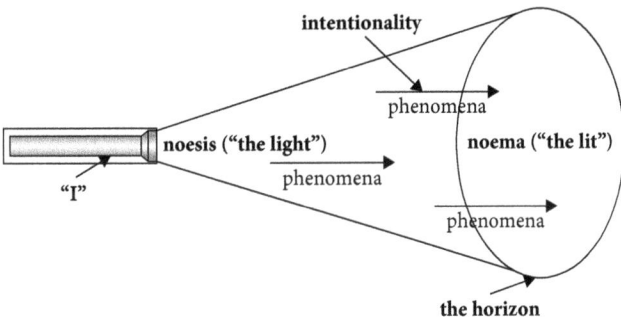

Figure 2: Basic Phenomenology Concepts. Source: Boeree (1998)

essense-description. The essence is description; the description is essence. This means that the description is directed to the essence, that is to say, the essence of what is described. The most important point of his philosophy is that Husserl establishes a connection between the description and the essence, or rather, the description of the desire to be a description towards the essence, an 'essence description'. What is important and sought is the 'essence', and this 'essence' is the one that will emerge after the outside world, nature, is parenthesized. (Öktem, 2005; Husserl, 2010). 'Selves' are derived from the directional analysis of the connection between the object as perceived and the subjective perception of that object or event. Husserl used the term orientation to describe the connection between noema and noesis (Sanders, 1982).

Noema-Noesis

The thing that consciousness is heading for is called noema in Husserl. Noema is what is perceived. Noesis is how things are perceived (Moustakes, 1994). Neoma is understood by noesis, and events are understood in different perceptions by this neosis. Therefore, the way the consciousness is directed is noesis, and the object defined by noesis is noema. Neoma is the irreal-oriented aspect of life, while noema is the real or objective aspect of life. All kinds of noesis as an irreal-oriented element carries noema as an objective meaning (Topakkaya, 2009).

Epoche and Eidetic Reduction

Epoche is a word used in ancient Greek to suspend judgment. It tries to get rid of all the natural attitudes that limit the person and provide a clear look. According to Boeree (1998), it is not heard if shouted too much. For the phenomenological attitude, it is necessary to temporarily suspend all personal prejudices, beliefs, preliminary ideas and assumptions. Husserl used this suspension as bracketing. Suspended, that is, those that are left out of brackets are fixed, they are not eliminated, what is done is not to take them into account (Sanders, 1982). Leaving aside the old knowledge, Epoche opens the doors of its personal and social dynamic resource along with new thinking. Through Epoche, phenomenology allows to exclude forms of thought that remain outside and predetermined goals, and to experience reality through a more authentic, responsible thinking and acting with others (Fay & Riot, 2007).

Eidetic reduction is the methodological path that leads from the concrete expression of a phenomenon to the implicit universal pure 'essence', a process that seeks a common ground by passing through, behind or underneath traditional ideas or structures (Sanders, 1982). Eidetic reduction tries to understand what the essence of the event actually means (Keller, 1999). With reduction, everything that is transcendent is excluded and this exclusion is achieved. The general assertion of natural behavior with reduction and the opinions that form the basis of this assertion, knowledge of the sciences, space-time etc. ids are subjected to epoche, avoiding judgment, they are bracketed. A phenomenological view cannot be obtained without reduction (Keskin, 2012).

Phenomenological Research

In phenomenological research in social sciences, the aim is to understand what the real experiences of people are. It is a matter of curiosity as to why and how events happen alongside 'what'. When a researcher adopting a phenomenological approach examines an event, he/she should make a detailed description as close to the original of all the components of that event as possible. Phenomenologically, the researcher is part of the social universe in which he / she researches. For example, if the subject of the research is tourist, it should be understood what the tourist life is and what the tourist experience is. It focuses on personal experiences such as what it means for tourists, how they define tourism, how they perceive their surroundings, how they interpret events (Denscombe, 2010).

The paradigms of phenomenological research can be listed as follows (Sanders, 1982):

World perception: The world is unstable and problematic. The phenomenon to be investigated is seen as a function of perceptions, orientations and personal meanings.

Phenomenon to be investigated: It is the experiences of people. Observed and perceived particular qualities are meanings for individuals.

Problem formulation: It starts with Epoche. All personal prejudices, beliefs or assumptions about the causal relationship are suspended.

Research method: The important thing is to describe the events from the perspective of those living and experiencing. All concepts and theories that emerge from the data formed in consciousness should not be repeated in the same way as they do and should be in an inductive approach.

Research objective and inferences: It is to find the universal pure essence. The logic of inferences is the direct comparison with the new meanings that are formed.

Generalization of results: The results can only be generalized to the group with the same experience. There is no generalization. Findings will be another research database.

Phases of phenomenological research can be listed as follows (Gray, 1997):
1. Identifying a specific phenomenon
   a. Phenomenological orientation
   b. Phenomenological analysis
   c. Phenomenological description
2. Review of general characteristics
3. Finding important relationships between qualifications
4. Track occurrence patterns
5. To follow the structure of the phenomenon in consciousness
6. Preventing and suspending beliefs about the phenomenon
7. Interpreting the meaning of the phenomenon

The phenomenological approach is important for understanding the questions and concepts such as what the tourist experience is, what it means to be a tourist, what is the feeling of being a tourist when it comes to interpreting and understanding human experiences. Phenomenology can be used in tourism research as a philosophical view that is aware of these differences in researches about people from different cultures and speaking different languages.

Results

The signs, which are the essence of the semiotic approach, do not make sense on their own. What makes them a sign is the person who perceives and interprets what it shows. At the core of human thought are symbols, shapes, signs, and it is also humans who create signs. The associations that occur in the human mind can be varied for a variety of reasons. However, the signs used to express

Table 1: Hermeneutic Phenomenology Preliminary Guide for Tourism Research

Research Purpose	Examining the experiences and understanding how the experiences are interpreted and understood (meaning of experiences in terms of the participants)
Ontology	We can reach the world and nature with our presence in the world. All our understanding of our existence in the world is perceptual and shaped by history, culture, language and past knowledge.
Epistemology	Main focus is on content, language and interpretation. The so-called truth is based on participation in interpretation, together structure and return to essence. Both the researcher and the participant are interpretive entities living in the real world. They play an important role in the process of understanding through dialogue and interpretation. The key is the player language.
Methodology	Interpretative and dialogue: the researcher wants to interpret the experience, seeks meaning, analyzes, discusses, and reconciles theory and data. The focus in the relationship is not the objectivity-subjectivity posture, but itself and others. Method: Interviews and participatory observation, writing detailed descriptions to understand meaning

Source: Pernecky & Jamal, (2010)

thoughts are limited in certain proportions. Therefore, the human mind sometimes has to use the same signs to create different meanings. The important point here is that the difference of meanings or interpretations used is person-based rather than sign-based and this reveals the importance of interpretation.

Hermeneutics is an approach used in social disciplines. Although the perspectives of hermeneutics may change, hermeneutics are interpreters of the unchanging feature throughout history, being aware of individual differences, and being aware that meanings change according to individuals and interpretations. Hermeneutic is a philosophy that is aware of the fact that humans are thinking social beings and that the differences in their lives will affect their thinking systems. For this reason, it is a philosophy that tells that the differences in thinking systems should be taken into consideration and that it should be seen as a wealth without being afraid of these differences. Hermeneutics is a human and interpretation-oriented philosophy that makes it valuable in research in the field of social sciences. For this reason, researchers working in the field of social sciences are more interested in hermeneutics than the researchers in science and continue to keep it alive.

The Hermeneutic paradigm should not be seen as a rival to other paradigms that shape science. It should enrich science with the perspective it brings, and it should be considered in the researches as an example to the impartiality of science. Whether science is related to non-human or human, the scientist is in need of understanding, explanation and interpretation. Hermeneutical philosophy, which has gone very far in the field of interpretation and has accumulated much since ancient times on how interpretive analysis and interpretive structure works, is one of the valuable source energies to be utilized. In phenomenology, consciousness is not a part of nature. Consciousness must be sought in human beings and human beings in consciousness. The fact that tourism is a human-centered social event makes hermeneutic phenomenology one of the important methods for tourism research.

References

Arnold, S. J. & Fischer, E. (1994). Hermeneutics and Consumer Research. *Journal of Consumer Research*, 21/1, p. 55–70.

Bayav, D. (2006). *Resimde Göstergebilim, Çocuk Resimlerinin Göstergebilimsel Çözümlenmesi*, Marmara University Institute of Educational Sciences, Doctoral Thesis, İstanbul.

Bernstein, R. J. (1983). *Beyond Objectivism and Relativism: Science, Hermeneutics, and Praxis*, Philadelphia: University of Pennsylvania Press.

Boeree, G. C. (1998). Qualitative Methods: Part One. Shippensburg University https://webspace.ship.edu/cgboer/qualmethone.html.

Chandler, D. (2007). *The Basics of Semiotics*, New York: Routledge.

Ciborra, C. U., Patriotta, G. & Erlicher, L, (1996). *Disassembling Frames on the Assembly Line: The Theory and Practice of the New Division of Learning in Advanced Manufacturing*, in W. J. Orlikowski, G. Walsham, M. R. Jones, and J. I. DeGross (eds), Information Technology and Changes in Organizational Work, London: Chapman and Hall, p. 397–418.

Denscombe, M. (2010). *Good Research Guide: For Small Scale Social Research Projects* (4th Edition), Berkshire: Open University Press.

Dervişcemaloğlu, B. (2008). *Göstergebilimin Tanımı ve Dalları*, http://www.ege-edebiyat.org/docs/493.pdf.

Fay, E. & Riot, P. (2007). *Phenomenological Approaches to Work, Life and Responsability*, Bradford: Emeral Group Publishing.

Felsefe.gen.tr (2019). *Hermeneutik Nedir?* http://www.felsefe.gen.tr/mustafa_gunay_hermeneutik.asp.

Gadamer, G. H. (2004). *Truth and Method*, London: Continuum Publishing.

Gottdiener, M. (1985). Hegemony and Mass Culture: A Semiotic Approach, *American Journal of Sociology*, 90/5, p. 979-1001.

Gray, M. J. (1997). Application of the Phenomenological Method to the Concept of Occupation, *Journal of Occupational Science*, 4/1, p. 5-17.

Heidegger, M. (1949). *Sein und Zeit*, Munich: Tübingen.

Husserl, E. (2010). *Fenomenoloji Üzerine Beş Ders* (Tr: Tepeh.)., Ankara: Bilge Su Publishing.

Johansen, D. J. & Larsen, E. S. (2002). *Signs in Use: An Introduction to Semiotics*, New York: Routledge.

Keller, P. (1999). *Husserl & Heidegger on Human Experience*, New York: Cambridge University Press.

Keskin, M. (2012). *Husserl'in Fenomenolojisine Giriş* www.20.uludag.edu.tr/~felsefet/kaygi/dergieski01/02.pdf.

Klein, H. K. & Myers, M. D. (1999). A Set of Principles for Conducting and Evaluating Interpretive Field Studies in Information Systems, *MIS Quarterly*, 23/1, p. 67-94.

Levine, H. G. & Rossmore, D. (1993). Diagnosing the Human Threats to Information Technology Implementation: A Missing Factor, Systems Analysis Illustrated in a Case Study, *Journal of Management Information Systems*, 10/2, p. 55-73.

Mick, D. G. (1986). Consumer Research and Semiotics: Exploring the Morphology of Signs, Symbols, and Significance, *Journal of Consumer Research*, 13, p. 196-213.

Morris, C. (1971). *Foundations of the Theory of Signs*. Chicago: University of Chicago Press.

Moustakas, C. (1994). *Phenomenological Research Methods*. California: Sage Publications.

Nelson, V. (2005). Representation and Images of People, Place and Nature in Grenada's Tourism, *Human Geography*, 2, p. 131-143.

Obenour, W., Patterson, M., Pedersen, P. & Pearson, L. (2006). Conceptualization of Meaning Based Research Approach for Tourism Service Experiences, *Tourism Management*, 27, p. 34-41.

Obenour, W. L. (2004). Understanding the Meaning of the Journey'to Budget Travellers. *International Journal of Tourism Research*, 6/1, p. 1-15.

Öktem, Ü. (2005). Fenomenoloji ve Edmund Husserl'de Apaçıklık (Evidenz) Problemi, *Ankara University The Journal of the Faculty of Languages and History-Geography*, 45/1, p. 27-55.

Palmer, E. R. (1969). *Hermeneutics*, USA: NorthWestern University Press.

Payne, K. R. (1999). At Midlife in Medieval Japan, *Japanese Journal of Religious Studies*, 26, p. 1-2.

Pernecky, T. & Jamal, T. (2010). (Hermeneutic) Phenomenology in Tourism Studies, *Annals of Tourism Research*, 37/4, p. 1055-1075.

Petr, C. (2002). Tourist Apprehension of Heritage: A Semiotic Approach to Behaviour Patterns, *International Journal of Arts Management*, 4/2, p. 25-38.

Pierce, C. S. (1955). Logic as Semiotic: the Theory of Signs', in JustusBucher (Ed.), Philosophical Writings of Peirce, New York: Dover, p. 98-119.

Radford, L. (2000). Signs and Meanings in Students' Emergent Algebraic Thinking: A Semiotic Analysis, *Educational Studied in Mathematics*. 42/3, p. 237-268.

Ryan, C. (2002). Tourism and Cultural Proximity: Examples from New Zeland, *Annals of Tourism Research*, 4/29. p. 952-971.

Ryder, M. (2004). Semiotics: "Language and Culture", ed. Carl Mitcham, Encyclopedia of Science, Technology, and Ethics, Detroit: Macmillan Reference USA.

Sanders, P. (1982). Phenomenology: A New Way of Viewing Organizational Research. *The Academy of Management Review*, 7/3, p. 353-360.

Saussure, F. (1916). *Course in General Linguistics*, ed. CharlesBally and AlbertSechehaye, trans. WadeBaskin. New York: Philosophical Library.

Schmidt, L. K. (2006). *Understanding Hermeneutics*, Acumen: Durham GBR.

Sfu.ca, (2019). *Hermeneutik Cycle*. http://www.sfu.ca/media-lab/cycle/presentation/design.html.

Sinha, D. (1963). Phenomenology and Positivism. *Philosophy and Phenomenological Research*, 23/4, p. 562-577.

Spychiger, B. M. (2001). Understanding Musical Activity and Musical Learning as Sign Processes: Toward a Semiotic, *Journal of Aesthetic Education*, 35/1, p. 53-67.

Thompson, J. C. (1997). Interpreting Consumers: A Hermeneutical Framework for Deriving Marketing Insights from the Texts of Consumers Consumption Stories, *Journal of Marketing Research*, 34, p. 438-455.

Topakkaya, A. (2007). Felsefi Hermeneutik, *FLSF (Journal of Philosophy and Social Sciences*,2007/4, p. 75-92.

Topakkaya, A. (2009). E. Husserl'de Noema ve Noesis Kavramları, *FLSF (Journal of Philosophy and Social Sciences)*, 2009/7, p. 121-136.

Young, E. D., Clifford, P., Denise, S. & Candler, C. (1997). A Hermeneutic Exposition of a Plains Healer's Concept of "The Grandfathers", *Antrophos*, 92/1-3, p. 115-128.

Mehmet CAN and Çağla ÜST CAN

17 Event Tourism

Introduction

Events that enable people to get together in line with their common values and beliefs (celebrations, commemorations, etc.) have been an indispensable part of societies since ancient times (Erden, 2014: 5). However, these events, which are conducted to a narrow extent, have changed dimensions and they were accepted as a type of event that supports the general policies of destinations and provides a significant contribution to achieving their social and cultural goals. They were also observed to transform into a strategic structure in economic terms (Raj et. al., 2013: 4).

Today, the increases in people's leisure time and disposable income have made events a core part of people's lifestyles, and the significance of events has elevated in terms of their number and demand (Etiosa, 2012: 4). In this context, events, as a significant factor of motivation in terms of tourism, have developed into an element of attraction that raises the awareness of destinations and increases tourists' demand for destinations (Getz, 2008: 403).

In terms of the needs tourism aims to meet and thanks to the social, cultural and economic benefits events provide for destinations, the orientation towards event tourism have elevated and this has led to a desire in countries to maintain the development of tourism with various local, national or international events (Saçılık & Çevik, 2017: 241).

1. The Concept of Event Tourism

An event, with the widest definition, is defined as the entirety of attractive events or activities that enable a certain group of people that shares a common belief to act together (Ekin, 2011: 3), and occur under certain circumstances (Getz, 1997: 4). While certain events are held in order to provide economic profits, others are conducted to attract attention to certain subjects (Yıldırım, 2014: 42).

The main point of the relationship between tourism and events is the transfer of the positive image created in the minds of individuals participating in the activities conducted within the event to the destination in question. Transferring an image of the event to a destination plays an important role in activating significant natural and physical points of attraction in the region in question (Jago et. al., 2002: 124;

Yıdırım et. al., 2016: 51). The income-creating influence of events in the region they are held has revealed that they are important tourism features, and events have recently become one of the most important elements of the tourism industry.

Event tourism is observed to be identified in the literature from two different perspectives: organizer and participant. In terms of participants, event tourism is defined as recreation activities that are held out of their daily lives in order to entertain people, increase their culture and gain new experiences (Tuna & Aknar, 2018: 266). In terms of the organizers, with the widest definition, event tourism is defined as the systematic planning, developing, marketing and conducting of events as tourist attraction elements that increase the tourism capacity of a destination and enable the creation or development of the image positively (Getz, 1997: 16; Tassiopoulos, 2005: 4).

Within the scope of the definitions mentioned, event tourism, for consumers, includes aims such as physical rest, mental relaxation, escaping from the routine activities of daily experiences and having a time that enables personal satisfaction (Karaküçük, 1997: 21). On the other hand, for the organizers, event tourism is observed to include aims such as creating a tourist attraction for the destination, creating an image, developing and marketing available values, encouraging and increasing tourist mobility of the destination, thus, gaining competitive advantage and gaining income in economic terms (Getz 1997: 53; Jago et. al., 2002: 124; Karagöz, 2006: 7; Raj et. al., 2013: 35).

Within tourism, in terms of the organizers of the events that are held in line with the expectations and needs of the participants, the economic, social and environmental benefits of the event to the region are of importance rather than the show aspect of the event. Accordingly, in order to gain more benefits from the events, which are regarded as a tourist attraction element, they are required to have certain features.

It is possible to state the main features of event tourism as the following (Getz, 1997: 68; Jago & Shaw, 1998: 28; Tassiopoulos, 2005: 11; Tuna & Aknar, 2018: 267):

- Events have a feature of being beyond daily experiences.
- Events are conducted within a certain period.
- Events are conducted for one or continuously with a certain frequency.
- Events do not generally require a building or a facility.
- All of the activities conducted within an event are practiced within the same region.
- Events are held in an open to public fashion (Except for scientific and sports events).

- Events are conducted for a celebration or a demonstration of a product or theme.
- Events include various activities.
- Events are conducted to raise awareness for the region or strengthen the image of the region.
- Events are desired to create a significant economic impact.
- Events are planned in a way to attract the attention of the media.

In order for events to be successful, not only the plans and investments, but also sound organizational skills, suitable advertisement and fitting weather conditions to conduct outdoor activities are required (Ispas & Hertanu, 2011: 127).

2. Types of Event Tourism

In terms of the organizers of events, to ease the measurement of the economic benefits of events to the region and make the benefits of events to the economy more apparent, it is important to categorize and classify events. In a general sense, events are analyzed within two titles, planned and unplanned events. Planned events, compared to unplanned events, cover events that have specific locations, schedules, target population and boundaries (Tassiopoulos, 2005: 10). Events that are held within tourism cover those that are conducted within a certain period and planned in all their parts. In the literature, the events that are conducted with a tourism-oriented plan are classified into two by tourism researchers. Within this framework, it is observed that certain researchers conduct their classification by considering the sizes of events while others consider the contents (themes) of the events in their classification (Allen et. al., 2002: 11).

2.1 Types of Events According to Their Sizes

These types of events are classified by scaling their demand and significance levels. Within this scope, events, according to their sizes, are divided into four topics: mega events, specialized events, regional events and local events (Getz, 2008: 407).

Mega Events: Mega events are generally held internationally at a certain time and for once, which causes a high level of tourism mobility in terms of their size and importance (Jago, 1997: 56). These types of events create a major impact in the media, have a considerable effect on the image of the destination and create the highest profit among other types of events (Allen et. al., 2002: 12). In order to classify an event as a mega event, it is needed to meet the criteria of a million visitors and a budget of at least 500 million dollars (Getz, 1997: 18). Olympic championships and world cup organizations can be stated as examples of mega events.

Specialized Events: These types of events are generally large-scale events, specific to the destination they are held in, conducted for once within a limited time or repeated within a certain frequency in order to raise awareness for a certain touristic destination and increase profitability (Ritchie, 1984: 2). These events are of importance in terms of awareness, attractiveness and profitability and enable significant amounts of competitive advantage in terms of prestige and the destination where they are held (Jago, 1997: 51; Tassiopoulos, 2005: 12). The Oktoberfest held in Munich and the Rio Carnival held in Rio de Janeiro can be examples of these types of events.

Regional Events: These types of events cover those that are held regionally and are mentioned with the name of the place where they are held. These events improve the national and regional attractiveness of the place and support the improvement of the touristic image and economic growth in the long term (Getz, 1997: 104). Regional events influence destinations in proportion to their extent, size, effect on the media, the number of visitors and economic impact (Allen et. al., 2002: 13). Special-themed festivals and exposition organizations conducted regionally can be mentioned as examples of these types of events.

Local Events: These events, which are placed at the lowest level in the classification made in terms of size, are those that are society-focused and held with the aim of strengthening the feelings of pride and belonging in societies (Yıldırım, 2014, 45). These events, which are embraced and supported by local governments and local communities, are conducted for the people inhabiting the region and for their socialization and entertainment (Tuna & Aknar, 2018: 266). Food festivals held at a local level can be mentioned as examples of these events.

2.2 Events According to Their Contents (Themes)

Events, in terms of their contents, are classified according to their themes. The events in this framework can generally be examined in eight sections as cultural, political, arts and entertainment, business and commerce, education and science, and sports, recreational and special events (Getz, 1997: 7; Allen et. al., 2002: 13).

Cultural Events: These events cover festivals, carnivals, commemorations and religious activities (Getz, 1997: 7). These events, which constitute a significant part of event tourism, have the power to attract a significant number of tourists (Allen et. al., 2002: 13). These types of events are local, regional, national and international events that are shaped around major building blocks of societies

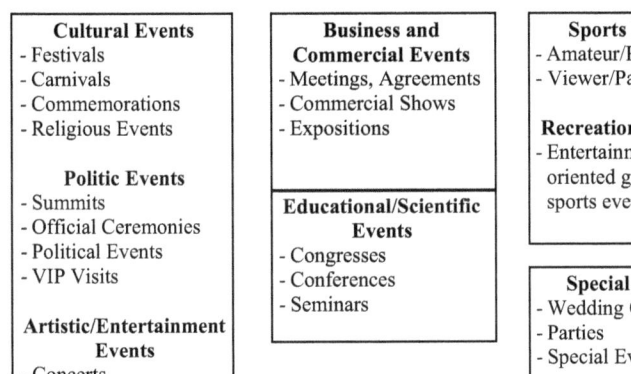

Figure 1: Events According to Their Contents: (Getz, 2008: 404)

(religion, social identity, ideology, etc.), and they are attached importance in terms of protecting, developing and introducing cultural identities in addition to their influence in enabling societies and participants to interact with each other (Tuna & Aknar, 2018: 275).

Political Events: These types of events cover summits, official visits, political meetings and VIP visits. Political events are generally conducted in well-known cities with developed infrastructure and they gather numerous politicians, large syndicates, unions, management levels of companies together (Çelik, 2009: 67). Although political events constitute the smallest category in events conducted within the framework of tourism, they are important events in terms of the attention of the media and viewers (Karagöz, 2006: 32). Davos Summit and G20 Summit can be mentioned among the examples of these types of events.

Artistic Events: These types of events cover concerts, performance shows, exhibitions and award ceremonies. Especially, due to the growth potential of the art market, it is observed that the interest for artistic events has increased and artistic events have become a tourist attraction. Thanks to the attention of the media and viewers to these types of events, they play significant roles in introducing destinations (Gilbert & Lizotte, 1998: 84; Karagöz, 2006: 31).

Business and Commercial Events: These types of events include meetings held with business and commercial aims, congresses, expositions and commercial shows (Getz, 1997: 8). Events with business and commercial aims, informing a certain type of occupational group on a certain subject or exhibiting and introducing products and services of businesses and creating demand for these products and gaining various commercial benefits are among these types of events (Tuna & Aknar, 2018: 275). These types of events, which are held at regional, national and international levels, are important in terms of creating a tourist attraction for the destination where they are held and bringing an economic liveliness thanks to the attention of participants and viewers.

Sports events: Sports events are one of the rapidly developing types of event industry (Allen et. al., 2002: 15). These types of events can be held in a specific country, region or destination within a certain time or they can be conducted in several different countries, regions or destinations synchronously (Karagöz, 2006: 22). Today, they are conducted for almost every branch of sports and every age group at different scales and sizes. The fact that sports events are held locally, regionally, nationally and internationally has increased the participation and attention to these events (Apaydın, 2011: 140). Sports events, which receive significant levels of participants, viewers and media interest, also contribute vital social and economic benefits in terms of tourism.

Scientific Events: These events cover congresses, seminars, conferences, symposiums, etc., which are conducted with the aim of sharing knowledge in a scientific field. The main goal of these types of events is learning and sharing knowledge among people (Getz, 1997: 9). These types of events, which are held for a certain time and within a certain program, are those that aim to gain knowledge in a professional or technical field or to share knowledge on a certain topic among experts in a professional or technical field (Ritchie, 1984: 6).

Recreation Events: These events cover all of the activities that the participants voluntarily choose to make use of their spare time that is left out from their jobs or meeting vital needs. Activities conducted within this framework are rather rich and various in terms of their content. Recreational activities can be gathered under general titles such as musical endeavors, sportive activities, games, artistic events, events that require skills, natural events, social and cultural events (Can, 2015: 4). Especially, individuals who live in metropolitan areas constitute an important factor in terms of the need for tourism-oriented recreational activities, which is due to their requirement for moving away from the crowded urban life, professional life and daily activities to have fun and gain new experiences (Şahin et. al., 2009: 63).

Special Events: These types of events cover weddings, parties or entertainments specific to an institution or person and social celebration activities (Getz, 197: 7). Because these activities are rather specific to individuals or societies, they are limited to the organizer and immediate surroundings. Therefore, the influence of these events on both organizers and participants is rather low. For this reason, the touristic activities that can be created by these events remain at a limited level (Erden, 2014: 17).

3. Effects of Event Tourism

Events that are held within the scope of tourism have numerous positive and negative effects on destinations. However, these effects depend on the type and size of the event and the features of the destination (Can, 2015: 8). It is possible to examine the effects of event tourism within four sections: social and cultural effects, physical and environmental effects, political effects and economic effects (Allen et. al., 2002: 32).

3.1 Social and Cultural Effects

The public is affected by tourism-oriented activities at the highest level. Events directly affect the social and cultural lifestyles of destinations. Events conducted within this framework have significant and positive social and cultural effects such as socializing and meeting the need for entertainment, sharing social experiences, reviving traditions that are about to be forgotten, strengthening the feeling of social pride and belonging, forming new ideas and increasing cultural experiences (Allen et. al., 2010: 35; Ekin, 2011: 44; Erden, 2014: 27). Furthermore, in addition to the positive social and cultural effect of event tourism, it can be stated that event tourism has possible negative effects such as disturbing social order and social peace, creating cultural conflicts and social alienation, commercializing cultural values, destroying cultural authenticity and damaging public image (Allen et. al., 2010: 35; Tuna & Aknar, 2018: 278).

3.2 Physical and Environmental Effects

Within the framework of event tourism, planned events have vital importance in terms of obtaining positive physical and environmental results. Events that are planned within this scope have various positive physical and environmental effects such as introducing the environment, raising environmental consciousness and improving environmental awareness, developing infrastructural and social services and enabling urban transformation and renewal (Erden, 2014: 28; Tuna & Aknar, 2018: 279). Events conducted within tourism provide significant physical and environmental contributions to the destination where they are held while unplanned events may lead to possible unwanted and negative situations that increase the physical and environmental cost of destinations. Due to the excessive busyness created in the destination of the event, traffic problems, increased in wastes that lead to environmental pollution, noise pollution, damaging historical and cultural structures can be mentioned among the possible negative physical and environmental effects (Çelik, 2009: 86).

3.3 Political Effects

A successful event that is supported by politicians has major importance in terms of the safety and support of society. Accordingly, politics have a significant position in event management. Events that are conducted within this framework have positive political effects such as improving recognition and providing prestige and enabling improved destination profile, marketing of investments and administrative experience. Negative political effects of events may include risks emerging from mistakes, misuse of funds, deficient and erroneous calculations, false propaganda, loss of social control, and disapproval of ideologies (Allen et. al., 2002: 32).

3.4 Economic Effects

In general terms, the success of an event is regarded as the equivalent of the economic yield of that event. This is due to the fact that the economic effects of events are regarded as materially visible and calculable. Karagöz (2006: 42) stated that the economic effects of events generally appear from three sources. These include the expenses made by the participants of the event, investment expenses made for the event and expenses made by organizers regarding the event.

Within this framework, the positive effects of event tourism include encouraging the local economy, increasing tourist expenses and tax income, improving jobs and employment opportunities and creating novel opportunities for investment. Furthermore, it can be stated that event tourism also has possible negative economic effects such as increases in prices in the local market and missing investment opportunities in different fields (Allen et. al., 2002: 39).

Conclusion

Today, the increases in leisure time and disposable income of people have made the events a core part of their lifestyle. Accordingly, events, which are important elements of motivation in terms of people's participation in tourism, have become a rapidly developing element for destinations as a significant factor of attraction.

Transformations compelled by globalization, problems caused by economic structuring and novel needs for creating an identity have activated many destinations to create alternative tourist attractions to benefit from available values and resources in social, cultural and economic terms at the highest level possible. Accordingly, significant levels of economic, social, environmental and political effects of events, which constitute a strong element of attraction, have led to a desire to create novel events or host existing events.

The nature of the effects of events on destinations has now revealed the reasons for destinations' interest in events. Accordingly, events are regarded as one of the most important factors that provide tourist mobility and enable a destination to introduce itself and gain an image. Furthermore, events do not only act as an element and supporter of tourism, but they also play many significant roles in building societies, urban transformation, strengthening national identity and social integration.

References

Allen, J., O'toole, W., Harris, R. & Mcdonnell, I. (2002). *Festival and Special Event Management*, Sydney: Wiley.

Apaydın, F. (2011). *City Marketing*, Ankara: Nobel Publications.

Can, E. (2015). The Relation among Leisure Time, Recreation and Event Tourism, *İstanbul Journal of Social Sciences*,10, p. 1–17.

Çelik, A. (2009). *Event Tourism and Perceived Impacts As A Destination Marketing Element the Case of İstanbul*, Gazi University Institute of Educational Sciences, Unpublished Master Thesis, Ankara.

Ekin, Y. (2011). *Festivals within the Context of Event Tourism and a Research about Social Impacts of Antalya Altın Portakal Film Festival on Residents*, Akdeniz University Social Sciences Institutes, Unpublished PhD Thesis, Antalya.

Erden, İ. Ö. (2014). *Perceptions of Local Community Towards the Impacts of Izmir International Fair*, Anadolu University Social Sciences Institutes, Unpublished Master Thesis, Eskişehir.

Etiosa, O. (2012). *The Impacts of Event Tourism on Host Communities. Case: The City of Pietarsaari*. Kokkola: Central Ostrobothnia University of Applied Sciences, Unit for Technology and Business.

Getz, D. (1997). *Event Management ve Event Tourism*. New York: Cognizant Communication Corporation.

Getz, D. (2008). Event Tourism: Definition, Evolution, and Research, *Tourism Management*, 29/3, p. 403–428.

Gilbert, D. & Lizotte, M. (1998). Tourism and Performing Arts, *Travel Tourism Analyst*, 1/1, p. 82–87.

Ispas, A. & Hertanu, A. (2011). Characteristics of Event Tourism Marketing. Case Study: the European Youth Olympic Festival, *Bulletin of the Transilvania University of Braşov*, 4/53, p. 127-134.

Jago, L. K. (1997). *Special Event and Tourism Behaviour: A Conceptualisation and an Empirical Analysis from a Values Perspective*, Victoria University Tourism and Marketing Faculty, Doctor of thesis, Sydney.

Jago, L. K. & Shaw, R. (1998). Special Events: A Conceptual and Definitional Framework, *Festival Management and Event Tourism*, 5/1, p. 21-32.

Jago, L. K., Chalip, L., Brown, G., Mules, T. & Ali, S. (2002). *The Role of Events in Helping to Brand a Destination*, Sydney: Australian Centre for Event Management, University of Technology.

Karagöz, D. (2006). *Event Tourism and Economic Impacts of Foreign Visitor's Expenditures: The Case of Formula 1 2005 Turkey Grand Prix*, Anadolu University Social Sciences Institutes, Unpublished Master Thesis, Eskişehir.

Karaküçük, S. (1997). *Recreation Leisure Time Evaluation Concept Scope and a Research*. Ankara: Seren Ofset.

Raj, R., Walters, P. & Rashid, T. (2013). *Events Management principles ve practice*. London: Sage Publications.

Ritchie, B. J. (1984). Assessing the Impact of Hallmark Events: Conceptual and Research Issues, *Journal of Travel Researh*, 23/2, p. 2-11.

Saçılık, Y. M. & Çevik, S. (2017). Determination of the Satisfaction Levels of "Erdek Time Festival" Participants in Terms of Benefits Provided By Event Tourism, *Route Educational and Social Science Journalss*, 4/7, p. 240-257.

Şahin, C. K., Akten, S. & Erol, U. E. (2009). A Study to Determine Recreational Participation Tendency of the Eğirdir Vocational School Students, *Artvin Çoruh University Faculty of Forestry Journal*, 10/1, 62-71.

Tassiopoulos, D. (2005). *Event Management: A Professional and Developmental Approach*. South Africa: Juta Academic.

Tuna, M. & Aknar, A. (2018). Event Tourism. (Editörler: Aydın, Ş. & Boz, M.). Current Issues and Trends in Tourism II within (p. 265-284). Ankara: Detay Publishing.

Yıldırım, H. M. (2014). *The Importance of Events on Destination Choice: Gallipoli Peninsula Historical National Park Sample*, Çanakkale Onsekiz Mart University Social Sciences Institutes, Unpublished PhD Thesis, Çanakkale.

Yıldırım, O., Karaca, O. B. & Çakıcı, A. C. (2016). A Research on the Perceptions and Satisfaction of Local People on "Adana, International Orange Blossom Carnival", *Journal of Travel and Hospitality Management*, 13/2, p. 50-68.

Mustafa Cüneyt ŞAPCILAR and Ahmet BÜYÜKŞALVARCI

18 The Effect of Nepotism[1]

Introduction

Since the phenomenon of tourism is multi-faceted and integrated with many fields, tourism education, tourism license or experience status of the persons engaged in tourism works can be ignored. Tourism business owners or managers may employ relatives or acquaintances in this field who they think are related to the tourism area in which they operate. This employment situation may create negative image for individuals who have tourism education or have tourism sector experience. Not only in the recruitment process, but also in wage, performance appraisal, promotion and etc. cases may also occur. This situation or the perception related to this situation can affect the loyalty, performance, job satisfaction and intention to leave the job. This situation is evaluated positively for individuals who are among the parties in the favoritism process and who create temporary opportunities and gain social and economic interest by using these opportunities, while a negative and undesirable situation for the employees other than these persons is evaluated.

The concept of favoritism is expressed as the preferential treatment of managers or power holders to their close environment or to the persons they are associated with (Saylı & Kızıldağ, 2007: 235). The concept of favoritism was used in the political and administrative literature for the first time in 1828 with the political appointments of US President Jackson (Tekiner & Aydın, 2015: 76) and the Asian-based financial crisis in 1997 was seen as the most important reason (Özkanan & Erdem, 2014: 181; Begley et. al., 2010: 282). Favoritism is the recruitment process or promotion of a person, taking into account the relationship of the kinship and friendship of the business owners without considering the factors such as knowledge, skills, abilities, success and education level (Pektaş, 1997: 90). Although the concept of favoritism is mostly seen in relation to official institutions, it is a common practice in private organizations and enterprises. Favoritism practices can be seen in all areas of the social system. Kinship, friendship, school/military friendship, professional association, being from the same sect, being from the same village, neighboring, being from the

---

[1] This study was produced from the PhD dissertation

same neighborhood, from being within the context of social belonging provide support for favoritism (Özkanan & Erdem, 2014: 180).

Favoritism practices continue in spite of the fact that it has negative impacts on economic, social and political growth and development in many countries (Loewe et. al., 2007: 27). Although favoritism is reduced to very low levels in developed countries, it is still one of the main problems of the organizational structure in developing countries (Tekiner & Aydın, 2015: 77). It is not necessary to have an interest that expresses only economic value for favoritism. Being a citizen, relative or voter is seen as sufficient for favoritism practices (Akalan, 2006: 113). Favoritism usually occurs during recruitment process, selection and career development (Fu, 2015: 4).

It is possible to see that favoritism is classified differently in the studies and related literature.

In the studies of Özüren (2017: 3), Demaj (2012: 13) and Avcı (2017: 4), the concept of favoritism is basically grouped into three types. In this categorization; *nepotism* is referred as the form of favoritism based on kinship between individuals, *cronyism* is defined as the favor of favoritism due to the friendship, and the type of favoritism based on politics is called *partisanship/patronage/political supporter* (Karacaoğlu & Yörük, 2012: 47).

Nepotism

In many studies, the concept of favoritism has been considered as nepotism and has a similar meaning. (Sarıboğa, 2017: 4; Karacaoğlu & Yörük, 2012: 47; Asunakutlu & Avcı, 2010: 96). Nepotism is defined as the employment or improvement of the position of persons who do not consider the ability, training, expertise, skill characteristics of a person, or who do not have the skills required by the job, solely on the basis of kinship relations. (Özler et. al., 2007: 438; Ansari et. al., 2015: 510). The reasons for the implementation of nepotism are considered as public, institutional and organizational (business). Lack of clear, fair, and transparent laws, policies and regulations which ensure that employee recruitment process and promotion, remuneration and performance procedures is the most important reason for the emergence of nepotism practices (İyiişleroğlu, 2006: 48).

It is a common practice to give priority to the recruitment of relatives or close contacts by new and small capital enterprises (Düz, 2012: 5). This can be explained by the efforts of the business owner to provide a sense of trust to him/her and benefit to the close environment. It seems natural to prominence favoritism in transferring management to second and third generations considering that working with acquaintances and relatives in the business is a condition that gives confidence and comfort in terms of management (Karacaoğlu & Yörük, 2012: 48). The close stance of the society to the concept of nepotism, gaining prestige in the family, in the work environment and in the society, seeking to improve the socio-economic situation of relatives, and providing job guarantee seem to be the other reasons of nepotism (Turhan, 2016: 108).

It is a weakness for people to use kinship relationships instead of their own knowledge, skills, efforts and abilities to get a certain job, to gain status and promotion. In terms of business, such a situation indicates a structure that supports favoritism. In all organizations and institutions where such structures exist and where success and talent are not taken into account in recruitment or promotion, job loss, labor turnover and institutional failure are generally considered to be inevitable outcomes (Asunakutlu & Avcı, 2010: 94).

Nepotism practices provide a number of advantages to institutions and businesses. These advantages explain the positive aspects of nepotism practices for institutions, businesses and employees (Turhan, 2016: 106; Karataş, 2013: 19; Asunakutlu & Avcı, 2009: 731; Asunakutlu & Avcı, 2010: 97; Tekiner & Aydın, 2015: 80; Ansari et. al., 2015: 510; Abdalla et. al., 1998: 555–556; Savaş, 2015: 12; Demaj, 2012: 39; Hudson et. al., 2017; İyiişleroğlu, 2006: 44–47):

➢ Working with relatives provides a material and spiritual reassuring and spiritually satisfactory working environment.
➢ Nepotism increases the commitment the job of family members or relatives.
➢ Employees who entering the job with nepotism can be trusted to than other employees.
➢ Employees who are members of the family, strong organizational commitment and family ties provide a significant competitive advantage.
➢ With the applications of nepotism, the people who will be employed in the senior management have the opportunity to get to know the business better and their intention to leave the job is reduced.
➢ Employees recruited with nepotism work with high performance in the business. It provides an environment of regular interaction and communication.
➢ It is believed that more robust steps have been taken in transferring business from generation to generation via nepotism and it provides a balanced transition.
➢ Employees who are hired with nepotism during a crisis they think the business as their own business and can make sacrifices.

➢ Employees can manage over the employees hired by nepotism. Nepotism practices can be done not only for emotional reasons but also for rational and strategic reasons.

In addition to providing some advantages to businesses, nepotism practices also have disadvantageous situations that affect the productivity, employee performance and even threaten the existence of business. These disadvantages reveal the negative aspects of nepotism practices for businesses and employees (Turhan, 2016: 106; Karataş, 2013: 19; Asunakutlu & Avcı, 2009: 731; Asunakutlu & Avcı, 2010: 97; Tekiner & Aydın, 2015: 80; Ansari, 2015: 510; Abdalla et. al., 1998: 555–556; Savaş, 2015: 12; Demaj, 2012: 39; Hudson et. al., 2017; Cabrera-Suárez, 2005: 92; Öztürk, 2008: 115):

➢ Nepotism is often seen in the human resources department of business. This department loses its operability and cannot fulfill its human resources functions.
➢ In business or countries where nepotism is widespread, the merit system is not taken into consideration and thus there is a loss of job and failure.
➢ Nepotism practices negatively affect organizational culture and organizational climate. There is a decrease in employees' organizational commitment and confidence in the business. These individuals become dissatisfied and pave the way for conflict.
➢ The performance appraisal system does not function properly in business with nepotism policies. The principle of equality is also ignored in wage policies. Working conditions is also differed. Family members and non-family members work under different conditions.
➢ Employees who are victims of nepotism practice are more thinking of leaving the jobs and are looking for a more respected job and business. Nepotism practices are one of the important factors affecting employee turnover.
➢ The biggest factor contributing to the success of the business and the sustainability of this success is the 'human factor'. Managers or individuals who ensure business success can be dismissed due to the employment of family members. This situation makes it difficult to employ qualified and competent employees. The business lacks intellectual capital as it prevents the employment of talented managers.
➢ Motivation decreases, an insincere organizational climate emerges, professionalism becomes difficult, and the principles of transparency and equality are violated in the business with nepotism practices.

The results of nepotism are dealt as social, organizational and individual results. (Erdem & İlhan, 2010: 154–158):

Social results; It is stated that the most important social result of nepotism policies and practices is 'brain drain'.
Organizational results; It is stated that it has an effect on the life span of business and organizations.
Individual results are evaluated in terms of people who benefit from nepotism practices and those who cannot benefit from nepotism practices

Job Satisfaction

The concept of job satisfaction is generally expressed as a result of the positive emotional perspectives of employees regarding various factors such as control, wage, working conditions, and opportunities for self-improvement, social relations and work environment (Çetin & Basım, 2011: 84). In another definition, job satisfaction is defined as a relaxing and soothing feeling that the individual endeavors to obtain from the work itself, managers, working group and work organization. (Eroğlu, 2013: 444). The most important feature of job satisfaction is that it is an emotional concept rather than a mental one (İşcan & Timuroğlu, 2007: 125). If the employee has positive attitudes towards his/her job, job satisfaction is achieved and in negative attitudes, there is job dissatisfaction. These attitudes emerge as a result of one's feelings, thoughts and evaluations about his/her job (Budak, 2006: 5).

The concept of job satisfaction has four important features (Luthans, 1994: 108):

- Job satisfaction is the emotional attitude towards situations that occur in the work environment. Job satisfaction is not seen by other people. It can be felt or expressed by the person himself/herself
- Job satisfaction in general is how much the job meets expectations.
- Job satisfaction represents different attitudes towards work. While the employee behaves a positive attitude towards one dimension of the job, he/she may be have a negative attitude towards the other dimension. Therefore, different dimensions of job satisfaction should be examined and general job satisfaction should be found.
- Job satisfaction can turn into job dissatisfaction faster like it can be achieved faster. Job satisfaction is a dynamic phenomenon.

The benefits of job satisfaction are examined in an individual, executive and organizational context. When the expectations of the employee are fulfilled, they are satisfied and this increases his loyalty and desire for the job. A positive development is achieved in the relations of the satisfied employees with the managers and this mutual satisfaction brings success to the business. The positive relationship between the employee and the manager contributes to the achievement of the desired level of development and resource use of the enterprises (Özpehlivan, 2015: 8).

The results of job satisfaction are so important that affect the physical and mental health of the employees, the peace and productivity of the organizations, the development and peace of the society. Job satisfaction results are not only individual but also organizational and social. Therefore, job satisfaction is very desirable, but not easy to obtain. The high level of job satisfaction is desirable by managers as it provides favorable working conditions. On the other hand, low satisfaction is seen as proof that things are not going well in an organization (Kök, 2006: 296–297). If the employee has the idea that his/her expectations from his/her job are not met sufficiently, job dissatisfaction occurs (Demir et. al., 2008: 140). Employees, who are dissatisfied with their job search for other job opportunities in relation to the unemployment rate, labor market conditions in their country, make a comparative assessment of their current job and decide to quit their jobs or continue to work in the same workplace (Kuzulugil, 2008: 130).

When the positive and negative effects of job satisfaction in terms of individual and organizational results are examined (Akıncı, 2002: 6):

Positive effects; Commitment to work and organization, absenteeism that occurs only under compulsory conditions, low employee turnover rate, steady increase in labor productivity, happy and healthy people, demonstrating talent and creativity, easier motivation to the objectives of the organization.

Negative effects; Lack of interest in work and organization, excessive absenteeism, increased job complaints, high employee turnover rate, low labor productivity and efficiency, unhappy and unhealthy people, lack of motivation (Akıncı, 2002: 6).

To learn and improve the level of satisfaction of employees in their job provides some organizational benefits for business. Job satisfaction surveys provide organizational benefits in terms of identifying employee problems, improving general attitude of employees, revising internal communication, identifying training needs, developing union activities, managing change and planning (Newstrom & Davis, 1997).

Intention to Leave the Job

Intention to leave the job is a cognitive process, which is listed as thinking, planning and wanting to quit. The intention to leave develops within a certain process and this process develops depending on many variables (Güzel & Ayazlar, 2014: 135). The concept of 'Intention to leave the job' refers to a conscious and cautious decision or intention in terms of quit the job (Barttlett, 1999: 70). In other words, the intention to leave the job is defined as the employee having a plan and thought of leaving until the time he/she leaves his/her current job or workplace (Yenihan et. al., 2014: 40).

Employees with the intention of leaving the job are unable to concentrate on their work and organizational goals, and their work result in lower productivity. Intention to leave the job is active action that the employee applies if employee is not enough satisfied with the employment conditions (Çekmecelioğlu, 2014: 26).

Intention to leave the job affects employees' behavior and leads to exhaustion. In terms of organization, the intention to leave the job means the loss of time and costs incurred for the purpose of training the employee. In addition, quit the job leads to loss of knowledge, skills and investment of the organization, and as a result of the recruitment of a new employee, the cost of adaptation and training is raised (Kaya, 2016: 51–52). Organizations have to find solutions to identify and eliminate the factors that affect the intention to leave in order preventing quit. In addition, it is important for organizations to try to reduce their intention to leave by offering various additional gains and training to their employees (Çetin, Güleç & Kayasandık, 2015: 21).

Methodology

This study was carried out in order to gain information about the favoritism practices faced or exposed by the people who work as employees in the tourism sector, and to contribute to the literature in this field. The main aim of the research is to determine the level and the application of nepotism practices in accommodation businesses.

When the studies carried out are examined, it is seen that nepotism is generally considered as a type of corruption at the level of public institutions, and these studies are directed to organizations in the following years. Especially in the tourism sector where service and human factor is more important than other sectors, it is determined that these studies are intensified. This study is discussed at the level of accommodation businesses. The fact that the study was conducted in resort hotels (Antalya) and city hotels (Konya) and the analysis of nepotism perceptions of employees working in these businesses emphasizes the importance of the study.

Job satisfaction and intention to leave are affected by many factors. The main focus of the study is to determine the relationship between nepotism, which is considered as a separate case, and job satisfaction and intention to leave.

After the literature review developed for this purpose, the following questions were sought answer in the scope of the study:

➢ Is there a relationship between nepotism and job satisfaction?
➢ Is there a relationship between nepotism and intention to leave?

The hotels are divided into city and resort hotels as of location and service differences. Konya is a city in Turkey that was chosen as the sample to measure nepotism perceptions of employees working in five star city hotels, and Antalya is a city in Turkey that was chosen as the sample to measure nepotism perceptions of employees working in five star resort hotels. The biggest limitation of the research is that the hotel businesses do not want to participate in the research.

Within the research, nepotism and job satisfaction and intention to leave the job were evaluated in relationally.

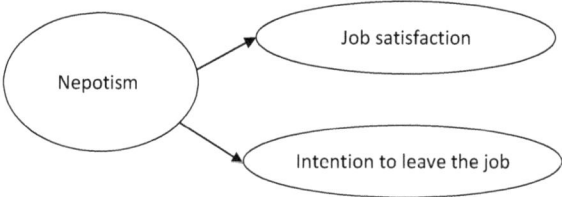

Figure 1: Research Model

The questionnaire was used as a data collection. The questionnaire method is a list of questions arranged according to a specific purpose and plan, and is the most preferred method of data collection (Yazıcıoğlu & Erdoğan, 2007: 75). In this study, 139 questionnaires from city hotels and 227 questionnaires from resort hotels, a total of 416 questionnaires were evaluated.

The questionnaire consists of three parts. In the first part, participants' demographic questions and statements about the reference are given. The second part consists of the nepotism scale. In the third part, the intention to leave and the job satisfaction are stated. Nepotism Scale; formed by Ford & McLaughlin (1985) and Abdalla et al. (1998), developed by Asunakutlu & Avcı (2009), and consists of 14 closed-ended 5-point Likert type expression; the questionnaire which included dimensions of nepotism in the promotion process, in the transaction, and in the recruitment process was used. Intention to leave; the intention to leave was adapted from the scale developed by Rosin & Korabick, 1995. Minnesota Job Satisfaction Scale; developed in 1967 by Weiss et. al.

Findings

The gender, age, education level, marital status and income levels of the employees were examined, the majority of the participants working in resort hotels consisted of male participants (66.4 %). When the gender of the participants working as employees in five-star hotels operating in Konya province is examined, it is seen that the ratio of men and women is close to each other (49.6 %–50.4 %). It is seen that 28.5 % of the participants working in resort hotels are in the 24–29 age range, and similarly, 36.7 % of the participants working in city hotels are in the 24–29 age range. The education level is taken into consideration, it is determined that the majority of the participants in the resort hotels are 'secondary education' graduates, and 28.1 % of the city hotels have 'bachelor degree'. The situation of the participants to get tourism training was examined, it was found that 41.5 % of the participants from the resort hotels got tourism training, and 33.1 % of the participants from the city hotels got tourism training. The income levels of the employees are examined, it is seen that the majority of the participants working in hotels operating in Konya and Antalya have 'minimum wage' or slightly higher income levels.

The working time of the participants in the tourism sector in Antalya is observed that 17.7 % work 7–8 years, and the working time of the participants in Konya is observed that 20.9 % ratio 3–4 years. This situation can be related to the fact that five star hotel businesses are newly opened enterprises in Konya. A similar situation is observed in the working conditions of the hotel businesses in which they operate. It is seen that 50.4 % of the participants from the province of Konya have worked in the enterprises for '11 months and less'. The working time of the employees working in hotel businesses in Antalya are taken into consideration, it has been determined that 33.2 % of the participants have been working in the same enterprises for ve '11 months or less'. The reason of the difference between the working time is the high turnover rate and seasonal characteristics in the hotel management sector which serves coastal tourism. The status of the participants in the research is examined, it is seen that 55.2 % of the participants working in resort hotels are permanent staff, and 86.3 % of the participants working in city hotels are permanent staff. It was determined that 32.5 % of the participants from the Antalya worked in the 'food and beverage' department, and 29.5 % of the participants from the Konya worked in the 'housekeeping' department. Table 1. shows the arithmetic mean and standard deviation values for determining the nepotism perceptions of the employees working in the resort hotels.

Table 1: Participant Opinions on Nepotism Perception

EXPRESSIONS	Resort Hotels (n=277)		City Hotels (n=139)	
Transaction Nepotism	$\bar{x}$	s	$\bar{x}$	s
1. Employees who have relatives in the management team of this business have more reputation than other employees.	2,69	1,39	2,15	1,29
2. It is very difficult for business owners or managers to dismiss or punish their relatives.	2,71	1,41	2,22	1,26
3. The authorities and responsibilities in this enterprise are primarily given to the relatives of the business owners / managers.	2,60	1,39	2,17	1,30
4. Employees who are relatives of business owners / managers benefit from opportunities of the enterprise more easily.	2,68	1,38	2,28	1,29
5. Employees who are relatives of business owners / managers in this enterprise have more privilege than the other employees.	2,53	1,35	2,32	1,30
6. I hesitate the employees who are relatives of business owners / managers in this enterprise.	2,68	1,36	2,04	1,30
Nepotism in Promotion				
7. Priority is given to employees who are relatives of business owners / managers in the promotion process.	2,55	1,35	2,20	1,29
8. It is easier to get promote employees who are relatives of business owners /managers in this enterprise.	2,68	1,32	2,20	1,29
9. I think that the knowledge, skills, experience and competencies of the employees in the promotion process are secondary importance.	2,51	1,41	2,22	1,24
10. Business owners /managers put their relatives in a better position regardless of their contribution to the enterprise.	2,58	1,39	2,12	1,27
11. No matter how successful I am in this enterprise, I cannot get ahead of the employees who are relatives of business owners /managers.	2,50	1,46	2,15	1,36
Nepotism in the Recruitment Process				
12. Relatives of business owners / managers have priority in recruiting personnel to this enterprise.	2,54	1,42	2,05	1,28
13. Relatives of business owners / managers are not forced in the selection process.	2,76	1,40	2,28	1,24
14. Relatives of business owners /managers is immediately hired when they applicant	2,71	1,43	2,20	1,21

In this context, transaction nepotism, nepotism in the promotion and nepotism in the recruitment process seem have a low level of participation.

The opinions of the employees working in city hotels on the perception of nepotism are given in Table 1. As can be seen from the arithmetic means, the opinions of the participants on nepotism practices are mainly 'disagree'. However, this does not mean that there are no nepotism practices in city hotels. The findings of the research can be considered as an indication that the five-star hotels operating in Konya are national and international chain hotel enterprises, and that there is an effort to avoid nepotist approaches in promotion, wages and general procedures.

Participants' opinions regarding intention to leave are examined in Table 2. When the opinions of participants who work as employees in city hotels regarding the intention to leave are evaluated, it can be stated that the intention of participants to leave is low than resort hotels.

Table 2: Participant Opinions on Intention to Leave

EXPRESSIONS	Resort Hotels (n=277)		City Hotels (n=139)	
	$\bar{x}$	s	$\bar{x}$	s
If I had a better alternative, I'd consider leaving this business.	2,64	1,41	3,08	1,34
If possible, I think of leaving this profession (hotel management).	2,87	1,48	2,89	1,38
I can accept alternative job offers.	3,00	1,49	3,20	1,35
I'm looking for a job in another business.	2,76	1,58	2,63	1,43

Participants' views on job satisfaction are given in Table 3. When the statements regarding the job satisfaction of the employees working in resort hotels operating in Antalya province are evaluated that was determined the participants were satisfied with the 'sense of success' in their work.

Table 3: Participant Opinions on Job Satisfaction Perception

EXPRESSIONS	Resort Hotels (n=277)		City Hotels (n=139)	
	$\bar{x}$	s	$\bar{x}$	s
Being able to keep busy all the time.	2,60	1,43	2,94	1,20
The chance to work alone on the job	3,22	1,43	3,33	1,15

(*continued on next page*)

EXPRESSIONS	Resort Hotels (n=277)		City Hotels (n=139)	
	x̄	s	x̄	s
The chance to do different things from time to time.	3,41	1,40	3,43	1,14
The chance to be 'somebody' in the community.	3,55	1,33	3,60	1,12
The way my boss handles his/her workers.	3,59	1,26	3,54	1,13
The competence of my supervisor in making decisions.	3,57	1,25	3,64	1,10
Being able to do things that go my conscience.	3,76	1,23	3,66	1,08
The way my job provides for steady employment	3,68	1,21	3,64	1,14
The chance to do things for other people.	3,75	1,25	3,61	1,08
The chance to tell people what to do	3,68	1,24	3,58	1,11
The chance to do something that makes use of my abilities.	3,70	1,34	3,73	1,04
The way company policies are put into practice.	3,68	1,27	3,59	1,14
My pay and the amount of work I do.	3,33	1,43	3,13	1,26
The chances for advancement on this job.	3,59	1,26	3,42	1,23
The freedom to use my own judgment.	3,59	1,29	3,54	1,07
The chance to try my own methods of doing the job.	3,75	1,23	3,51	1,15
The working conditions.	3,44	1,41	3,46	1,22
The way my co-workers get along with each other.	3,71	1,24	3,58	1,22
The praise I get for doing a good job.	3,74	1,28	3,57	1,22
The feeling of accomplishment I get from the job.	3,85	1,22	3,76	1,18

When the job satisfaction of the employees in city hotels is examined, the item with the highest average is 'the feeling of accomplishment I get from the job'. This is followed by the item that 'the chance to do something that makes use of my abilities' and 'being able to do things that go my conscience'.

According to the findings in Table 4, no significant relationship was found between nepotism, and job satisfaction and job satisfaction dimensions. Although there is no significant relationship, it is noteworthy that the relationships are negative.

When the relationship between nepotism, and job satisfaction and job satisfaction dimensions is considered in terms of city hotels, a negative relationship is found between nepotism, and facility and chance satisfaction dimension (r= -0.234, p= 0.005). There is a negative correlation between nepotism, and

Table 4: Correlation Analysis of Nepotism and Job Satisfaction Dimensions (Resort Hotels)

		Sum of nepotism	Internal satisfy dimension	Manager and facility dimension	Chance and application dimension	Sum of job satisfaction
Sum of nepotism	Pearson p	-				
internal satisfy dimension	Pearson p	-,068 ,259	-			
Manager and facility dimension	Pearson p	-,086 ,153	,606** ,000	-		
Chance and application dimension	Pearson p	-,084 ,164	,591** ,000	,783** ,000	-	
Sum of job satisfaction	Pearson p	-,090 ,134	,764** ,000	,886** ,000	,956** ,000	-

**. Correlation is significant at the 0.01 level (2-tailed).

application and wage job satisfaction dimension (r= -0.194, p= 0.022). When the relationship between nepotism, and manager and success dimension was examined, a negative and low level relationship was determined. When the nepotism and dimension of internal satisfaction are examined, it is possible to say that the relationship between the two variables is negative and low level (r= -0.246, p= 0.003). It is possible to say that there is a negative relationship between nepotism practices and perception of job satisfaction.

Table 5: Correlation Analysis of Nepotism and Job Satisfaction Dimensions (City Hotels)

		Sum of nepotism	Facility and chance dimension	Application and wage dimension	Manager and success dimension	Internal satisfaction dimension	Sum of job satisfaction
Sum of nepotism	Pearson p	-					
Facility and chance dimension	Pearson p	-,234** ,005	-				
Application and wage dimension	Pearson p	-,194* ,022	,706** ,000	-			
Manager and success dimension	Pearson p	-,277** ,001	,728** ,000	,639** ,000	-		
Internal satisfaction dimension	Pearson p	-,246** ,003	,641** ,000	,517** ,000	,667** ,000	-	
Sum of job satisfaction	Pearson p	-,279** ,001	,905** ,000	,819** ,000	,900** ,000	,805** ,000	-

**. Correlation is significant at the 0.01 level (2-tailed).
*. Correlation is significant at the 0.05 level (2-tailed).

Table 6: Correlation Analysis of Nepotism and Intention to Leave (Resort Hotels)

		Sum of nepotism	Sum of intention to leave
Sum of nepotism	Pearson	-	
	p		
Sum of intention to leave	Pearson	,222**	-
	p	,000	

**. Correlation is significant at the 0.01 level (2-tailed).

Table 7: Correlation Analysis of Nepotism and Intention to Leave (City Hotels)

		Sum of nepotism	Sum of intention to leave
Sum of nepotism	Pearson	-	
	p		
Sum of intention to leave	Pearson	,393**	-
	p	,000	

**. Correlation is significant at the 0.01 level (2-tailed).

Correlation analysis of nepotism and intention to leave was evaluated in Table 6. and Table 7. in terms of resorts and city hotels.

According to Table 6, there is a positively low-level significant relationship between the perceptions of nepotism and intention to leave the job (r= 0.222, p= 0.00).

According to the results of the correlation analysis, which measures the relationship between nepotism and intention to leave in terms of city hotels; There is a positive and moderate relationship between nepotism and intention to leave the job (r= 0.393, p= 0.00).

Table 8 shows the results of simple linear regression analysis to explain the relationship between nepotism perceptions of employees, and the perceptions of job satisfaction and job satisfaction dimensions. The established model is statistically significant (p<0.05). Accordingly, for a simple linear regression model, the regression coefficients are not equivalent to zero. t statistics indicating the significance of regression coefficients; chance and application satisfaction dimension (t= 1,396; p>0.05), internal satisfaction dimension (t= -1,332; p>0.05) and sum of satisfaction (t= -1,502; p>0.05) are statistically meaningless. Perceptions of nepotism have a very low and negative effect on the manager and facility satisfaction dimension

Table 8: Results of Regression Analysis on Nepotism and Job Satisfaction Dimensions (Resort Hotels)

	Dependent Variable				Independent Variable		
Job Satisfaction Dimensions	Manager and facility dimension				Nepotism		
	R^2	Adjusted R^2	F	p	β	t	p
	,007	,004	2,057	,000	-,086	-1,434	,000
	Dependent Variable				Independent Variable		
	Chance and application dimension				Nepotism		
	R^2	Adjusted R2	F	p	β	t	p
	,007	,003	1,949	,000	-,084	-1,396	,164
	Dependent Variable				Independent Variable		
	Internal satisfaction dimension				Nepotism		
	R^2	Adjusted R2	F	p	β	t	p
	,005	,001	1,281	,000	-,068	-1,132	,259
	Dependent Variable				Independent Variable		
	Sum of job satisfaction				Nepotism		
	R^2	Adjusted R2	F	p	β	t	p
	,008	,005	2,256	,000	-,090	-1,502	,134

Table 9 shows the results of a simple linear regression analysis to explain the relationship between nepotism perceptions of city hotels employees, and perceptions of job satisfaction and job satisfaction dimensions with a mathematical model. Firstly, the relationship results between nepotism and job satisfaction sub-dimensions were evaluated. According the results of simple linear regression analysis to explain the relationship between nepotism and job satisfaction' sub-dimension 'facility and chance' with a mathematical model, the model was found to be statistically significant (F=7,963; t=-2,822; p<0,05). For simple linear regression model, the regression coefficient is different from zero. R^2 value, which is the dependent variable explanation ratio of the independent variable, was calculated as 0,055. Although the R2 value was found to be very low, and independent variables were found to be insufficient to explain the perception of nepotism, this result shows that 5.5 % of the change in the 'facility and chance' job satisfaction dimension is explained by nepotism. According to Table 9, nepotism contributes negatively to the 'facility and chance' sub-dimension of job satisfaction. One unit increase in perceptions of nepotism reduces the 'facility and

Table 9: Results of Regression Analysis on Nepotism and Job Satisfaction Dimensions (City Hotels)

	Dependent Variable				Independent Variable			
	Facility and chance dimension				Nepotism			
	R²	Adjusted R2	F		p	β	t	p
	,055	,048	7,963		,000	-,234	-2,822	,005
	Dependent Variable				**Independent Variable**			
	Application and wage dimension				Nepotism			
Job Satisfaction Dimensions	R²	Adjusted R2	F		p	β	t	p
	,037	,030	5,337		,000	-,194	-2,310	,022
	Dependent Variable				**Independent Variable**			
	Manager and success dimension				Nepotism			
	R²	Adjusted R2	F		p	β	t	p
	,077	,070	11,381		,000	-,277	-3,374	,001
	Dependent Variable				**Independent Variable**			
	Internal dimension				Nepotism			
	R²	Adjusted R2	F		p	β	t	p
	,061	,054	8,845		,000	-,246	-2,974	,003
	Dependent Variable				**Independent Variable**			
	Sum of job satisfaction				Nepotism			
	R²	Adjusted R2	F		p	β	t	p
	,078	,071	11,533		,000	-,279	-3,396	,001

chance' job satisfaction dimension by 0.234 units. In the simple linear regression analysis conducted to explain the relationship between the 'Application and wage', 'Manager and success' and 'Internal' dimensions which are the other sub-dimensions of job satisfaction and nepotism, the model was found to statistically significant. It is seen that nepotism contributes negatively to the expressed job satisfaction sub-dimensions. The model, established to explain the relationship between nepotism perceptions and job satisfaction perceptions of employees working in city hotels, is statistically significant (F=11,533; t= -3,396; p<0,05). Accordingly, for a simple linear regression model, the regression coefficient is different from zero. The value of R^2, which is the dependent variable explanation ratio of the independent variable, was calculated as 0.078. This result shows that 7.8 % of the change in job satisfaction is explained by nepotism. According to Table 9, nepotism has a negative effect on job satisfaction. One unit increase in perceptions of nepotism reduces job satisfaction by -0,279 units. As a result of the analysis, simple regression model; Job satisfaction = 3,979 + (- 0,210) (nepotism) was determined.

Table 10: Results of Regression Analysis on Nepotism and Intention to Leave (Resort Hotels)

Dependent Variable				Independent Variable		
Intention to Leave the Job				Nepotism		
R^2	Adjusted R2	F	p	β	t	p
,049	,046	14,254	,000	,222	3,775	,000

Table 10 shows that nepotism positively contributes to the intention to leave the job. One unit increase in perceptions of nepotism increases intention to leave the job by 0.222 units.

Table 11: Results of Regression Analysis on Nepotism and Intention to Leave (City Hotels)

Dependent Variable				Independent Variable		
Intention to Leave the Job				Nepotism		
R^2	Adjusted R2	F	p	β	t	p
,154	,148	24,953	,000	,393	4,995	,000

As a result of the analysis, simple regression model; Intention to leave the job 2,168 + (0,250) (nepotism) was determined.

Table 11 shows the results of simple linear regression analysis to explain the relationship between nepotism perceptions and intention to leave the job with a mathematical model. The established model is statistically significant in terms of nepotism. Accordingly, for a simple linear regression model, the regression coefficient is different from zero. R^2 value which is the dependent variable explanation ratio of the independent variable was determined as 0,154 for nepotism. This result shows that 15.4 % of the intention to leave the job is explained by nepotism. Table 11 shows that nepotism positively contributes to the intention to leave. One unit increase in perceptions of nepotism increases the intention to leave by 0.393 units. As a result of the analysis, simple regression model; Intention to leave 2,037+ (0,419) (nepotism) was determined.

Conclusions

The participants were evaluated from the perspective of city and resort hotels in the study. When the results related to the perception of nepotism which serve the purpose of the study is taken into consideration, it is determined that the perception of nepotism is lower in city hotels than in resort resorts. When the findings of intention to leave are examined, the level of intention to leave is higher in city hotels. 'Looking for a job in another business' was determined more intense in resort hotels than city hotels. When the resort hotels and city hotels were compared in terms of job satisfaction, it was found that the participants who work in resort hotels had higher job satisfaction than those who worked in city hotels. When the job satisfaction is evaluated in general, it is seen that the satisfaction level is due to the sense of accomplishment.

Correlation and regression analysis were applied to the data obtained in order to solve the problem of the research. In the correlation analysis, the relationship between nepotism and job satisfaction of the participants working in city hotels is statistically significant and negative. This is an indication that participants' perceptions of nepotism have a negative effect on job satisfaction. On the other hand, no significant relationship was found between nepotism and job satisfaction in resort hotels. When the relationship between perception of nepotism and intention to leave the job is examined, it is seen that both sample groups have a different structure. This situation can be expressed 'as the perception of nepotism increases, increase of the intention to leave'.

Simple linear regression analysis was performed to explain the relationship between nepotism perceptions and job satisfaction and intention to leave, and to determine the effect and direction on the variables, which differed in terms of hotels' location. Simple linear regression analysis was performed to test the linear relationship between nepotism (independent variables) and job satisfaction (dependent variables), to determine how much of the change in perception of nepotism was explained by job satisfaction and to express the relationship between variables statistically. T statistic indicating the significance of regression coefficients for resort hotels was statistically insignificant for job satisfaction dimension (t= -1,502; p>0.05). For city hotels, 7.8 % of the change in job satisfaction was explained by nepotism. One unit increase in perceptions of nepotism reduces job satisfaction by -0,279 units.

The results of a simple linear regression analysis to explain the relationship between the perception of nepotism and intention to leave the employees working in the resort hotels; 4.9 % of the intention to leave is explained by nepotism. It shows that 15.4 % of the intention to leave for city hotels is explained by nepotism.

Evaluation was made in terms of city and hotel businesses in this study. These evaluations were limited to Konya and Antalya provinces. In the other studies, the perception of nepotism can be search for accommodation business that serve different concepts and the results can be compared. Furthermore, it is considered that the evaluation of nepotism applications in terms of corporate travel agencies and airlines operating in the sector for many years will contribute to the literature.

References

Abdalla, H. F., Maghrabi, A. S. & Raggad, B. G. (1998). Assesing the Perceptions of Human Resource Managers Toward Nepotism. A Cross-Cultural Study. *International Journal of Manpower*,19, 554–570.

Akalan, A. R. (2006). *Türk Kamu Hizmetinde İyi Yönetim ve Yolsuzlukla Mücadele*. Basılmamış Doktora Tezi. Selçuk Üniversitesi, Sosyal Bilimler Enstitüsü, Konya.

Akıncı, Z. (2002). Turizm Sektöründe İşgören İş Tatminini Etkileyen Faktörler: Beş Yıldızlı Konaklama İşletmelerinde Bir Uygulama. *Akdeniz İ.İ.B.F. Dergisi*, 4, 1–25.

Ansari, Y., Merdasi, A. & Aliabad, F. A. (2015). The Impact of Nepotism in the Organization on the Attitude of Employees to the Organization: A Case Study in a Private Bank. *International Journal of Review in Life Sciences*. 5(5), 509–515.

Asunakutlu, T. & Avcı, U. (2009). *Nepotizm-İş Tatmini İlişkisi: Aile İşletmelerinde Bir İnceleme*. Ulusal Yönetim ve Organizasyon Kongresi Bildiri Kitabı.21-23 Mayıs 2009 Eskişehir. 17, 728–734.

Asunakutlu, T. & Avcı, U. (2010). Nepotizm İş Tatmini İlişkisi: Aile İşletmelerinde Bir İnceleme. *Süleyman Demirel Üniversitesi İktisadi ve İdari Bilimler Fakültesi Dergisi*, 15, 2–9.

Avcı, A. (2017). *Şirketlerde Nepotizm Uygulamasının Çalışanların İş Tatminive Tükenmişlik Düzeylerine Etkisi*. Basılmamış Yüksek Lisans Tezi. İstanbul SabahattinZaimÜniversitesi, SosyalBilimlerEnstitüsü, İstanbul.

Bartlett, K. R. (1999). *The Relationship Between Training and Organizational Commitment in the Health Care Field*. The Degree of Doctor of Philosophy, The University of Illinois, Urbana.

Begley, T. M., Khatri, N. & Tsang, E. W. K. (2010). Networks and Cronyism: A Social Exchange Analysis. *Asia Pacific Journal of Management*, 27(2), 281–297.

Budak, A. (2006). *Kamu Sektöründe Çalışanların İş Tatmin Düzeyi: Milli Savunma Bakanlığı Akaryakıt İkmal ve Nato Pol Tesisleri'nde BirUygulama.*

Basılmamış Yüksek Lisans Tezi. Anadolu Üniversitesi, Sosyal Bilimler Enstitüsü, Eskişehir.

Cabrera-Suárez, K. (2005). Leadership Transfer and the Successor's Development in the Family Firm. *The Leadership Quarterly*, 16, 71-96.

Çekmecioğlu, H. G. (2014). Göreve ve İnsana Yönelik Liderlik Tarzlarının Örgütsel Bağlılık, İş Performansı ve İşten Ayrılma Niyeti Üzerindeki Etkileri. KONBED, 28, 21-34.

Çetin, A., Güleç, R. & Kayasandık, A. E. (2015). Etik İklim Algısının Çalışanların İşten Ayrılma Niyetine Etkisi: Tükenmişliğin Aracı Değişken Rolü. *Electronic Journal of Vocational Colleges*- Ekim, (26)2 18-31.

Çetin, F. & Basım, H. N. (2011). Psikolojik Dayanıklılığın İş Tatmini ve Örgütsel Bağlılık Tutumlarındaki Rolü. *İŞ GÜÇ Endüstri İlişkileri ve İnsan Kaynakları e-Dergisi*, 13(1), 79-94.

Demaj, E. (2012). *Nepotism, Favoritism and Cronyism and Their Impact on Organizational Trust and Commitment; the Service Sector Case in Albania.* Basılmamış Yüksek Lisans Tezi. Epoka University, Business Administration Department, Faculty of Economics and Administrative Sciences, Epoka.

Demir, H., Usta, R. & Okan, T. (2008). İçsel Pazarlamanın Örgütsel Bağlılık ve İş Tatminine Etkisi. *H.Ü. İktisadi ve İdari Bilimler Fakültesi Dergisi*, 26(2), 135-161.

Düz, S. (2012). *Konaklama İşletmelerinde Nepotizm İle Örgütsel Bağlılık Arasındaki İlişkinin İncelenmesi.* Basılmamış Yüksek Lisans Tezi, Afyon Kocatepe Üniversitesi, Sosyal Bilimler Enstitüsü, Afyonkarahisar.

Erdem, R. & İlhan, T. (2010). *Yönetim ve Örgüt Açısından Kayırmacılık.* R. Erdemve T. İlhan (der.) Akraba Kayırmacılığı (Nepotizm). İstanbul: Beta Yayınları, 135-166.

Eroğlu, F. (2013). *Davranış Bilimleri.* İstanbul: Beta Yayınları.

Fu, P. (2015). Favoritism: Ethical Dilemmas Viewed through Multiple Paradigms. *The Journal of Values-Based Leadership*, 8(1), 1-7.

Güzel, B. & Ayazlar, G. (2014). Örgütsel Adaletin Örgütsel Sinizm ve İşten Ayrılma Niyetine Etkisi: Otel İşletmeleri Araştırması. *KMÜ Sosyal ve Ekonomik Araştırmalar Dergisi*, 16(26), 133-142.

Hudson, S., González-Gómez, H. V. & Claasen, C. (2017). Legitimacy, Particularism and Employee Commitment and Justice. *Journal of Business Ethics*, (157)3, 1-15.

İşcan, Ö. M. & Timuroğlu, M. K. (2007). Örgüt Kültürünün İş Tatmini Üzerindeki Etkisi ve Bir Uygulama. *İktisadi ve İdari Bilimler Dergisi*, 21(1), 119-135.

İyiişleroğlu, S. C. (2006). *Aile Şirketleri: Adana ve Çevresinde Faaliyet Gösteren Aile Şirketlerinde Nepotizm Uygulamasının Tespitine Yönelik Bir Araştırma.* Basılmamış Yüksek Lisans Tezi. Çukurova Üniversitesi, Sosyal Bilimler Enstitüsü, Adana.

Karacaoğlu, K. & Yörük, D. (2012). Çalışanların Nepotizm ve Örgütsel Adalet Algılamaları: Orta Anadolu Bölgesinde Bir Aile İşletmesi Uygulaması. *İŞ, GÜÇ Endüstri İlişkileri ve İnsan Kaynakları Dergisi*, 14(3), 43–64.

Karataş, A. (2013). *Otel İşletmelerinde Kronizmin İş Tatminive İşten Ayrılma Niyeti Üzerindeki Etkileri: Muğla İlinde Bir Araştırma.* Basılmamış Yüksek Lisans Tezi. Balıkesir Üniversitesi, Sosyal Bilimler Enstitüsü, Balıkesir.

Kaya, N. (2016). *Mobbingin Örgütte Adalet Algısı ve Örgütsel Bağlılık İlişkisi İle Çalışanların İş Performansına ve İşten Ayrılma Niyetine Etkisi.* Basılmamış Doktora Tezi. Beyken Üniversitesi, Sosyal Bilimler Enstitüsü, İstanbul.

Kök, S. B. (2006). İş Tatmini ve Örgütsel Bağlılığın İncelenmesine Yönelik Bir Araştırma. *Atatürk Üniversitesi İktisadi ve İdari Bilimler Dergisi*, 20(1), 291–317.

Kuzulugil, Ş. (2012). Kamu Hastaneleri Çalışanlarında İş Tatminini Etkileyen Faktörlerin İncelenmesine Yönelik Bir Araştırma. *İstanbul Üniversitesi İşletme Fakültesi Dergisi*, 41(1), 129–141.

Loewe, M., Blume, J., Schönleber, V., Seibert, S., Speer, J. & Voss, C. (2007). *The Impact of Favouritism on the Business Climate: A Study on Waste in Jordan.* Bonn: Deutsches Institutfür Entwicklungspolitik.

Luthans, F. (1994). *Organizational Behavior.* (1. Edition). USA: Mcgraw-Hill Inc.

Newstrom, J. & Davis, K. (1997). *Human Behaviour at Work, Organizational Behaviour.* New York: Mcgrawhill.

Özkanan, A. & Erdem, R. (2014). Yönetimde Kayırmacı Uygulamalar: Kavramsal Bir Çerçeve. *Süleyman Demirel Üniversitesi Sosyal Bilimler EnstitüsüDergisi*, 20, 179–206.

Özler, H., Özler, D. E. & Gümüştekin, G. E. (2007). Aile İşletmelerinde Nepotizmin Gelişim Evreleri ve Kurumsallaşma. *Selçuk Üniversitesi Sosyal Bilimler Enstitüsü Dergisi*, 17, 437–450.

Özpehlivan, M. (2015). *Kültürel Farklılıkların İşletmelerde Örgüt İçi İletişim, İş Tatmini, Bireysel Performans ve Örgütsel Bağlılık Kavramları Arasındaki İlişkiye Etkileri: Türkiye-Rusya Örneği.* Basılmamış Doktora Tezi. Okan Üniversitesi, Sosyal Bilimler Enstitüsü, İstanbul.

Öztürk, T. (2008). Değişen Çağın Aile İşletmelerinde Kurum Kültürünün Yerleştirilmesinde Profesyonel Yöneticilerden Beklentiler. *Çankaya Üniversitesi Fen-Edebiyat Fakültesi*, 10, 109–116.

Özüren, Ü. (2017). *Tekstil İşletmelerinde Nepotizm Uygulamalarına Bağlı Olarak Üretkenlik Karşıtı Davranışlar ve Sonuçları*. Basılmamış Yüksek Lisans Tezi. İstanbul Kültür Üniversitesi, Sosyal Bilimler Enstitüsü, İstanbul.

Pektaş, E. K. (1997). *Büyük Kent Belediyelerinin Eğitim ve Kültür Hizmetlerine Siyasal Parti İdeolojilerinin Yansıması*. Basılmamış Yüksek Lisans Tezi. Dokuz Eylül Üniversitesi, Sosyal Bilimler Enstitüsü, İzmir.

Rosin, H. & Korabik, K. (1995). Organizational Experiences and Propensity to Leave: A Multivariate Investigation of Men and Women Managers. *Journal of Vocational Behavior*, 46, 1–16.

Sarıboğa, M. (2017). *Nepotizmin Örgütsel Bağlılık ve İş Doyumuna Etkisi Ve Otel Çalışanları Üzerine Bir Araştırma*. Basılmamış Yüksek Lisans Tezi. Doğuş Üniversitesi, Sosyal Bilimler Enstitüsü, İstanbul.

Savaş, Y. (2015). *Nepotizmin Yenilik ve Yetenek Yönetimi Üzerine Etkisi*. Basılmamış Yüksek Lisans Tezi. Aksaray Üniversitesi, Sosyal Bilimler Enstitüsü, Aksaray.

Saylı, H. & Kızıldağ, D. (2007). Yönetsel Etik ve Yönetsel Etiğin Oluşmasında İnsan Kaynakları Yönetiminin Rolünü Belirlemeye Yönelik Bir Analiz. *Sosyal Bilimler Dergisi*, 9(1), 231–251.

Tekiner, M. A. & Aydin, R. (2015). Analysis of Relationship Between Favoritism and Officer Motivation: Evidence from Turkish Police Force. *Inquiry-Sarajevo Journal of Social Sciences*, 1(2), 75–97.

Turhan, R. (2016). *Nepotizm, Kronizm ve Patronaj Eğilimlerinin Kurumsallaşma Algısı Bağlamında Analizi*. Basılmamış Yüksek Lisans Tezi. Ege Üniversitesi, Sosyal Bilimler, İzmir.

Weiss, D. J., Dawis, R. V., England, G. W. & Lofquist, L. H. (1967). *Manual for the Minnesota Satisfaction Questionnaire*. Minneapolis: University of Minnesota, Work Adjustment Project Industrial Relations Center.

Yazıcıoğlu, Y. & Erdoğan, S. (2007). *Spss Uygulamalı Bilimsel Araştırma Yöntemleri*.(2. Baskı). Ankara: DetayYayıncılık.

Yenihan, B., Öner, M. & Çiftyıldız, K. (2014). İş Stresi ve İşten Ayrılma Niyeti Arasındaki İlişki: Otomotiv İşletmesinde Bir Araştırma. *Çalışma İlişkileri Dergisi*, 5(1), 38–49.

Serdar ÇÖP

19 Sharing Economy for Sustainability in Tourism

Introduction

Thoughtless consumption of natural resources worldwide and following the industrial revolution, accompanying environmental and cultural deterioration necessitated the emergence of the concept of sustainability (Kasli et. al., 2015: 29). Ciocoiu (2011: 33) defined the term 'sustainability' as, not ignoring the needs of future generations while meeting today's needs.

Sustainability in the tourism sector depends on economic, social, technological, political and environmental impacts. All these impacts should be regulated by tourism sector stakeholders through policies, strategies, directives and regulations. (OECD, 2018: 1). Environmental, cultural, social and economic sustainability can be ensured through the policies, strategies, directives and regulations. It is obvious that the sharing culture and the effects of today's sharing economy coincide in ensuring sustainability.

While it was once impossible for people to share their cars, music, machines, houses and meals, this has become possible with the change of life as a result of technological development (Beutin, 2018: 5). Particularly, the developments in mobile phone technology and mobile applications have made this situation more widespread.

As a result of not only the technological developments in the world but also population growth, environmental problems, and the desire to use money more efficiently, the sense of sharing has been integrated with today's economic system and revealed the sharing economy. Thanks to the sharing economy; consumption of goods and services may be offered to others, whether paid or unpaid, although ownership is still not changed. The fact that people meet their needs over the internet rather than the markets where the buyer and seller come together defines the existence of a new economic system (Kiraci, 2017: 52; Kurt & Ünlüönen, 2017: 13). This new economic system is significant in terms of reaching out to the masses easily, paying attention to social welfare rather than personal, allowing using and returning things instead of ownership and comprising rather of economically more efficient applications (Kiracı, 2017: 52).

Owing to the sharing economy, significant changes have occurred in the tourism distribution system. The sharing economy is also growing rapidly in transportation (Blabla), accommodation (Airbnb), catering (Eatwith) sectors (Özdemir & Çelebi, 2018: 28). Considering the effects of a stronger sharing economy on sustainability in tourism, environmental sustainability comes into mind first, followed by social and economic sustainability. Due to the increasing importance of sustainability in today's society, the concepts of sustainability and sustainability in tourism should be examined firstly in order to better understand the role of sharing economy in sustainable tourism.

1. Sustainability in Tourism

The terms 'sustainability' and 'sustainable' stand for durability and durable in French, permanence in German and have the same meaning with maintain, pursue, provide and support in Dutch and other languages (Pisani, 2006: 85). Ulrich Grober, investigating the term 'sustainability', stated that this concept was used for the first time in '*Sylvicultura Oeconomica*' when narrating the significance of forests in raw material production (Schmandt, 2010: 11). Sustainability, in general terms, is the ability of a process or activity to protect natural resources and to leave the environment in good order for future generations while meeting current needs (Collin, 2004: 265). In another study, sustainability is defined as a concept that aims to meet today's needs and desires without compromising the ability to meet the needs of the future (United Nations, 1987). At the 'Development Summit' organized by the United Nations, phenomena such as poverty, health, education, clean water, hygienic living conditions, demographic equality, responsible consumption, economic growth and biodiversity have been determined as targets of sustainability and/or sustainable development, to be realized by 2030 (Özgen et. al., 2016: 33–34; Yavuz, 2010: 66). The role, contribution and opportunities of tourism have a significant role in the realization of these objectives. Tourism has also become an important tool to meet the demands of human beings such as the search for innovation and change. The devastation of mass tourism developed by globalization (excessive use of resources, degradation of natural balance, forest slaughter, damage to cultural heritage, etc) (Ayaş, 2007: 61) and the estimation that tourism will be the fastest and most steady growing sector in the next forty years (WTO, 1999: 12) have increased the need to implement sustainability in the tourism sector. Sustainable tourism, which has gained more importance after the Rio World Conference (Weaver, 2006: 11) is defined as a form of development that utilizes environmental resources optimally, helps to preserve natural resources and biological diversity by maintaining the necessary ecological processes, respects the socio-cultural reality of communities, preserves established and living cultural heritage as well as traditional values, contributes to intercultural understanding and tolerance, provides long-term economic operations and stable employment, generates income, hosts communities, contributing to poverty reduction by providing socio-economic benefits including social services and shares to all stakeholders (UNEP & UNWTO, 2005: 11). The concept of sustainable tourism which attaches importance to nature, living creatures and social culture should be ecologically and environmentally sustainable, economically viable and socially acceptable (Kuter & Unal, 2009: 148).

There are tools that contribute to the implementation of tourism policies (Countryside Commission, 1995: 2) that can sustain the local economy without damaging the environment to which it is connected or dependent. Transportation capacity, environmental impact assessment, ecological footprint factors are tools that contribute to future sustainability of tourism (Çeken, 2016: 75-83). In order to achieve sustainable tourism, a touristic potential should be exhibited by determining an effective touristic policy, appointing the team to realize the planning and creating a local organization in addition to these tools (Çeken, 2016: 118-136).

There are examples of sustainable tourism practices available in various regional, local accommodation, travel, airline, food and beverage businesses both in Turkey and in the world.

'Calm City (Citta slow)' movement which has emerged in the Tuscany region of Italy, spread to the world and became a trend is an example of local sustainable tourism practices. Seferihisar district of Izmir province is the first settlement in Turkey to participate in this movement (Özgen et. al., 2016: 270-271). In a study by Karadeniz (2014: 104), Persembe district of the province of Ordu was appointed as a calm city and it was pointed out that Persembe has continuously contributed to the economy and development without giving harm to the nature, the inhabitants, the abstract and concrete cultural heritages, while fulfilling objectives related with sustainable tourism. This study also stated that the principles of sustainable tourism are in parallel with the principles of calm city where the historical texture in the settlement area and human labor have been preserved, the agricultural awareness as well as recognition of local tastes and handicrafts have been increased through social participation and the use of clean energy has become widespread so that a sustainable tourism strategy can be followed (Karadeniz, 2014: 104). Seferihisar district and Ephesus ancient city were also evaluated within the context of same tourism principles. Some of the practices featured following the evaluations in Seferihisar were seed swap festivals organized to maintain biological diversity, counting trees and identifying historic trees, to promote recycling and separation of waste for environmental cleaning, practices to reduce image and noise pollution, peasant markets established to promote local welfare, supporting local product sales of women producers to promote social equity; and regarding the ancient city of Ephesus these were preserving cultural values and improving historical sites (Özgen et. al., 2016: 272; Eser, 2010: 32–33).

Awareness regarding sustainability is constantly spreading. In this way, a conscious mass in the field of tourism is increasing day by day and the supply of tourism activities is directed in line with the wishes of the goal of sustainable environment. A visitor who has his/her attitude channeled towards sustainability in tourism should pay attention to reduce energy, water, plastics consumption, not to buy things that are related to endangered species, to use public transport, to support the local economy, to learn about the company's tourism policies before traveling with it and to prefer entities or regions that implement sustainable tourism (Çeken, 2016: 83). As stated in a Native American Proverb, we do not inherit the earth from our ancestors; we borrow it from our children (Sezgin & Karaman, 2008: 436); and that we should support the efforts in raising awareness in order to ensure sustainability in tourism.

2. Sharing Economy

Sharing is the most rational way of survival, observed throughout human existence (Yakin & Kazancoglu, 2018: 2; Belk, 2014: 1595). Although the historical background of behavior of sharing goes back to the existence of mankind, the concept of sharing economy was initially discussed by Felson & Speath (1978: 642) under the title 'Collaborative consumption', and since then it has frequently appeared in the literature as 'Sharing economy'.

Sharing economy is making use of goods and services in the most economical way by sharing them with others (Özdemir & Çelebi, 2018: 25). Development of information technologies, thus facilitating the exchange of goods or services over the internet, resulted in the emergence of a concept like sharing economy (Hamari, 2016: 2048). Sharing economy, which aims to acquire value and benefits, causes a change in the structure of production and consumption (Doyduk & Tiftik, 2017: 140). In this structure, people have the chance to use the product or service instead of buying it. The process in this economy is based on providing the opportunity to use a product or service in return for a fee or even without a fee (Özdemir & Çelebi, 2018: 28). In addition, the desire to gain new practices and experiences comes to the forefront instead of ownership (Özdemir & Çelebi, 2018: 25). On the other hand, this economy is expressed as a system based on creating value through innovative initiatives and enterprises (Kacar & Yakin, 2018: 736).

Thanks to the advantages of mobile internet technology and social networks, the ability to reach crowds without being dependent on space and time has influenced sharing economy (Demirer & Hassan, 2016: 44). It is estimated to reach a sector of 335 billion dollars by 2025 (PWC, 2015: 5). The impact of companies as well as individuals on the growth of the market is quite high. Following a period of brand experience in which companies communicate with their customers through their websites, today there is sharing economy, which is dominated by customer-orientation and satisfaction, promoting using and giving back (leasing) rather than buying (Kişi, 2018: 58). Sharing economy platforms are means that provide economic opportunity to their beneficiaries, include sustainable consumption, have unstructured markets, serve the neo-liberal economic paradigm, and provide benefits to parties besides sharing (Martin, 2016: 153).

Apart from these advantages, sharing economy is associated with some disadvantages such as perceiving these platforms as means of informal economy lacking necessary legal regulations (Kurt & Unluonen, 2017: 12). Sharing economy may have negative effects on the institutions and organizations within the tourism system as well as individuals and society as reflection of customer and guest relations (Dredge & Gyimothy, 2015: 295). Considering the advantages; we can mention that it leads individuals to consume less and thus preventing waste and promoting savings through leasing opportunities (Yakin & Kazancoglu, 2018: 1). In this way, sharing economy contributes to the protection of resources, strengthening of the local community, reducing expenses, meeting the expectations of low-income groups, the emergence of new business opportunities and the ability of people to travel more freely (Beutin, 2018: 5).

Github and SourceForge are examples of resources of open source software, Wikipedia, Youtube and Instagram are examples of content sharing web sites and The Pirate Bay is an example for end-to-end file sharing web sites within the sharing economy. Kiva which is used as a financial system and Kickstarter which is used as a mass funding system are other examples in the sharing economy. The most sophisticated player in the tourism sector is the Airbnb platform. Since its foundation in 2008 in San Francisco (Kacar, 2018: 40), Airbnb has been an integrated system of high economic value, in which the service provider and consumer come together throughout the online accommodation network within the tourism industry (Özdemir & Çelebi, 2018: 30).

The number of platforms, which are referred to as the integrated systems within the sharing economy are increasing day by day. Best known and popular of these platforms in the field of accommodation and travel are Couchsurfing,

Homestay; Uber, Blablacar, Godrive, Olacabs, Jumpinstudent, Getaround for car rental; Taskrabbit, Skillshare, Fverr, Skillpages, Etsy, Fubles in the field of skills and services; Vizeat, Mealsharing, Travelspoon, Talktochef.com, Foodswapnetwork in the field of wining and dining; Kiva, Indiegogo, Kickstarter in the field of finance; Wework, Workinton, Sharedesk, Loosecubes in the field of office spaces (Kacar, 2018: 38).

When examining different industries where we take the advantage of the sharing economy, the popular industries and rates are as follows: media and entertainment 28 %, accommodation 20 %, transportation 19 % and Retail and Consumer Goods 19 % (Beutin, 2018: 9–13). The industrial size attained by the sharing economy reveals the necessity of a better and more rational management.

3. Sharing Economy for Sustainability in Tourism

Excessive and unplanned consumption of resources worldwide following the industrial revolution have raked up universal problems such as environmental pollution, global warming, income injustice, excessive use of energy, erosion of human values through individualization, population growth and hunger. Elimination of these problems, controlling or mitigating their impacts forms the basis of sustainability (Madran & Yakin, 2018: 61).

The United Nations and non-governmental organizations develop various policies and strategies for sustainability. Sustainable Development Objectives, which was discussed under chapter 17 were put into force by United Nations in January 2016 (Madran & Yakin, 2018: 64; UNDP, 2018). Sharing economy within the context of sustainable tourism can have a profound impact on clean water and clean energy, decent work and economic growth, cleaner industry, innovation and infrastructure, reduction of inequalities, sustainable cities and communities, conscious production and consumption.

Sharing economy can provide significant advantages for sustainability. Resources are used more effectively; business models can also be organized within this framework. With the development, systemization and legalization of the sharing economy over time people's desire to own will be minimized. This will also minimize the carbon footprint for a sustainable environment. In addition, it can be set forth that urban planners will tend to gravitate towards sharing economy oriented practices over time with the effect of global pressures (Sundararajan, 2014: 4).

Shared travelling and accommodation are the services where sharing economy is the most intense in the tourism sector. It is known that the negative effects of the road vehicles, most preferred means of journey, on the sustainable environment are quite high. Reducing the negative effects of the land vehicles via shared travel in the tourism sector is a promising indicator. It is particularly important in terms of sustainability that vehicles such as bicycles take part in sharing economy platforms (Madran & Yakin, 2018: 77). Utilizing idle resources in the accommodation sector other than travelling will ensure that surplus houses are also put into use. This will prevent the establishment of more facilities or businesses. In addition, it can be stated that accommodation realized through platforms related to the sharing economy serves sustainability in tourism since it enables less water, detergent and energy consumption and exhibiting more environmentalist behaviors.

Results

The new economic system created by sharing goods and services among consumers is shaped through sociological and cultural transformations, economic and environmental concerns and the consumption behavior of the new generation (Yakın & Kazancaglu, 2018: 14).

Rapidly adapting to the movement of sharing on the web and on the internet, and realizing that sharing can take place smoothly and efficiently are two indicators of a growing sharing economy in the future (Kiraci, 2017: 52). It is estimated that this economic culture that grows with the development of internet technologies will become widespread and shall standardize over time.

People who offer their homes, cars, meals to other people as an entrepreneur through the sharing economy, outside their daily routines, serve to strengthen the local economy and to increase touristic mobility and competition (Özdemir & Çelebi, 2018: 33). Airbnb, one of the supporting platforms, is more environmentally friendly than other accommodation platforms in terms of economic and social development as well as reducing carbon footprint. From a sustainability perspective, it is also known to contribute to reducing the carbon footprint (Özdemir & Çelebi, 2018: 30). Other than platforms used for transportation purposes such as Uber and Blablacar, the use of smaller, hybrid and electric vehicles as well and reduced energy consumption and greenhouse gas emissions will ensure sustainability in the sharing economy (Chen & Kockelman, 2016: 283). From this point of view it can be said that, the sharing economy can limit unsustainable customer behavior and completely change the existing practices with destructive innovation (Martin, 2016: 153).

Non-clarity of legal status of companies in the sharing economy, tax applications and limitations will determine the future position of the sharing economy. However; there are also attempts by larger enterprises to take place in the sharing economy with the influence of capitalism. The sharing economy culture is estimated to become more widespread and particularly pursue its activities as a new sector; therefore, it is essential to support policies, programs and environment friendly, green investments to ensure sustainability in tourism (OECD, 2018: 2).

The organization of consumption in co-operation with sharing economy to ensure sustainability in tourism will affect the sustainable use of resources. Increasing awareness of producers and consumers within the sharing economy, the preparation and implementation of the policies will have a significant impact on leaving tourism destinations unharmed to future generations.

References

Ayaş, N. (2007). Çevresel Sürdürülebilir Turizm Gelişmesi. *Gazi Üniversitesi İktisadi ve İdari Bilimler Fakültesi Dergisi*, 9(1), p. 59–69.

Belk, R. (2014). You Are What You Can Acces: Sharing and Collaborative Consumption Online. *Journal of Business Research*, 67, p. 1595–1600.

Beutin, N. (2018). Share Economy 2017 The New Business Model. Londra: PricewaterhouseCoopers GmbH Wirtschaftsprüfungsgesellschaft International Limited (PwCIL).

Chen, T. D. & Kockelman, K. M. (2016). Carsharing's Life-Cycle Impacts on Energy Use and Greenhouse Gas Emissions. *Transportation Research Part D*, 47, p. 276–284.

Ciocoiu, C. N. (2011). Integrating Digital Economy and Green Economy: Opportunities for Sustainable Development. *Theoretical & Empirical Researches in Urban Management*, 6(1), p. 33–43.

Collin, P. H. (2004). Dictionary of Environment & Ecology. Londra: Bloomsbury Publishing Plc.

Countryside Commission. (1995). Sustaining Rural Tourism. Cheltenham: Countryside Commission (CCP 483).

Çeken, H. (2016). Sürdürülebilir Turizm, Temel Kavramlar ve İlkeler. Ankara: Detay Yayıncılık.

Demirer, D. & Hassan, A. (2016). Değiş Tokuş ve Kiralama Uygulamalarının Konaklama İşletmelerin Üzerindeki Olası Etkileri. *Anatolia: Turizm Araştırmaları Dergisi*, 27(1), p. 41–63.

Doyduk, H. B. B. & Tiftik, C. (2017). Nesnelerin İnterneti: Kapsamı, Gelecek Yönelimi ve İş Fırsatları. *Üçüncü Sektör Sosyal Ekonomi*, 52(3), p. 127–147.

Dredge, D. & Gyimothy, S. (2015). The Collaborative Economy and Tourism: Critical Perspectives, Questionable Claims and Silenced Voices. *Tourism Recreation Research*, 40(3), p. 286–302.

Eser, S. (2010). Sürdürülebilir Kültür Turizmi: Efes Örneği. *Ege Coğrafya Dergisi*, 19(2), p. 27–34.

Felson, M. & Spaeth, J. L. (1978). Community Structure and Collaborative Consumption. A Routine Activity Approach. American Behavioral Scientist, (21)4, p. 614–624.

Hamari, J., Sjöklint, M. & Ukkonen, A. (2016). The Sharing Economy: Why People Participate in Collaborative Consumption. *Journal of the Association for Information Science and Technology*, 67(9), p. 2047–2059.

Kacar, A. İ. (2018). Toplulukların Gücü: Paylaşımcı Markalar, A. Canan, V. Yakın, A. İdil Kacar (Ed.), Paylaşım Ekonomisi A'dan Z'ye Tüm Boyutlarıyla içinde (p. 1–18). Ankara: Akademisyen Yayınevi.

Kacar, A. İ. & Yakın, V. (2018). Paylaşım Ekonomisi ve Değer Yaratmak: Kanvas İş Modeli Örneği. *Üçüncü Sektör Sosyal Ekonomi*, 53(3), p. 724–739.

Karadeniz, C. B. (2014). Sürdürülebilir Turizm Bağlamında Sakin Şehir Perşembe. Uluslararası Sosyal Araştırmalar Dergisi, 7(29), p. 84–107.

Kaşlı, M., Cankül, D., Köz, E. N., & Ekici, A. (2015). Gastronomik Miras ve Sürdürülebilirlik: Eskişehir örneği. *Eko-Gastronomi Dergisi*, 1(2), p. 27–46.

Kiracı, H. (2017). Ortak Tüketim Ekseninde Paylaşılan/Paylaşılmayan Varlıklar ve Bireylerin Paylaşım Davranışlarını Etkileyen Faktörler Üzerine Bir Araştırma. *International Journal of Economic and Administrative Studies*, 16. UİK Özel Sayısı, p. 51–70.

Kişi, N. (2018). Paylaşım Ekonomisinin Ulşsım Sektörüne Yansımaları: UBER Örneği. *Uluslararası Yönetim ve Sosyal Araştırmalar Dergisi*, 5(10), p. 57–68.

Kurt, S. & Ünlüönen, K. (2017). Paylaşım Ekonomisi Kapsamında Turizm Sisteminin Değerlendirilmesi. *Gazi Üniversitesi Turizm Fakültesi Dergisi*, 1, p. 1–21.

Kuter, N. & Ünal, H. E. (2009). Sürdürülebilirlik Kapsamında Ekoturizmin Çevresel, Ekonomik ve Sosyo-kültürel Etkileri. *Kastamonu Üniversitesi Orman Fakültesi Dergisi*, 9(2), p. 146–156.

Madran, C. & Yakın, V. (2018). Sürdürülebilirlik ve Paylaşım Ekonomisi, A. Canan, V. Yakın, A. İdil Kacar (Ed.). Paylaşım Ekonomisi A'dan Z'ye Tüm Boyutlarıyla içinde (p. 62–85). Ankara: Akademisyen Yayınevi.

Martin, C. J. (2016). The Sharing Economy: A Pathway to Sustainability or a Nightmarish Form of Noliberal Capitalism?*Ecological Economics*, 121, p. 149–159.

Özdemir, G. & Çelebi, D. (2018). Paylaşım Ekonomisi: Airbnb Örneği. *İş, Güç Endüstri İlişkileri ve İnsan Kaynakları Dergisi*, 20(2), p. 21–38.

Özgen, H. K. Ş., Dilek, S. E., Türksoy, S. S., Çelebi, S. K. (2016). Sürdürülebilir Turizm Yönetimi. Ankara: Detay Yayıncılık.

Pisani, J. A. D. (2006). Sustainable Development – Historical Roots of the Concept. *Environmental Sciences*, 3(2), p. 83–96.

PricewaterhouseCoopers (PWC) (2015). Sharing or Paring? Growth of the Sharing Economy. https://www.pwc.com/hu/en/kiadvanyok/assets/pdf/sharing-economy-en.pdf. (11.10.2019).

Schmandt, J. (2010). George P. Mitchell and the Idea of Sustainability. Teksas: Texas A & M Press.

Sezgin, M. & Karaman, A. (2008). Turistik Destinasyon Çerçevesinde Sürdürülebilir Turizm Yönetimi ve Pazarlaması. *Selçuk Üniversitesi Sosyal Bilimler Enstitüsü Dergisi*, (19), p. 429–438.

Sundararajan, A. (2014). Peer-to-peer businesses and the sharing (collaborative) economy: overview, economic effects and regulatory issues. Written testimony for the hearing title, the power of connection: peer-to-peer businesses, held by the committee on small businesses of the U.S. House of Representatives. https://republicans-smallbusiness.house.gov/uploadedfiles/1-15-2014_revised_sundararajan_testimony.pdf.

The Organisation for Economic Co-Operation and Development (OECD) (2018). OECD Tourism Trends and Policies 2018. https://www.oecd-ilibrary.org/urban-rural and-regional-development/oecd-tourism-trends-and-policies-2018_tour-2018-en (10.10.2019).

United Nations. (1987). Report of the World Commission on Environment and Development Our Common Future. https://www.are.admin.ch/are/en/home/sustainable-development/international-cooperation/2030agenda/un-_-milestones-in-sustainable-development/1987--brundtland-report.html. (18.11.2019).

United Nations Development Programme (UNDP). (2018). https://www.tr.undp.org/content/turkey/tr/home/sustainable-development-goals.html (10.10.2019).

United Nations World Tourism Organization (UNWTO). (2005). Making Tourism More Sustainable: A Guide for Policy Makers, Paris ve Madrid: UNEP & UNWTO Publishing.

Weaver, D. (2006). Sustainable Tourism. Londra: Routledge, Taylor & Francis.

World Tourism Organization (WTO). (1999). Tourism: 2020 Vision. Executive Summary Updated.

Yakın, V. & Kazançoğlu, İ. (2018). Paylaşım Ekonomisin Gelişimi. A. Canan, V. Yakın, A. İdil Kacar (Ed.), Paylaşım Ekonomisi A'dan Z'ye Tüm Boyutlarıyla içinde (p. 1–18). Ankara: Akademisyen Yayınevi.

Yavuz, A. (2010). Sürdürülebilirlik Kavramı ve İşletmeler Açısından Sürdürülebilir Üretim Stratejileri. *Mustafa Kemal Üniversitesi Sosyal Bilimler Enstitüsü Dergisi*, 7(14), p. 63–86.

Serdar EREN

20 Sustainable Tourism Criteria and Turkish Restaurants

Introduction

There is a thriving attendance on sustainability within the restaurant industry in Turkey. Managers and staff became aware of sustainable practices and their importance in the Restaurant industry. Moreover a growing number of patrons of the restaurants started to demand sustainable healthy food and environment-friendly business forms throughout the Turkish restaurants (Temizkan et. al., 2017). Hence, sustainability involves the economic, social and environmental impacts, and restaurant managers are keen on creating value through sustainability.

Sustainability criteria for global tourism are a toolbox of rules defined by many associations and researchers. Although most of the criteria studies cover particular areas of tourism as destinations, hotels, and tour operators, various research concerning restaurant sustainability is accelerating (Levy & Duverger, 2010). In this manner, it is essential to review and reveal the relationships with sustainability criteria and the conditions of the restaurant industry.

Sustainable Tourism Indicators

Sustainability is the development that fulfills the needs of today without getting along with the competence of prospective peers to fit their manner of use and commitments (WCED, 1987: 49). Sustainable tourism is a sub-division of sustainable development. Sustainable tourism is a structure that boosts approximate progress, with a concentrate on the measures as the standard of living and the positive outcome that is meaningful for people (Hall et. al., 2015). Such a focus encourages companies to evolve within the finite resources available to society and future generations (Legrand et. al., 2010).

By naming as 'Global Sustainable Tourism Criteria (GSTC)', United Nations established a significant development in this area. Over thirty criteria organized around the practices of the united nations to support tourism sustainability (Bricker & Schultz, 2011; GSTC, 2015). GTSC criteria mainly focus on four dimensions such as; (1) minimizing impact on destinations' environments (waste reduction, re-use, recycling), (2) preservation of local culture and customs, heritage conservation and communities, (3) minimizing negative climate impacts (carbon emissions, etc.), and (4) Maximizing financial benefits for local economy. Even though global sustainable tourism criteria focus on destinations, hotels and tour operators, some of the criteria can be adapted to restaurant industry (Bristow & Jenkins, 2018). Several attempts have been accomplished to fill the gap of indicators needed for restaurant industry. For instance; Green restaurant association managed the development of sustainable restaurant indicators. Lorenzini (1994) determines restaurants specified as green are the currently used or refurbished businesses drafted, established, conducted and annihilatedwith the purpose of environmentally attentive and energy-saving in practice. The central spotlight of a green restaurant differs from traditional restaurants in five essential dimensions, such as: Reduce, Reuse, Recycle, Efficiency, and Energy (Hu et. al., 2010; Bandyopadhyay & Munjal, 2014).

Apart from associations, a few researchers conducted studies to clarify the sustainability in restaurant industry. Even though sustainability studies spotlighted on four leading restaurant indicators as food, water, energy and waste (Freeman, 2011), Legrand et. al., (2010) proposed seven performance indicators that revolve around the concept of the sustainable restaurant. With the standard of the living scope of restaurant sustainability, Cavagnaro (2015), proposed five measures as culture, health, nature, quality, and profit. On the other hand Teng et. al., (2014) support the view of guest intention factors on determining the sustainability of restaurants. Baldwin et. al., (2009) recommend that environmentally friendly and accredited products are the most prominent part of restaurant sustainability. Moreover, Jenkins & Bristow (2013) stated that creativity, management, and service attributes, food sourcing and preparations shall be the factors recognized through the successful management of restaurant sustainability.

Turkish Restaurant Industry and Sustainability

The Turkish restaurant industry can be classified into two categories within the tourism network: Coastal tourism-based restaurants and City tourism-based restaurants. Coastal tourism has a seasonality effect which creates overloading to destination capacity during summer months (Corluka et. al., 2016: 73). The intensity of tourist combined with residents generates many obstacles to beachfront restaurants including environmental sustainability, socio-economic stability and restaurant-quality (Yamkovaya et. al., 2019: 3). On the other hand, City restaurants distressed by the conditions of cities, dense population, pollution, living stress, environmental deterioration and so on (Mori & Christodoulou, 2012). Such conditions influence restaurant industry to reach natural resources such as local food (Duram & Crawley, 2012) and factors turning around it as carbon footprint, waste management, innovative water, and energy production (Freeman, 2011; Hall & Gössling, 2013; Sharma, 2014).

Turkish tourism literature has put shallowed attention to the sustainability criteria within the subject of the restaurant industry. The studies conducted in the area focused variety of subjects including green restaurants, waste management, local food, innovation, and development. In a more detailed study, Yazıcıoğlu et. al., (2018) suggested that sustainable food practices focus on energy saving, water saving, waste management, and chemical reduction. Even though the managers would like to implement such practices, there is a concern about knowledge and costs (Kurnaz & Özdoğan, 2017). This apprehension and the pressure from the customers let the restaurants show their interest in sustainability providing more efficient use of restaurant resources (Atay et. al., 2013; Mesci, 2014). Restaurants usually show their certificates to green practices and sustainability on the web sites; however, they do not explain clearly about their practices and developments linking with global tourism sustainability criteria (Ertas et. al., 2018; Şahingöz & Güleç, 2019).

A few researchers have identified the problem of food waste in Turkish restaurants. According to a study comparing the restaurant's sustainability practices between Rome and Alanya, food waste seemed to be a massive issue in restaurants (Doğan et. al., 2015). Almost same results have been reached by Tekin & Ilyasov (2017), verifying that Turkish guests had a very high waste of food and this varies according to personal and demographic factors. Özdemir & Güçer (2018), concluded that such high food waste rate is a result of the perception and attitudes of the guests.

There is a considerable amount of literature regarding the role of local food in the sustainability of the restaurant business. Kargiglioğlu & Ayyıldız (2018), found that most family restaurants serve local food in a destination. Restaurants operating in coastal tourism destinations have problems reaching the local food because the supply side within the local producers is weak and the relationships among retail stakeholders are not very active (Eren, 2017). On the other hand restaurants serve in rural tourism destinations have a strong tendency to search and use the local delicacies with the heritage-based dishes (Erdem et. al., 2018). Business owners give importance to local foods in terms of cost, demand, and necessity. Besides, food and supplier-related issues are useful in regional food purchases (Erdem et. al., 2017; Güripek & Usta, 2018).

A growing body of sustainability literature about restaurants concentrated on the topics of innovation and development. For instance Birdir & Kale (2014) remind that restaurant businesses apply innovative services in terms of food quality, service innovations, and environmental awareness. In some of the practices of goods/services, process, marketing, and social responsibility innovation, restaurants were found to keep pace with innovation (Albayrak, 2017). For example in the process of a new product development named 'living kitchen', sustainability practices affect the audience in a very positive way, and they had very positive thoughts about the contributions of this new kitchen flow. Similarly, Eren (2018) found that agricultural production systems can be created with permaculture designs of different sizes to be installed in the restaurant or garden. As appropriate inventory levels are identified, these systems will operate sustainably.

Conclusion

Sustainable tourism is a structure that boosts approximate progress, with a concentrate on the measures as the standard of living and the positive outcome that is meaningful for people (Hall et. al., 2015). Such a focus encourages companies to evolve within the finite resources available to society and future generations (Legrand et. al., 2010).

There are a few criteria related to tourism and restaurant sustainability in general. Among them, GTSC criteria and Green Restaurant Criteria are the most important documents which correspond to the sustainable practices in Turkish restaurants in a partial manner. Turkish restaurant studies have focused more on the variety of subjects including green restaurants, waste management, local food, innovation, and development. Among the studies' most significant findings demonstrate that most of the restaurant managers have a willingness to adapt their business to sustainability criteria, biggest obstacle is the 'lack of knowledge' concern (Sünnetçioğlu & Yılmaz, 2015). Most of the restaurant managers think that sustainable practices create extra costs to the operations but when the project partners inform them about the benefits and costs, managers agree to follow the sustainability criteria (Yazıcıoğlu & Yılmaz, 2018).

Although there are studies compatible with the documents of sustainable criteria, some subjects of sustainability, as stated in both criteria in the Turkish restaurant literature, are missing. For instance, the relationship of cultural heritage and culinary practices, the sustainability of traditional food production, biodiversity and sustainability in Turkish restaurants are the emerging research that should be investigated as crucial points of convergence. Further studies should target to create and develop more restaurant based general sustainability criteria. They may carry out more research on the correlations between culture and sustainability within the restaurant industry.

References

Albayrak, A. (2017). Restoran işletmelerinin yenilik uygulama durumları: İstanbul'daki birinci sınıf restoranlar üzerine bir çalışma. Journal of Tourism and Gastronomy Studies, 5(3), 53-73.

Atay, L. ve Dilek, S. E. (2013).„Konaklama İşletmelerinde Yeşil Pazarlama Uygulamaları: Ibis Otel Örneği", Süleyman Demirel Üniversitesi İktisadi ve İdari Bilimler Fakültesi Dergisi, 18(1), ss. 203-219.

Baldwin, C., Wilberforce, N., & Kapur, A. (2009). Restaurant and food service life cycle assessment and development of a sustainability standard. International Journal of Life Cycle Assessment, 13(1), 1-10.

Bandyopadhyay, R. & Munjal, S. (2014). Sustainable Restaurant: Current Perspectives and Way Forward. In Managing Sustainability in the Hospitality and Tourism Industry (pp. 417-460). Oakville: Apple Academic Press.

Birdir, S., & Kale, E. (2014). Restoran işletmelerinde yenilik uygulamaları: Mersin ve Adana örneği. Seyahat ve Otel İşletmeciliği Dergisi, 11(3), 57-72.

Bricker, K. S. & Schultz, J. (2011). Sustainable tourism in the USA: A comparative look at the Global Sustainable Tourism criteria. *Tourism Recreation Research*, 36(3), 215-229.

Bristow, R. S. & Jenkins, I. (2018). Restaurant assessment of local food and the Global Sustainable Tourism criteria. *European Journal of Tourism Research*, 18, 120-132.

Cavagnaro, E. (2015). Sustainable restaurant concepts focus on F & B. In P. Sloan, W. Legrand, and C. Hindley (Eds.), The Routledge handbook of sustainable food and gastronomy (pp. 245-252). London, New York: Routledge.

Corluka, G., Mikinac, K., & Milenkovska, A. (2016). Classification of tourist season in coastal tourism. *UTMS Journal of Economics*, 7(1), 71-83.

Culha, O., & Dagkiran, S. (2016). Restoran işletmelerinde üst düzey çalışanlar açısından yöresel yiyecekler: faydalar, engeller ve satın alma ölçütleri. *Anatolia: Turizm Araştırmaları Dergisi*, 27(2), 195-212.

Doğan, H., Nebioğlu, O. & Demirağ, M. (2015). A comparative study for green management practices in Rome and Alanya restaurants from managerial perspectives. *Journal of Tourism and Gastronomy Study*, 3(2), 3-11.

Duram, L. & Cawley, M. (2012). Irish chefs and restaurants in the geography of local food value chains. *The Open Geography Journal*, 5, 16-25.

Erdem, B. & Akyürek, S. (2017). Yeni Bir Mutfak Akımı: Yaşayan Mutfaklar. *Journal of Tourism and Gastronomy Studies*, 5(2), 103-126.

Erdem, O., Bayram, F., Ciftci, B., & Kemer, A. K. (2018). Mutfak Şeflerinin Yöresel Mantarları Tanıma ve Kullanım Durumlarına İlişkin Keşifsel Bir Araştırma. *Journal of Tourism and Gastronomy Studies*, 6/Special issue3, 225-239.

Erdem, O., Mizrak, M. & Kemer, A. K. (2017). Yöresel yemeklerin bölge restoranlarında kullanılma durumu: Mengen örneği. *Uluslararası Türk Dünyası Turizm Araştırmaları Dergisi*, 3(1), 44-61.

Eren, S. (2017). The Consumption of Local Food in Restaurants: A Study in Kas. *Uluslararası Güncel Turizm Araştırmaları Dergisi*, 1(2), 55-64.

Eren, S. (2018). Ekolojik Restoranlar ve Perma-Kültür Uygulamaları: Ekbiçyeiç Restoranı Üzerine Bir Araştırma. *Güncel Turizm Araştırmaları Dergisi*, 2(Ek1), 534-552.

Ertas, M., Kirlar Can, B., Yesilyurt, H. & Kocak, N. (2018). Konaklama İşletmelerinin Yeşil Yıldız Uygulamaları Kapsamında Çevre Duyarlılığının Değerlendirilmesi, *Seyahat ve Otel İşletmeciliği Dergisi*, 15(1):102-119.

Freeman, E. M. (2011). Restaurant industry sustainability: barriers and solutions to sustainable practice indicators. Unpublished Masters Thesis, Arizona State University.

Gstccouncil (2015). Global Sustainable Tourism Criteria, available online: (accessed on 27.10.2019), https://www.gstcouncil.org/wpcontent/uploads/2015/11/GSTCIndustry_Needs_TOR_and_Development_Process_v1_1_11-12-2015.pdf.

Güripek, E. & Usta, O. (2018). Turizm Destinasyonlarının Rekabet Gücünün Artırılmasında Stratejik Destinasyon Yönetimi: Çeşme Alaçatı Destinasyonu Üzerine Bir Uygulama. *Journal of Tourism and Gastronomy Studies*, 6(4), 496-523.

Hall, C. M. & Gossling, S. (Eds.). (2013). *Sustainable culinary systems: Local foods, innovation, and tourism & hospitality*. Abingdon: Routledge.

Hall, C. M., Gossling, S. & Scott, D. (Eds.). (2015). *The Routledge handbook of tourism and sustainability*. Abingdon: Routledge.

Hu, H. H., Parsa, H. G. & Self, J. (2010). The Dynamics of Green Restaurant Patronage. *Cornell Hospitality Quarterly*, 51(3), 344-36.

Jenkins, I. & Bristow, R. S. (2013). Global Sustainable Tourism Criteria: An International Perspective on Restaurant Sustainability Model Development. *Innovative Business Practices: Prevailing a Turbulent Era*, chp. 16. 371-392.

Kargiglioglu, O. G. S. & Ayyıldız, O. G. S. (2018). Mutfak Kültürünün Sürdürülebilirliği Bakımından Yöresel Yiyeceklerin Menülerde Yer Alma Düzeyi: Sinop Ölçeğinde Bir Araştırma, *Asos Journal*, 6(86), 345-355.

Koroglu, O., Buzlukcu, C., Ulusoy Yildirim, H. & Oflaz, M. (2019). Ekolojik Tarım Turizm Faaliyetlerine Katılan Ziyaretçilerin Ekolojik Çiftliklere Yönelik Beklenti ve Algılarının Tespit Edilmesi. *Journal of Tourism and Gastronomy Studies*, 7(1), 25–45.

Kurnaz, A., & Özdoğan, O. N. (2017). İstanbul'da Yer Alan Yeşil Restoran İşletmeleri Hizmet Kalitesinin Grserv Modeli İle Değerlendirilmesi. *Dokuz Eylül Üniversitesi İşletme Fakültesi Dergisi*, 18(1), 75–99.

Legrand, W., Sloan, P., Simons-Kaufmann, C. & Fleischer, C. (2010). A review of sustainable restaurant indicators. *Advances in Hospitality and Leisure*, 6, 167–183. DOI: 10.1108/ S1745-3542(2010)0000006013.

Levy, S. E. & Duverger, P. (2010). Consumer perceptions of sustainability in the lodging industry: Examination of sustainable tourism criteria. In Proceedings of the International CHRIE Conference-Refereed Track, Event 31, Amherst, MA, USA, 28–31 July; pp. 2–8.

Lorenzini, B. (1994). The green restaurant, part II: Systems and service. *Restaurant & Institutions*, 104(11), 119–136.

Mesci, Z. (2014). Otellerin çevreci uygulamalarının değerlendirilmesi: yeşil yıldızlı bir otel işletmesinde örnek olay çalışması. *Seyahat ve Otel İşletmeciliği Dergisi*, 11(1).

Mori, K. & Christodoulou, A. (2012). Review of sustainability indices and indicators: Towards a new City Sustainability Index (CSI). *Environmental Impact Assessment Review*, 32(1), 94–106.

Özdemir, G. & Güçer, E. (2018). Food Waste Management within Sustainability Perspective: A Study on Five Star Chain Hotels. *Journal of Tourism and Gastronomy Studies*, 6, 280–299.

Özkan, C. & Aydın, S. (2018). Yerel Yiyecekler Aracılığı İle Sürdürülebilir Destinasyonlar: Ayvacık Örneği. *Journal of Tourism and Gastronomy Studies*, 335, 349.

Şahingöz, S. A. & Güleç, E. (2019). Restoran mutfaklarında yeşil nesil restoran hareketi: "La Mancha Restoran" örneği. *Journal of Tourism Theory and Research*, 5(2), 292–300.

Searcy, C., Karapetrovic, S., & McCartney, D. (2005). Designing sustainable development indicator: Analysis for a case utility. Measuring Business Excellence, 9(2), 33–41.

Selvi, M., & Demirer, D. (2012). Ekolojik Tatil Çiftliklerinin TATUTA Projesi Deneyimine İlişkin Örnek Olay İncelemesi. *Anatolia: Turizm Araştırmaları Dergisi*, 23(2), 187–202.

Sharma, S. (2014). *Sustainable Culinary Practices*. In: Jauharı, Vinnie. Managing Sustainability in the Hospitality and Tourism Industry. Oakville: Apple Academic Press, pp. 303–334. ISBN 978-1-926895-72-7.

Sünnetçioğlu, S. & Yılmaz, B. S. (2015). İzmir'deki restoran yöneticilerinin sürdürülebilir restoran işletmeciliği üzerine yaklaşımlarının değerlendirilmesi. *Karabük Üniversitesi Sosyal Bilimler Enstitüsü Dergisi*, 5(1), 94–114.

Tekin, O. A. & Ilyasov, A. (2017). The Food Waste in Five-Star Hotels: A Study on Turkish Guests' Attitudes. *Journal of Tourism and Gastronomy Studies*, 13, 31.

Temizkan, R., Temizkan, S. P. & Sever, Y. (2017). Development of Green Kitchen Quality (G-KITCHQUAL) Scale. *Journal of Tourism and Gastronomy Studies*, 3, 16.

Teng, Y. M., Wu, K. S. & Huang, D. M. (2014). The influence of green restaurant decision formation using the VAB model: The effect of environmental concerns upon intent to visit. *Sustainability*, 6, 8736–8755.

Yamkovaya, M., Arcila, M., Cardoso Martins, F. & Izquierdo, A. (2019). Sustainable Development of Coastal Food Services. *Sustainability*, 11(13), 3728.

Yazıcıoğlu, I. & Aydin, A. (2018). Yeşil Restoran Uygulamaları Üzerine Nitel Bir Araştırma: İstanbul Örneği. *Gazi Üniversitesi Turizm Fakültesi Dergisi*, 21(1), 55–79.

Yazıcıoğlu, I., Ozata, E. & Yaris, A. (2018). Sürdürülebilir yiyecek ve içecek işletmeciliği: Ankara İlinde bir araştırma, *Journal of Tourism and Gastronomy Studies*, 6(2), 350–368.

Uğur CEYLAN
21 City Tourism and Kütahya

Introduction

In 2016, approximately 55 % of the world's population lived in cities. This ratio is expected to rise to 66 % by 2050. In parallel with this increasing urbanization, cities often compete to increase the number of overnight stays and become increasingly tourist attractions. City tourism has gained popularity in recent years, especially in Europe, where overnight stays have increased by 14.2 % between 2012 and 2016. This trend is reflected in other statistics. For example, there was a sharp increase in the number of overnight stays between 2012 and 2014, especially in urban environments (up to + 10.0 %). This increased interest is mainly due to increased and improved flight connections, increases in consumers leisure time, changes in tourists' holiday preferences, and an increase in weekend trips (Boivin & Tanguay, 2019: 67).

In recent years, tourists are moving away from mass tourism and prefer alternative tourism types instead. The primary reason for this is the changes in consumers' lifestyles and behaviors. This change led the tourists to move for a short time for many purposes such as social, cultural, artistic etc. Tourists visit many cities around the world for the purpose of seeing the architecture, recognizing the socio-cultural structure of the local people, visiting museums, participating in activities such as concerts and theaters, shopping, participating in sports activities. During these activities, they spend money in the city they visit and contribute to the economies of the cities (Şarkaya İçellioğlu, 2014: 38).

The perishable nature of tourism products increases the importance of future planning and forecasting. Loss of revenue from an unsold hotel room, the number of tourists expected to a destination, the preparation of substructure and superstructures according to the expected number of tourists, making adequate hotel investments for this and employing staff. Therefore, planning for the future and predicting what will happen in the future is crucial to the success of the entire tourism industry. Long-term and short-term forecasts are both important for different managerial purposes. Long-term forecasts for tourism demand for the next few years may help infrastructure planning a destination, while short-term forecasting of demand forecast over the next two or three months can help the destination have greater operational flexibility. For example, the number of buses from the airport to the city center (Frechtling, 2001).

Kütahya is a city that stands out with its historical, cultural and social elements but it is not a city that has demonstrated its potential until today. Although the cities around which it has competed within the scope of city tourism (Eskişehir, Afyon etc.) have made significant progress and investments in recent years, the city of Kütahya could not make a difference in this point. In this section, city tourism potential of Kütahya city center is examined. Strengths and weaknesses in terms of internal dynamics of Kütahya province and opportunities and threats in terms of external dynamics were tried to be determined.

City Tourism and Kütahya

City tourism is regarded as an important form of tourism to promote local economic and cultural development of the urban landscape. Despite its importance, city tourism is not clearly defined and continues without a systematic understanding structure (Ashworth & Page, 2011). This is due in part to a wide range of tourism-related activities in an urban environment ranging from visiting friends and relatives, attending cultural events and visitor attractions, going to the beach, shopping, living the nightlife and studying in the city environment, and leisure activities related to it (Law, 1994). Subsequently, there is a lack of recognition of the possible relationships and interactions between tourism and the multifaceted presence of the city. The main focus of city tourism, cities and large cities, are large multifunctional enterprises that can absorb a wide variety of tourist types and as a result undergo many changes to meet the relevant demands of the market (Page & Hall, 2003).

The attractiveness of a city is not limited to heritage ideas. Instead, action should be taken to utilize long-term planning and sustainable resources for tourism. As a matter of fact, cities are trying to improve their facilities as a means of tourism development; pedestrian areas, specially themed areas and planned environment. In addition, some city features may affect the appearance of the destination positively or negatively. For example, transport potential for tourists, signs, green / public spaces could potentially be satisfactory (Mikulić et. al., 2016: 75). For tourists, street noise, street litter and insufficient transport network are known to be serious threats to regional tourism competitiveness. The environmental quality of the city and the accessibility of a city seem to be important factors that attract tourists (Provenzano, 2015: 443). Managing city tourism through strategic planning can contribute positively to the social and economic life of a destination by alleviating negative externalities, focusing on investments involving multiple players for many years and providing a co-ordination mechanism to communicate common vision for the future. (Van der Berg & Russo, 2004).

The requirements of city tourism in terms of demand are listed as follows: (Emekli, 2011: 33-34):

- City tourism is a type of tourism that has attracted the attention of tourists of almost all ages, including young tourists.
- Third age tourist groups participate in city tourism and related cultural activities and spend a lot during city tours.
- Culture is an important factor that motivates tourists in city tourism. Tourists also want to experience the cultures they want to get to know during their trips. Therefore, tourists attach great importance to the texture and atmosphere of the city. When planning museums, historical buildings, shopping centers, events, the city's natural structure and should not be spoiled and authenticity should be preserved as much as possible.
- The impact of internet and social networks in city tourism is increasing day by day. In addition, friend-relative advice is still important.
- The frequency and number of tourists traveling is constantly increasing. Especially with weekend tours, tourists participate in city tourism and want to satisfy different interests.
- Environmental awareness is one of the important elements observed in tourists in recent years. Tourism enterprises also support environmental awareness with green practices. This support also affects the preferences of tourists. Tourists enjoy the use of recyclable and energy-saving products in hotels. Hotel managers state that many tourists are willing to participate directly in the hotel's recycling efforts.

Law (1996) mentions some of the characteristics of tourist destinations. In big cities, people visit more relatives and friends to avoid population density. Tourists visit more places of interest due to the development of large cities. Airports and scheduled shuttles are easily accessible when traveling to big cities. There are many accommodation facilities built to serve business travelers. Big city destinations can cater to a range of different tourist markets while providing communication, transport, services and facilities to meet tourist needs. These markets include a more educated population attracted to the cultural heritage of cities and towns, elderly people with a greater interest in cultural and historical heritage, young people engaged in the excitement of the city urban environment (entertainment, nightlife and sports events etc.), business travelers, meetings and fairs. It is possible to develop various parallel and common forms of tourism in the city area. The main reasons for visiting the city at leisure times are: (Hall & Page, 2006):

- Participation in cultural and sports activities,
- Participation in religious ceremonies,
- Entertainment,
- Shopping,
- Personal work
- Visiting family and friends,
- Participation in training and business meetings,
- Fairs and conferences.
- Travel

City tourism is a destination as a whole. Although not all destinations have favorable conditions for city tourism, many destinations have favorable conditions for city tourism. A significant number of tourists in urban areas visit for a primary purpose apart from leisure time, including business, conferences, shopping and visiting friends and relatives. Local people are important users of attraction centers and infrastructures, which are generally developed for non-tourism purposes. While a destination offers different alternatives to different tourists in the field of tourism, it also competes with other sectors. During this competition, investments in the city should be planned together with both tourism and other sectors in mind. For this reason, changing expectations of tourism should be taken into consideration when planning and policy making in destinations. All public and private sector managers should be involved in this process. Finally, in urban areas, there is a complex restriction on development; natural environmental factors are often less prominent and cultural heritage and housing factors are more important than other types of tourism. (Edwards et. al., 2008: 1033). Çeken (2008) explained the factors that should be considered in order to develop a destination in terms of tourism and increase the level of development as follows:

- The cultural values of the tourism region (archaeological monuments, folklore and festivals) should be analyzed,
- A preliminary analysis of the region's infrastructure (climate, natural coverings and land suitable for tourism) should be carried out,
- The current facilities (hotels, motels and pensions) in the region and the availability of additional investments should be analyzed,
- Businesses in the tourism region (businesses that offer direct / indirect private services; businesses that producing and marketing private goods),
- After realistically analyzing factors such as the perception of local people in the region in a tourism region, the impact of tourism on regional development will also increase.

The most important area of tourism activities in cities is generally central areas and often includes historical areas. The positive elements are the main functions of the city, which include commercial and administrative facilities, as well as architectural monuments and cultural facilities concentrated in these areas, and well-developed food, transport and accommodation services. The location of the tourist areas is also related to the distribution of hotel facilities in these areas. (Lapko, 2014: 208). In addition to these factors, it is possible to classify the factors affecting the city tourism as follows (Manente, 2005):

- Media; It is important for the promotion and marketing of the city's domestic and foreign tourism. For example, a successful television program, commercial or a film in the city,
- In terms of city tourism, ethnicity, cultural elements and niche markets can make the city more advantageous than other cities,
- There is increasing interest in traditional food culture, slow city and slow food practices. Italy and France stand out in this regard.
- Cool capitals, where design stands out, have gained importance in recent years. In particular, hotels, architectural structures, art, fashion and food are the main elements of attraction in the perception of cities. There is growing interest in well-known architectural structures, such as the Guggenheim Museum in Bilbao.
- Interest in local cultural activities and festivals is increasing, and the number of such activities is increasing in proportion to this.

When the domestic literature is examined, it is seen that many studies have been done on city tourism. Tutal Cheviron (2008) explores how İstanbul is transforming with the developments in tourism, culture and congress sectors. Başarangil (2014), has examined the phenomenon of city tourism in the province of Kırklareli and put forward the requirements for structuring sustainable city tourism. Zeğerek & Ortaçeşme (2017) investigated the contribution of green areas to city tourism in Antalya. Çiftçi & Çolak (2017) examined the image of Mardin within the scope of city tourism.

In 2017, Kütahya was awarded the title of 'Creative City' by joining the UNESCO Creative Cities Network consisting of literature, music, design, gastronomy, cinema/film, craft, and folk and media arts. This award was given in the field of 'Craft and Folk Arts'. Adopted to the UNESCO Creative Cities Network, Kütahya is registered as one of the 37 Craft and Folk Art cities in the world (Governorship of Kütahya, 2017). Kütahya has many elements that will attract tourists in the tourism area. At the point of accessibility, almost all roads are connected to neighboring cities, and also have a railway network and an airport.

Table 1: Overnight Statistics of Tourism and Municipality Certified Enterprises

	2014	2015	2016	2017	2018	2019 (January–June)
Overnight Statistics	448.234	435.974	440.793	434.086	443.382	241.261

Source: Kütahya Provincial Directorate of Culture and Tourism

Table 2: Number of Hotels and Beds in Kütahya (June, 2019)

Enterprisess	Number	Number of Rooms	Number of Beds
Tourism Operation Certificated	11	581	1.167
Tourism Investment Certificated	5	366	910
Municipality Certificated	65	2.239	6.389
Total	81	3.186	8.466

Source: Kütahya Provincial Directorate of Culture and Tourism

It also has enough facilities to accommodate tourists coming to the city. The current tourism potential of Kütahya is given below:

Kütahya was visited by 241,261 tourists in the first six months of 2019. When we examine the number of tourists since 2014, there is not a significant increase or decrease in the number of tourists. This means the number of tourists going on their own course in the last five years (www.kutahyakultur.gov.tr).

There are 81 hotels operating in Kütahya with tourism, tourism investment and municipality certificated. The total number of rooms in these hotels is 3186 and the total number of beds is 8466. These data show that Kütahya can accommodate approximately three million people throughout the year (www.kutahyakultur.gov.tr).

The number of visitors of the three museums in the city center of Kütahya is given in Table 3. They were visited by 22,768 tourists in the first six months of 2019. In addition to these museums, other private museums in the city center under the supervision of the Museum Directorate;

- Kütahya Municipality City History Museum
- Kütahya Municipality Geological Museum
- Kütahya Municipality Sıtkı Olçar Tile Museum

- Dumlupınar University Museum
- Tugay Anatolian Culture Art and Archeology Museum

Table 3: Number of Visitors of Museums in Kütahya City Center (January–June 2019)

Museums of the Museum Directorate	2019 (January–June) Number of Visitors
Archaeological Museum	10.988
Tile Museum	3.453
Kossuth Museum	8.327
Total	22.768

Source: Kütahya Provincial Directorate of Culture and Tourism

Evaluation of Kütahya City Center Within the Scope of City Tourism

SWOT analysis is one of the basic methods used in the case analysis. The interactions of institutions and organizations with their environment are systematically examined by SWOT analysis. In this respect, internal strengths and weaknesses of institutions and organizations are analyzed and external opportunities and threats are analyzed. In this way, it helps organizations or businesses to develop plans and strategies for the future. The method to be determined in SWOT analysis should be handled in a very comprehensive and versatile manner (Çoban & Karakaya, 2010: 348). In this context, the city center of Kütahya will be examined by SWOT analysis within the scope of city tourism.

Strengths

- Kütahya province is located in the east of Aegean Region and it is in the transition point in the east-west and north-south direction. In particular, there is a crossing point of the İstanbul-Antalya and Izmir-Ankara highway routes (Taşkın & Şener, 2013: 255).
- Kütahya has been home to many civilizations throughout history. The province of Kütahya was home to Phrygian, Roman, Seljuk and Ottoman states. Bears cultural traces from all periods. Phrygian Valley, Aizanoi Ancient City, Kütahya Castle, Germiyan Street and so on. This cultural wealth is an attraction element for tourists (Taşkın et. al., 2014: 144).

- Yoncalı is an important center in terms of thermal tourism and has social, cultural and historical values that can be used for the leisure time of the tourists coming here. Thus, the rich historical and cultural opportunities of Kütahya province will be provided to tourism through thermal tourism. Cultural and historical heritage, which is not prominent here, will be evaluated through thermal tourism (Önder et. al., 2010: 53).
- There are many historical places in the city center of Kütahya that will attract tourists to the city such as Kütahya Castle, Germiyan Street, Kossuth Museum, Ulucami (Taşkın et. al., 2014: 144).
- The presence of both road, air and rail transportation systems in the city center of Kütahya makes it an easily accessible city for tourists.
- The climatic conditions of Kütahya are suitable for tourists for four seasons. Particularly in the summer, the city does not experience a lot of heat and will allow tourists to visit the city easily.
- The hospitable nature of the public will ensure that tourists leave satisfied. In this way, the promotion of the city will increase in proportion to this.
- Handicraft items which have an important place in our culture such as tiling, ceramics and manuscripts can be offered to tourists as an alternative tourism product (Taşkın & Şener, 2013: 255).
- Zafer Airport is located in Kütahya. Especially the fact that the airport is located in Kütahya is important in the competition with Afyon. In the competition in thermal tourism, having air transport is an important element.
- Tourists coming to the city center of Kütahya will not be trapped inside the hotel and in the city center by including a wide package tour together with the tourism elements in the districts (Aizanoi Ancient City, Phrygian Valley, Dumlupınar Martyrdom etc.). In this way, tourists can both extend their stay and have more enjoyable time

Weaknesses

- The city center of Kütahya does not have a proper city plan. As in developed cities, there is no broad streets and alleys parallel to each other and planned geometric city planning (Taşkın & Şener, 2013: 255).
- Unhealthy structures (slums, etc.) in certain regions that disrupt the general appearance of the city caused distorted urbanization and distorted the image of the city.
- The buildings constructed by the construction companies without considering the architectural features of the city aesthetically distort the image of the city and do not reveal a good image.

- Art activities are important for both locals and tourists visiting the city. Cities such as Milan, Vienna and Istanbul are visited by tourists only for artistic activities. In Kütahya, venues for art activities should be prepared in this direction and art activities should be organized in such a way that public participation can be ensured. Ensuring that art is encouraged, developed and respected will be an important tool both for the quality of life of the locals living in the city and for attracting tourists.
- Kütahya province has many historical and touristic values dating back to ancient times. First of all, these values should be transferred to future generations such as cultural heritage within the scope of sustainable tourism. Rather than seeing historical and touristic values as income generating elements, they should be adopted as national values. There are people living in Kütahya who have not visited the Kossuth Museum, Aizanoi and Kütahya Castle. These people should also be informed by organizations to be organized by public institutions and these sources of information should be transferred to future generations.
- Kütahya province has many tourist elements such as tiling, thermal tourism and historical sites. Any of these elements could not become the symbol of the city. For example, the Vase (Vazo) in the center of the city is generally known to tourists visiting the city. The Vase can be given a larger and more specific style, symbolizing the city like the Eiffel tower in Paris.
- Due to the lack of advertising and promotion activities of Kütahya province, it is still unknown by many target groups. Increasing the promotion and marketing activities will increase the duration and expenditure of tourists in the city (Taşkın & Şener, 2013: 255).
- There are insufficient signs, documents and offices to guide tourists in the city. Implementation of this issue will increase the promotion of the province.
- There is a lack of qualified staff in the city. Inability to provide good service to tourists coming to the city will also adversely affect the promotion of the province.
- There are no parking spaces for tour buses in the historical and touristic places in the city center (especially in the area of Ulucami, the Dönenler Mosque and the Archaeological Museum).

Opportunities

- On 31 October 2017, Kütahya was granted the title of 'Creative City' of the UNESCO Creative Cities Network. As a result, Kütahya has become one

of the 180 cities in the world that is accepted a creative city by UNESCO. However, the promotion and marketing of the province in the international arena can be achieved easily.
- The tourists coming to our country from developed countries spend more because of the value of their money. In fact, they receive cheaper products and services in our country than other countries. Tourists prefer regions with a historical and cultural richness, such as Kütahya province, which will less force their budgets. While they will be able to stay less in the cities of developed countries, they will be able to have more holidays in Kütahya with the same budget.
- Urban regeneration is an important issue in terms of building a healthier structure and attracting more tourists.
- Increasing tourism activities, albeit slowly, reveal new investments and create new employment opportunities.
- Local people and city managers are keen on tourism. This will cause tourism to develop faster in the city.
- It is an important opportunity for Kütahya to be located on the Istanbul-Antalya and Ankara-İzmir routes and to complete the roads on this route.
- The province of Kütahya is connected to major cities such as Balıkesir, Ankara, Eskişehir and İzmir by railway.
- Culinary culture that has been going on since the Seljuk and Ottoman periods (Göveç, cimcik, goose rag, etc.) still continues. Tourists coming to the city will be able to see the richness of the culinary culture of the province. This wealth will enable the city to develop in gastronomic tourism.

Threats

- With the acceleration of tourism activities and the increase in the number of tourists visiting the city, the infrastructure of the city will be insufficient. If infrastructure services are not increased in parallel with the development of tourism, problems such as traffic density, environment and noise pollution will arise within the city. This will have a negative impact on the quality of life of both local people and tourists. Therefore, city managers should take precautions against these problems that may arise in the future.
- There are not enough qualified personnel trained in the field of tourism. In particular, there are not enough staff or tourist guides in the city. This may create dissatisfaction for the tourists visiting the city and may have negative effects on the promotion of the city (Taşkın & Şener, 2013: 255).

- Terrorism, social events and problems with neighboring countries in our country may make it difficult for tourists to prefer our country even if these events occur in certain regions. Tourists do not prefer such regions and countries unless they feel safe. Although such promotion and marketing activities are carried out, it does not affect tourists' preferences (Morgan, 2012).
- Kütahya is in competition with the neighboring provinces of Afyon, Eskişehir and Uşak. Tourism activities and strategies that may make a difference from these provinces should be put forward. Especially Afyon and Eskişehir are among the primary competitors of Kütahya province in competition.
- A significant level of financing is needed for the development of tourism activities in the province of Kütahya. Improvement of infrastructure, restoration of historical mansions and increasing service quality require significant budgets. Failure to meet these budgets or lack of adequate support from the private sector will result in the failure to implement the plans and strategies to be introduced.
- Not planning the city in an environmentally sensitive way, destroying the green areas and not making the restorations in accordance with the original will harm the historical, cultural and natural environment in the long term even if it is not noticed in the short term. Building areas where people will breathe, preserving nature and carrying cultural and historical monuments to future generations should be identified as priorities in planning (Beritelli et. al., 2013).
- There are many historical mansions in the province of Kütahya dating from the Seljuk and Ottoman periods. These mansions must be restored by qualified personnel. Poor restorations will disrupt the original state of the building and prevent the transfer of cultural assets to the future in a healthy way.
- Eskişehir and Afyon provinces show important developments in the field of city tourism. The fact that tourists prefer these provinces as a priority in this regard will put Kütahya in the second place in the competition.

Results

City tourism in the world and in Turkey in recent years has emerged as an increasingly important forms of tourism. While cities like Paris, London, New York and Amsterdam have become prominent in the world in terms of city tourism, in Turkey the cities such as İstanbul, Ankara, İzmir and Antalya are also made improvements in urban tourism. Historical and touristic attractions, traditions and festivals in the city centers are extensively used to attract tourists to the city.

In this section, the swot analysis of historical, touristic and cultural attraction elements located in the city center of Kütahya province was analyzed.

The city of Kütahya is in competition with the neighboring provinces such as Eskişehir and Afyon. While investments in these two cities continue rapidly, there is a slower progress in the province of Kütahya. The tourists coming to this region primarily visit these two provinces and according to their length of stay they visit Kütahya. It is known that mobility from Kütahya province to these two cities is quite high especially on weekends. In this respect, it is necessary to establish areas where local people can spend time in Kütahya. Increasing the number of social areas such as theme parks and activity areas emerges as priority planning.

Considered as a creative city by UNESCO, Kütahya should produce touristic products that will attract and increase the interest of tourists in this area. In the field of crafts and folk arts, the city should prioritize the persons and elements that symbolize the city and should carry out promotional and marketing activities on these issues. Craftsmen and folk artists trained by the city should be promoted worldwide and exhibitions should be organized for their works in certain countries. In addition, one of the important arts of the province of Kütahya tiles in this context should be developed and introduced.

The fact that Kütahya is located on the Ankara-İzmir and İstanbul-Antalya routes is an important advantage for the province. In the summer months, it is an important opportunity for millions of people to travel on this line. Building areas that may be of interest to people traveling through the city, placing promotional signs along the way will guide people to both stop and visit the city. With these arrangements, the number of tourists visiting the province is expected to increase significantly. Especially the number of daily tourists will increase by this way.

The infrastructure and superstructures of Kütahya city center should be improved by the related public institutions. The high traffic density in the city center and the serious parking problem is an important problem for the tourists who will visit the city. The fact that there is no parking place for tour buses and cars on the square of Ulucami, Archaeological Museum, the Dönenler Mosque, the old mansions and especially Germiyan Street is a problem for tourists. Traffic arrangements should be made in these areas, new parking areas should be created and construction should not be allowed. Arrangements to be made in these areas will allow tourists visit in a spacious way and easily.

Culinary culture of the province of Kütahya is one of the elements that can attract many tourists to the city. The food culture, which has been going on since the Seljuk and Ottoman periods, can be offered to tourists in special places (mansions). Although a few mansions are already trying to continue this work, they do not receive enough attention due to lack of promotion and marketing. Providing the promotion of the local food of Kütahya in gastronomic activities will increase the number of tourists coming to the city center in this area to a certain extent.

The tourists coming to the city center of Kütahya can be directed to the touristic products in the districts with various tours. In this way they will have a better time and increase their overnight stay in the city. Touristic products such as Aizanoi Ancient City, Phrygian Valley, Dumlupınar Martyrdom, Domaniç Plateaus will be integrated with city tourism.

The managers of Kütahya province should make the necessary investments simultaneously with the development of city tourism in the coming years. They should minimize the problems that will arise with the growth of the city. In line with this growth, they should provide the necessary support to encourage the hotel investments that the city needs. The infrastructure and superstructure investments that the city will need should be made immediately and these investments should be arranged in a way to ease the city tourism.

Hotel enterprises and travel agencies operating in the field of tourism in the province of Kütahya should support the city in advertising and promotion. In particular, travel agencies should organize package tours to promote city tourism. The addition of charm elements in the districts to the package tours to be organized will extend the overnight stay of tourists in the city. Hotel enterprises should organize activities to ensure that tourists have a good time rather than just meeting their accommodation needs. They should include more local foods of Kütahya province in their menus.

The studies to be carried out by academicians in the following years should measure the expectations of the tourists coming to the city. The results of these researches should be shared with both public institutions and managers of tourism enterprises and plans should be made in line with these expectations. Finally, together with city tourism, it is important to carry out researches to develop other tourism types of Kütahya province.

References

Ashworth, G. & Page, S. J. (2011). Urban Tourism Research: Recent Progress and Current Paradoxes, *Tourism Management*, 32, p. 1–15.

Başarangil, İ. (2014). Kent Turizminin Sürdürülebilir Geleceği: Turizm Potansiyeli Açısından Kırklareli İli'nin Değerlendirilmesi, *Uluslararası Hakemli Pazarlama ve Pazar Araştırmaları Dergisi*, 3/1, p. 82–99.

Beritelli, P., Bieger, T. & Laesser, C. (2013). The New Frontiers of Destination Management: Applying Variable Geometry as a Function-Based Approach, *Journal of Travel Research*, 53/4, p. 403-417.

Boivin, M. & Tanguay, G. A. (2019). Analysis of the Determinants of Urban Tourism Attractiveness: The Case of Québec City and Bordeaux, *Journal of Destination Marketing & Management*, 11, p. 67-79.

Çeken, H. (2008). Turizmin Bölgesel Kalkınmaya Etkisi Üzerine Teorik Bir İnceleme, *Afyon Kocatepe Üniversitesi, İ.İ.B.F. Dergisi*, 10/2, p. 293-306.

Çiftçi, H. & Çolak, O. (2017). Turistik Destinasyonların Pazarlamasında Kent İmajı Algısı: Mardin İli Üzerine Bir Uygulama, *International Journal of Academic Value Studies (Javstudies)*, 3/14, p. 224-236.

Çoban, B. & Karakaya, Y. E. (2010). Geleceği Planlamada Stratejik Yönetim ve Swot Analizi: Kavramsal Yaklaşımlar, *e-Journal of New World Sciences Academy, Social Sciences*, 3C0052, 5/4, p. 342-352.

Edwards, D., Griffin, T. & Hayllar, B. (2008). Urban Tourism Research, Developing an Agenda, *Annals of Tourism Research*, 35, p. 1032-1052.

Emekli, G. (2011). Öğrenen Turizm Bölgeleri, Kentler ve Kent Turizmine Kuramsal Yaklaşım, *Aegean Geographical Journal*, 20/2, p. 27-39.

Frechtling, D. C. (2001). *Forecasting Tourism Demand: Methods and Strategies*. Burlington, MA: Butterworth Heinemann.

Hall, C. M. & Page, S. J. (2006). *The Geography of Tourism and Recreation*. Environment, Place and Space (3rd ed.), London: Taylor & Francis e-Library, p. 248-250.

Lapko, A. (2014). Urban Tourism in Szczecin and Its Impact on the Functioning of the Urban Transport System. *Procedia – Social and Behavioral Sciences*, 151, p. 207-214.

Law, M. C. (1994). *Urban Tourism: Attracting Visitors to Large Cities*. London: Mansell Publishing.

Law, M. C. (1996). *Tourism in Major Cities*. International Thompson Business Press: Routledge.

Manente, M. (2005). New *Paradigms for City Tourism Management, Session 2: Market Opportunities and Competitiveness of the City Tourism Key Challenges for City Tourism Competitiveness*. Turkey: WTO Forum Istanbul, 1-3 June.

Mikulić, J., Krešić, D., Miličević, K., Šerić, M., & Ćurković, B. (2016). Destination Attractiveness Drivers among Urban Hostel Tourists: An Analysis of Frustrators and Delighters, *International Journal of Tourism Research*, 18/1, p. 74-81.

Morgan, N. (2012). Time for 'Mindful' Destination Management and Marketing, *Journal of Destination Marketing & Management*,1(1–2), November 2012, p. 8–9.

Önder, H., Özçelik, Ö. & Odabaşı, Y. (2010). Termal Turizm Bölgelerinde İkincil Konutların Turizme Kazandırılması: Kütahya - Yoncalı Termal Turizm Bölgesi Örneği, *Ticaret ve Turizm Eğitim Fakültesi Dergisi*, 2, p. 40–57.

Page, S. J. & Hall, C. M. (2003). *Managing Urban Tourism*. London: Pearson Education.

Provenzano, D. (2015). A Dynamic Analysis of Tourism Determinants in Sicily, *Tourism Economics*, 21/3, p. 441–454.

Şarkaya İçellioğlu, C. (2014). Kent Turizmi ve Marka Kentler: Turizm Potansiyeli Açısından İstanbul'un Swot Analizi, *İstanbul Üniversitesi Sosyal Bilimler Dergisi*, 1, p. 37–55.

Taşkın, E., Söylemez, C. & Baran, A. (2014). *Turizm Pazarlamasında Üniversite Öğrencilerinin Rolü: Kütahya DPÜ Örneği*, VII. Lisansüstü Turizm Öğrencileri Araştırma Kongresi, 04–05 Nisan 2014, Kuşadası, Aydın, p. 140–146.

Taşkın, E. & Şener, H. Y. (2013). Sağlık Turizm Markası: Kütahya, *Dumlupınar Üniversitesi Sosyal Bilimler Dergisi*, 36, p. 253–260.

Tutal Cheviron, N. (2008). İstanbul'da Kent Turizminin Yeni Biçimleri, *Galatasaray Üniversitesi İletişim Dergisi*, 9, p. 203–223.

Van der Berg, L. & Russo, A. P. (2004). *The Student City: Strategic Planning for Student Communities in EU Cities*. London: Ashgate Publishing.

Zeğerek, P. & OrtaçeşmeV. (2017). Yeşil Alanların Kent Turizmine Katkısının Antalya Örneğinde İncelenmesi, *Mediterranean Agricultural Sciences*, 30/3, p. 205–212.

http://www.kutahya.gov.tr/kutahya-unesco-yaratici-sehirler-aginda

Kütahya Provincial Directorate of Culture and Tourism (2019).www.kutahyakultur.gov.tr

Yeliz PEKERŞEN

22 An Overview of Creative Tourism Concept

Introduction

Creativity is an effective strategy used to reveal the different and individual talents of tourists. This strategy is not only preferred for the promotion of personal talents, but is also regarded as the most important tool for development, globalization and competition acquisition by many destinations in the world (Al Ababneh, 2017: 282). The drastic changes in tourism supply and demand and the creation of experiential environments have led the tourism sector to creativity (Bastenegar et. al., 2018: 541).

The fact that tourism has escaped from its characteristics for the mass and gained individuality (Zoğal & Emekli, 2017: 22) and that creativity has become more and more important for the development of tourism has increased the interest in new and different tourist tendencies (Aşık, 2014: 786). The concept of creative tourism (Altınay & Dinçer, 2017: 344), which includes more education, social, emotional and interactive communication with different cultures, people and destinations is a new generation of tourism enabling the tourists to recognize different cultural endeavors, authentic experiences, discover and experience cultural and local values and offering culture, travel and entertainment together (Gülüm, 2015: 87). Destinations can engage in creative tourism activities to attract more visitors, reinforce their brand image and strengthen their position in local development (Jelinčić & Žuvela, 2012: 78).

Creative tourism is regarded as a 'form or extension' of cultural tourism (Richards & Marques, 2012: 1). However, these two concepts have common as well as different effects. While activities in cultural tourism are mostly based on visuality, production and processing of cultural products is essential in creative tourism (Aşık, 2014: 787). Active participation, real experiences, development of individual creative potential development skills and close relations with local people constitute the main features of creative tourism. These experiences are mostly related to daily life. Traditional handicraft making, clay pot painting, gastronomy, porcelain painting, carpet weaving and dancing are the most active applications of this type of tourism (Tan et. al.,, 2013: 153).

The essence of creative tourism lies in the integration of ideas, science and technology (Tang, 2014: 15642). Today, tourists are looking for more interactive and satisfying experiences (Tan et. al., 2014: 248). It is the responsibility of tourists to learn about the environment in creative tourism and practice this knowledge to improve their skills. In this sense, tourists are actively directed to improve themselves. The 'need for learning' that arises in tourists is often addressed by focusing on local activities. In this way, the host community can both gain and establish positive relationships with the guests (Lemy, 2016: 95).

In this section, the concept of creative tourism is emphasized; its emergence and development process are examined and an attempt has been made to explain creative tourists and their characteristics. The studies in the literature have been examined in detail and evaluations have been made on the types of tourism in which creative tourism activities can be applied, the relationship between sustainable tourism and creative tourism, positive and negative aspects of creative tourism.

Concept of Creativity and Creative Tourism

Creativity means developing new ideas and using them to design scientific discoveries, technological innovations and authentic cultural products (Lindroth et. al., 2007: 54). It is a cognitive skill that enables the emergence of a new product in a unique and talent-based way by using the intelligence of the individual in a unique and productive manner (Aslan, 2016: 20). It provides personal gains to individuals and is regarded as a sign of mental health and emotional well-being (Simonton, 2000: 152). It is also thought that the term creativity attracts consumers and creates a cool association (Ali et. al., 2016: 87). Sacco & Segre (2009: 2) described creativity as the process of redirecting the value-added forms of production processes towards intangible values. Jarábková & Hamada (2012: 6) described creativity as the ability to innovate, apply innovation from one geographic area/sector to another, or bring innovation to one geographic area and sector. Creativity is not only a profit-making tool for the tourism sector, but also a strategy that encourages innovation and the development of personal skills (Ray, 1998: 5). In this respect; it includes a form of activity where tourists can learn by experiencing the family relationships in the destination, homemade food, the way of life, the natural and cultural heritage of the region (Ali et. al., 2016: 88). Creativity also provides branding for a creative city and atmosphere. Tourists can gain experience and learning about the region and culture thanks to the atmosphere created by the local people (Rabazauskaitė, 2015: 125).

In the tourism industry, with a flow from mass tourism to special interest travels and tourists getting bored of the holiday concept of sea, sand, and sun and seeking new experiences, especially with the interaction with local cultures and desire for experiencing creative activities in the region, the concept of creative tourism started to become prominent (Bastenegar et. al., 2018: 542). Creative tourism 'is a type of tourism that offers tourists the opportunity to develop their creative potential actively participating in learning experiences specific to a tourism destination' (Richards & Raymond, 2000: 20). The concept of creative tourism creates a common exchange between visitors and locals and includes detailed knowledge of how products, services and experiences are created together by producers and consumers (Richards & Wilson, 2006: 12). In this knowledge, there are common aspects such as active participation, authentic experiences, personal development and skill development and these are mostly related to daily life (Tan et. al., 2013: 154). Creative tourism, which offers tourists the opportunity to participate in creative activities such as arts, crafts and cookery workshops and to be part of the region's unique culture, encourages individuals to learn about the cultural accumulation, artistic activities and characteristics of the region, and to gain experience through intensive communication with the people living in the region (UNESCO, 2006: 3). In short, creative tourism can be defined as a sustainable form of tourism that provides the opportunity to experience and learn the region's cultural identity, concrete and intangible cultural heritage, art products through active participation by establishing close relations with the local people in the region (Zoğal & Emekli, 2017: 25).

Creative tourism is a tool for providing creative travel zones to tourists, a creative method that enriches the existing resources, a resource for creating new places in the regions and a strategic method used for revitalizing the regions (Aşık, 2014: 788). Activities related to creative tourism enable tourists to learn more about the local skills, specialties, traditions and unique qualities of the places they visit (Tan et. al., 2013: 155). It includes a learning process that consists of experiences that enable tourists to develop their creative potential skills by communicating with the people living in the region and learning the culture of the region (Richards, 2011: 1237). In addition to all these, in terms of process and activity; creative tourism facilitates learning by educating about the environment and ecosystem by creating awareness and awareness on tourists, community, local people (Singsomboon, 2014: 36). This type of tourism focuses on a specific tourism destination or creative activities (Ohridska-Olson & Stanislav, 2010). Tourists enter into a social, participatory, educational and emotional interaction with the destination, local people and the culture of that region. In this way, they feel like citizens of the region (UNESCO, 2006: 2). Creative tourism, which makes the local an attraction area for tourism (Gülüm, 2015: 92), is an important alternative for the development of destinations that have the potential to meet the individual expectations of tourists (Rogerson, 2006: 150).

Emergence and Development of the Concept of Creative Tourism

While people's search for difference, interest on different cultures and desire to live with them increase cultural travels, it also mobilized creative tourism demand (Chhabra et. al., 2003: 703). The tourism industry is experiencing a shift in demand from traditional tourism to creative tourism, where tourists try to participate in local cultures and tend to experience creative activities (Bastenegar et. al., 2018: 542). Therefore, creativity is vital in tourism. Creative tourism has become increasingly important as a result of tourists wanting to have interesting experiences after raising their awareness (Richards, 2009: 78). The rise of qualified consumption, the importance of identity formation and cultural capital gain in post-modern society have also contributed to the emergence of a new type of tourism based on wider use of culture, intangible cultural assets, stronger relations with local people and effective information transfer (Carvalho, 2014: 27).

The term 'creative tourism' was first used by Raymond and Richards, who were inspired by impressive travel experiences during their trips to Thailand, Indonesia and Australia between 1999–2000 and defined as the type of tourism by the visitors in order to develop their creative potential via active participation to travel region-specific courses and learning experiences (Richards & Raymond, 2000: 18). Given the growing role of tourism in the urban economy, it is not surprising that the physical restructuring of cities for creative purposes is closely related to tourism and entertainment (Richards & Wilson, 2006: 17). Creative tourism is used as a new revitalization strategy for cities that use cultural planning and creative industries as a tool in urban transformation and plays an active role in the emergence of creative city concepts (Booyens & Rogerson, 2015: 408). The activities offered must be compatible with each other and be relevant to history, culture and lifestyle in terms of learning and experience. The aim is that tourists are not only tourists but also active citizens of the community (UNESCO, 2006: 3). Creative tourism has emerged as an extension of cultural tourism and has started to develop in response to cultural tourism as a result of active participation, self-discovery and the aim of visitors to be a part of the region (Tan et. al., 2013: 155). Thus, although it was initially considered as a potential tourism form, it was later adopted as a type of tourism that offers tourists experiences of discovering their creative potential, actively participating in courses and learning about the regional culture (Ali et. al., 2016: 87). Depending on this development, creative tourism destinations have been regarded as a strategy to contribute to local economies physically, culturally and socially and to encourage tourists to learn the social and cultural features of the spaces (Alvarez et. al., 2010). Creative tourism includes economic approaches, environment, culture, sociology, anthropology and interdisciplinary activities that can be studied according to the sciences of history, art and physiology (Bastenegar et. al., 2018: 543). At the same time, it tends to create a more people-oriented society from a sociological point of view, transforming into more specific activities with travelers' symbols, values, knowledge, experience and participation (Richards, 2016: 32). As a result, creative tourism has emerged as a set of activities that provide individuals with the opportunity to learn and develop their personal skills and discover their creativity, and to give them the opportunity to become part of the local traditions (Richards, 2011: 1227). Therefore, creative tourism not only helped to develop the ties between the visitor, the host and the guest, but also encouraged tourists to discover and realize themselves (Chang et. al., 2014: 403).

Creative tourism is nourished by the overall growth in cultural tourism, while addressing people who want more than cultural experiences (Richards, 2014: 130). Instead of manipulating and exploiting it as a new generation of tourism, it wants to evaluate and enrich cultural, personal and natural resources, to use creativity as the main source for originality and sustainability while preserving concrete and intangible cultural heritage, defending the recovery of intangible assets (Juzefovic, 2015: 73). Individuals who want to improve themselves desire a range of activities to paint, draw, design, photograph, implement cultural activities, be part of the life of the region, experience local food and drink in the regions they visit (Richards & Wilson, 2006: 10). Today's creative tourism includes artists, designers, commercial or event producers, creative product organizations and networks, global foundations and blockbuster exhibitions, educational tourists such as students, and creative business tourists such as cultural tourists that motivate economic transfer (Kostopoulou, 2013: 4587). In the context of responding to these demands and providing alternative approaches to tourism development in many places, interest in creative tourism is increasing internationally (Duxbury & Richards, 2019: 1).

Creative Tourists and Its Properties

Creative tourists are individuals that aim to experience the cultural features of a destination, gaining natural and authentic experience and therefore increasing their personal experiences and creativity (Jelinčić & Žuvela, 2012: 81). The most significant characteristics of creative tourists are curiosity, high production power and versatility of their interests (Çellek, 2003: 7). The basic needs of creative tourists whose expectations are different from other tourists during their travels are to participate in cultural and creative activities (Aşık, 2014: 790). The creative tourist often wants to get to know a culture on the spot, meet real people with whom s/he can interact and experience their daily life. In this direction, learning from local producers a craft and skill is the most effective way to achieve this (Richards, 2010: 14). The basis of the creative tourism experience is the interaction and communication with the other people in the destination they visit (Selstad, 2007: 25). During the creation of a creative product, creative tourists are both participants and producers, and the content of their own travel is the determining factor in the formation of the demand for creative tourism (Prahalad &

Ramaswamy, 2004: 51). Nobody has control over the experiences gained by the tourists participating in creative tourism activities (Anderson, 2007: 46). Creative tourists who participate in activities that are the main factor of creative tourism (Richards & Raymond, 2000: 18) are participants who enjoy developing new skills and prefer applied learning (Tan, Luh & Kung, 2014: 250). Tourists participating in creative tourism activities are mostly; they are more interested in abstract culture, recognize different cultures and live as an opportunity for personal development, and have already been informed about their destination (Aşık, 2014: 789).

Creative Tourist Types (Profile)

The experiences and perceptions of individuals participating in creative tourism activities vary according to the profiles of tourists (Tan et. al., 2014: 248). Raymond (2003: 3) states that creative tourism mainly addresses 'third age tourists, young tourists and tourists interested in traditional culture'. Third age group tourists are the advanced and mature age group of tourists that arise as a result of increasing life expectancy (Kılıçlar et. al., 2017: 80). This tourist profile wants to carry out activities and experiences that they could not perform during their years of employment during their retirement (Ceylan, 2018: 671). Another creative tourist profile consists of young tourists. Young tourists travel to different destinations for adventure and education in order to experience new cultures (Polat, 2017: 245). This group, which is under thirty years of age, is open to new ideas and is interested in travel that will contribute to the development of mental and physical aspects (Aşık, 2014: 789). The creative tourist profile, which is interested in the traditional culture, has a structure that integrates with the culture of the region and wants to see all the elements of the local culture (Yüncü, 2010: 28). These tourists experience a specific destination for reasons such as personal satisfaction and personal development, learning (Sigala & Leslie, 2005: 28).

Creative Tourism in Terms of Sustainable Tourism and Tourism Types Where Creative Tourism Activities Can Be Applied

One of the most basic characteristics of creative tourism is its sustainability (Bastenegar et. al., 2018: 542). In the classical tourism approach, tourism products should be evaluated in certain times and places and tourism should be sustainable in an increasing competitive environment (Akgöz et. al.,2016: 397). However, since living with the people of the region is at the forefront in creative tourism, there is no risk of damaging the nature and natural destination resources and the arrangements to prevent this are always taken into consideration. Like other types of tourism, creative tourism, which is not affected by seasonality, can maintain its continuity throughout the year (Stipanović & Rudan 2014: 510; Yılmaz & Alçin, 2017: 221).

Creative tourism is not just a combination of creativity and tourism. It is also a new form of tourism in providing the cultural needs of tourists and the sustainability of tourism. In this context, creative tourism has a structure that can be updated and improved at all times (Zhang, 2013: 177–179) and it can be a tool to develop sustainable tourism models in the renewal and revitalization of cultural resources (Richards & Marques, 2012: 7).

Creative tourism is closely related to cultural tourism. In this sense, it includes many creative activities such as agricultural tourism, gastronomy, culinary arts, music, ecotourism, crafts, theater, dance, etc. (Zoğal & Emekli, 2017: 31). Unlike cultural tourism, the fact that tourists provide experiences that contribute to their development is the most fundamental distinguishing feature (Tan et. al., 2014: 249). Creative tourism practices are frequently seen in congress tourism activities. Participants can take part in in creative activities by including appropriate activities (workshops, soap making, etc.) in the congress program (Yozcu & İçöz, 2010: 111).

Table 1: Activities That Can Be Used for Creative Tourism

Traditional culture and historical and cultural heritage	Crafts, antiques and restoration	Jewelry, textile, ceramics, wood processing, architectural restoration, etc.
	Traditional cuisine	Food making, local cuisine, etc.
	Historical, cultural and natural heritage	Museums, archives, libraries, monuments, natural parks, etc.
Art	Visual arts	Painting, sculpture, photography, literature, etc.
	Performance arts	Theater, dance, opera, circus, puppet, etc.
Cultural Industries	Audiovisual	Film, television, video, radio, documentary, etc.
	Music	Live music, album, record, CD, etc.
	Publishing	Books, magazines, newspapers, etc.
Creative Activities	Design	Fashion design, graphic design, interior architecture, etc.
	Creative services	Architecture, advertising, etc.

Source: (INTELI, 2011: 20).

As can be observed in Table 1, all branches of art and works, creative activities, traditional culture and cultural heritage elements can be used as a source of supply in creative tourism. While natural and cultural attractions constitute the source of tourism types such as city tourism, rural tourism and nature tourism; creative tourism, which is an extension of cultural tourism and which can bring together different types of tourism in this sense, plays an active role on the attractiveness and life course of touristic destinations (Güven, 2016: 327).

Positive and Negative Aspects of Creative Tourism

Creative tourism allows tourists to spend a more satisfying holiday. They assist the manifestation of the local creativity features of destination for those hosting tourism. Development of innovative tourism products are significant in terms of small- and medium-scale businesses. For destinations, it is an effective tool to highlight themselves in a competitive environment. It helps develop positive relationships between the visitors and local community (Richards, 2015). In creative tourism, tourists discover different traditions and lifestyles and can see daily life in all its colors ((Juzefovič, 2015: 73). Creative tourism promotes intercultural and individual interaction by preserving local culture and values (Ohridska-Olson & Stanislav, 2010: 10).

If creative tourism cannot be developed with sustainability, society culture may be adversely affected. The changes that occur in the regions where tourism mobility is intense may arise as a social problem. As the level of interaction between tourists and local people increases, local people are affected by the culture they are exposed to and they can shape their lives according to the culture they are influenced by moving away from their habits such as clothing, lifestyle and eating and drinking routines. In addition, the traditional culture may be in danger of extinction, while it may move away from its moral values (Uygur & Baykan, 2007: 18).

The cultural values and local traditions of the destinations are the main sources of creative tourism. In the order of preference of people, the fact that they prioritize Italy for the highest level of opera, Russia or Navajo for religious icon painting art and Santa Fe for pot/clay courses indicates that these activities, both in domestic and international travels, represent the supply factors of creative tourism (Ohridska-Olson & Stanislav, 2010: 12). Thailand is also one of the countries that recently started to use the concept of creative tourism to attract more tourists during off-season periods (Wattanacharoensil & Schuckert, 2016). Creative tourism has developed over several years in many countries such as New Zealand, Austria, Spain, Canada, the United States and Taiwan (Tan et.

al., 2013: 155). Traditional handicrafts encourage the sustainability of tourism activities in countries such as Turkey that are rich in terms of creative working areas and developing in terms of tourism industry. In the destination, tourists can experience the traditional handicrafts of the region and buy these products as souvenirs and have an unforgettable experience. In this way, they add cultural motifs to their tourist experiences. The fact that these values are unique to the region is the main starting point of creative tourism (Altınay & Dinçer, 2017: 346).

Conclusion and Evaluation

People's expectations and desires change constantly. These different desires and expectations have led to the tourism activities and have resulted in the diversification of touristic products. In this context, creative tourism is a type of tourism realized by wondering different cultures, discovering the cultural and regional activities of the destination and experiencing these activities directly. Recently, it has become an accepted approach all over the world in order to attract tourists to destinations by influencing them with creative activities and to increase economic development. Creative tourism can be used as an element of attraction due to reasons such as increasing the number of tourists, ensuring the sustainability of cultural heritage and providing personal development opportunities to individuals.

Creative tourism is also a key part of the current competitive environment. Apart from encouraging the development of cultural tourism supply sources, it has a structure that constitutes a source of income and is able to meet the demands and needs of tourists. It provides the diversity of cultural tourism, enables the development of concrete and intangible cultural heritage elements, and provides the opportunity to participate in their lives by establishing a bridge between tourists and local people.

With the shaping of sustainability principles and new alternative tourism forms, seasonal density can be controlled in the destinations, the damages caused by mass tourism can be prevented and areas can be developed for tourists seeking different experiences. Original and participatory creative approaches, such as creative tourism, will stimulate the creation of new tourism experiences and innovative forms of consumption by reforming traditional tourism products. Creative tourism will be able to complement itself perfectly as tourists interact with destinations and the environment. The key element here is to encourage the participation of local people in the work process and enable tourists to experience the unique culture of the region with active participation.

References

Akgöz, E., Göral. R. & Tengilimoğlu, E. (2016). Turistik Ürün Çeşitlendirmenin Sürdürülebilir Destinasyonları Açısından Önemi, *Akademik Bakış Dergisi*, 55, pp. 397–407.

Al Ababneh, M. M. (2017). Creative Tourism, *Journal of Tourism & Hospitality*, 6/2, pp. 282.

Ali, F., Ryu, K. & Hussain, K. (2016). Influence of Experiences on Memories, Satisfaction and Behavioral Intentions: A Study of Creative Tourism, *Journal of Travel & Tourism Marketing*, 33/1, pp. 85–100.

Altınay, M. & Dinçer, İ. F. (2017). Geleneksel El Sanatlarının Yaratıcı Turizm Kavramında Değerlendirilmesi, *Journal of Recreation and Tourism Research*, 4/(Special Issue 1), pp. 343–352.

Alvarez, M. D., Salman, D. & Uygur, D. (2010). Creative Tourism and Emotional Labor: An Investigatory Model of Possible Interactions, *International Journal of Culture, Tourism and Hospitality Research*, 4/3, pp. 186–197.

Anderson, D. T. (2007). The Tourist in the Experience Economy, *Scandinavian Journal of Hospitality and Tourism*, 7/1, pp. 46–58.

Aslan, A. E. (2016). Kavram Boyutunda Yaratıcılık, *Türk Psikolojik Danışma ve Rehberlik Dergisi*, 2/16/, pp. 15–20.

Aşık, A. N. (2014). Yaratıcı (Kreatif) Turizm, *Uluslararası Sosyal Araştırmalar Dergisi*, 7/31, pp. 786–795.

Bastenegar, M., Hassani, A. & Khakzar, B. M. (2018). Thematic Analysis of Creative Tourism: Conceptual Model Design, *Amazonia Investiga*, 7/17, pp. 541–554.

Booyens, I. & Rogerson, C. M. (2015). Creative Tourism in Cape Town: An Innovation Perspective, *Urban Forum*, 26/4, pp. 405–424.

Carvalho, R. (2014). A Literature Review of the role of Cultural Capital in Creative Tourism, In J. A. C. Santos, M. Correira, M. Santos, & F. Serra (Eds.), TMS 2014: Management Studies International Conference (pp. 17–28), University of the Algarve, School of Management, Hospitality and Tourism.

Ceylan, U. (2018). Üçüncü Yaş Turistlerin Termal Konaklama İşletmelerinden Beklentileri Üzerine Bir Araştırma, *MANAS Sosyal Araştırmalar Dergisi*, 7/4, pp. 671–685.

Chang, L. L. F., Backman, K. & Huang, C. Y. (2014). Creative Tourism: A Preliminary Examination of Creative Tourists' Motivation, Experience,

Perceived Value and Revisit Intention, *International Journal of Culture, Tourism and Hospitality Research*, 8/4, pp. 401–419.

Chhabra, D., Healy, R. & Sills, E. (2003). StagedAuthenticity and Heritage Tourism, *Annals of Tourism Research*,30/3, pp. 702–719.

Çellek, T. (2003). Sanat ve Bilim Eğitiminde Yaratıcılık. *Pivolka*,2/8, pp. 4–11.

Duxbury, N. & Richards, G. (2019). Towards a Research Agenda for Creative Tourism: Developments, Diversity, and Dynamics. Duxbury, N. ve Richards, G. (Ed.) A Research Agenda for Creative Tourism. Edward Elgar Publishing Limited, Cheltenham.

Gülüm, E. (2015). Yaratıcı Turizm-Halk Kültürü İlişkisi ve Yerelin Popülerleşmesi. *Milli Folklor*, 105, pp. 87–98.

Güven, A. (2016). Yaratıcı Turizm Kapsamında Antalya İli Turizm Kaynaklarının Değerlendirilmesi. *3rd Congress on Social Sciences, China to Adriatic.* (pp. 326–336). Antalya: Akdeniz Üniversitesi.

INTELI (2011). Creative-based Strategies in Small and Medium-sized Cities: Guidelines for Local Authorities, (pp.1-124). Final output of the URBACT project "Creative Clusters in urban areas of low density", INTELI Technical Action Plan, Portugal.

Jarábková, J. & Hamada, M. (2012). Creativity and Rural Tourism, *Creative and Knowledge Society*, 2/2, pp. 5–15.

Jelinčić, D. A. & Žuvela, A. (2012). Facing the Challenge? Creative Tourism in Croatia, *Journal of Tourism Consumption and Practice*, 4/2, pp. 78–90.

Juzefovič, A. (2015). Creative Tourism: The İssues of Philosophy, Sociology and Communication, *Creativity Studies*, 8/2, pp. 73–74.

Kılıçlar, A., Aysen, E. & Küçükergin, F. (2017). Demografik Değişimlerin Turizm Türleri Üzerindeki Belirleyici Etkisi: Üçüncü Yaş Turizmi, *Gazi Üniversitesi Turizm Fakültesi Dergisi*, 2, pp. 80–100.

Kostopoulou, S. (2013). On the Revitalized Waterfront: Creative Milieu for Creative Tourism, *Sustainability*, 5/11, pp. 4578–4593.

Lemy, D. (2016). Creative Tourism Activities in Ecotourism. A Study in Ujung Kulon National Park Banten Indonesia. Asia Tourism Forum 2016 – The 12th Biennial Conference of Hospitality and Tourism Industry in Asia, (ATF-16) (pp. 95–98), Atlantis Press.

Lindroth, K., Ritalahti, J. & Soisalon-Soinien, T. (2007). Creative Tourism in Destination Development, *Tourism Review*, 62,3/4, pp. 53–58.

Ohridska-Olson, R. & Stanislav, I. (2010). Creative Tourism Business Model and its Application in Bulgaria Proceedings of the Black Sea Tourism Forum

'Cultural Tourism – The Future of Bulgaria', September 24. (pp. 23–39), Varna, Bulgaria.

Polat, E. (2017). Genç Bireylerin Turistik Tercihlerinin Belirlenmesi Üzerine Balıkesir Kent Merkezinde Bir Araştırma, *Uluslararası Kültürel ve Sosyal Araştırmalar Dergisi*, 3/2, pp. 234–250.

Prahalad, C. K. & Ramaswamy, V. (2004). *The Future of Competition: Co-Creating Unique Value With Customers*. Boston: Harvard Business Press.

Rabazauskaitė, V. (2015). Revitalisation of Public Spaces in the Context of Creative Tourism, *Creativity Studies*, 8/2, pp. 124–133.

Ray, C. (1998). Culture, Intellectual Property and Territorial Rural Development, *Sociologia Ruralis*,38, pp. 3–20.

Raymond, C. (2003). Case Study- Creative Tourism New Zealand. Retrieved October 30 2019 from https://www.yumpu.com/en/document/read/27688957/case-study-creative-tourism-new-zealand.

Richards, G. (2009). Creative Tourism and Local Development. In *Creative Tourism: a Global Conversation* (pp. 78–90). Santa Fe.

Richards, G. (2010). Tourism Development Trajectories: From Culture to Creativity?, *Encontros Científicos–Tourism & Management Studies*, 6, pp. 9–15.

Richards, G. (2011). Creativity and Tourism: The State of the Art, *Annals of Tourism Research*, 38, pp. 1225–1253.

Richards, G. (2014). Creativity and Tourism in the City, *Current Issues in Tourism*, 17/2, pp. 119–144.

Richards, G. (2015). Creative Tourism: New Opportunities for Destinations Worldwide?' Presentation at the World Travel Market Conference on 'Creative Tourism: All that You Need to Know about this Growing Sector', Retrieved October 30 2019 from https://www.academia.edu/17835707/Creative_Tourism_New_Opportunities_for_Destinations_Worldwide.

Richards, G. (2016). The Challenge of Creative Tourism. *Ethnologies*, 38/1–2, pp. 31–45.

Richards, G. & Marques, L. (2012). Exploring Creative Tourism: Editors Introduction, *Journal of Tourism Consumption and Practice*, 4/2, pp. 1–11.

Richards, G. & Raymond, C. (2000). Creative Tourism. *ATLAS News*, 23, pp. 16–20.

Richards, G. & Wilson, J. (2006). The Creative Turn in Regeneration: Creative Spaces, Spectacles and Tourism in Cities. In M. K. Smith (Ed.), Tourism, Culture and Regeneration (pp. 12–24). Wallingford: CABI.

Rogerson, M. C. (2006). Creative Industries and Urban Tourism: South African Perspectives, *Urban Forum*, 17/2, pp. 149–166.

Sacco, P. L. & Segre, G. (2009). Creativity, Cultural Investment and Local Development: A New Theoretical Framework for Endogenous Growth. In *Growth and Innovation of Competitive Regions* (pp. 281–294). Berlin: Springer Heidelberg.

Selstad, L. (2007). The Social Anthropology of the Tourist Experience Exploring the Middle Role, *Scandinavian Journal of Hospitality and Tourism*, 7/1, pp. 19–33.

Sigala, M. & Leslie, D. (2005). *International Culturaş Tourism Management: İmplication an Cases*. Elsevier/Butterworth. Retrieved October 30 2019 from https://books.google.com.tr/books?hl=tr&lr=&id= lmEeJb9rySYC&oi=fnd&pg=PR3&dq=igala,+M.,+%26+Leslie,+D.,+(200 5).+International+cultural+tourism:+Management,+implications+and+- cases.+Heinemann:+Elsevier/Butterworth.&ots=BlMPotDNws&sig=UMK4 9Kp4T-DVtKddn9EERWipoLY&redir_esc=y#v=onepage&q&f=false.

Simonton, D. K. (2000). Creativity: Cognitive, Personal, Developmental and Social Aspects, *American Psychologist*, 55/1, pp. 151.

Singsomboon, T. (2014). Tourism Promotion and the use of Local Wisdom through Creative Tourism Process, *International Journal of Business Tourism and Applied Sciences*, 2/2, pp. 32–37.

Stipanović, C. & Rudan, E. (2014). Development Concept and Strategy for Creative Tourism of the Kvarner Destination. 22nd Biennial International Congress Tourism & Hospitality Industry 2014, University of Rijeka, Faculty of Tourism & Hospitality Management, pp. 507–517.

Tan, S. K., Kung, S. F. & Luh, D. B. (2013). A Model of "Creative Experience" in Creative Tourism, *Annals of Tourism Research*, 41, pp. 153–174.

Tan, S. K., Luh, D. B. & Kung, S. F. (2014). A Taxonomy of Creative Tourists in Creative Tourism, *Tourism Management*, 42, pp. 248–259.

Tang, Z. (2014). Development Stratejies for Cultural Creative Tourism of Heilongjiang, China, *Bio Technology An Indian Journal*,10/24, pp. 15641–15646.

UNESCO. (2006). Towards Sustainable Strategies for Creative Tourism. *Discussion Report of the Planning Meeting for 2008International Conference on Creative Tourism*, Santa Fe, New Mexico, U.S.A., October 25–27, Retrieved October 28 2019, from https://unesdoc.unesco.org/ark:/48223/ pf0000159811.

Uygur, M. S. & Baykan, E. (2007). Kültür Turizmi ve Turizmin Kültürel Varlıklar Üzerindeki Etkileri, *Ticaret ve Turizm Egitim Fakültesi Dergisi*, 2, pp. 30-49.

Wattanacharoensil, W. & Schuckert, M. (2016). Reviewing Thailand's Master Plans and Policies: Implications For Creative Tourism?*Current Issues in Tourism*, 19/10, pp. 1045-1070.

Yılmaz, Ö. G. & Alçin, R. (2017). Yaratıcı Turizm Kapsamında NazarBoncuğu El Sanatının Değerlendirilmesi: Nazarköy Örneği, *Journal of Current Researches on Social Sciences*,7/4, pp. 213-224.

Yozcu, K. Ö. & İçöz, O. (2010). A Model Proposal on the Use of Creative Tourism Experiences in Congress Tourism and the Congress Marketing Mix, *PASOS. Revista de Turismo y Patrimonio Cultural*, Special Issue 8/3, pp. 105-113.

Yüncü, H. R. (2010). Sürdürülebilir Turizm Açısından Gastronomi Turizmi ve Perşembe Yaylası. 10.Aybastı-Kabataş Kurultayı, 11 (27-34). Ankara. Retrieved October 28 2019, from https://docplayer.biz.tr/2975841-10-aybasti-kabatas-kurultayi.html.

Zhang, Y. (2013). Study on the Main Characteristics and Development Countermeasures of Creative Tourism. International Conference on Education, Management and Social Science, ICEMSS, Paris-France, pp. 7-180.

Zoğal, V. & Emekli, G. (2017). Yaratıcı Turizme Kavramsal ve Coğrafi Bir Yaklaşım, *Ege Coğrafya Dergisi*, 26/1, pp. 21-34.

Yılmaz SECİM

23 Gastronomy Festivals in Turkey

Introduction

The concept of gastronomy festival developed after the increase of gastronomy and effectiveness in the world. The general purpose of gastronomy festivals is to introduce products belonging to a region or region at national and international level. Rather than pick one or a few individuals in the study festival, all festivals held in Turkey is aimed to make a general assessment basis. Before the gastronomic festivals in Turkey are required to explain the concept of tourism and festivals.

As a result of the increase in welfare level in the world, tourism activities have improved. Tourism; It is possible to diversify according to accommodation, duration of travel, average age of tourists and means of transportation used (Atalay, 2017: 4). Turkey, when evaluated touristic point of sea, sand and sun has come to the fore with the trio. In addition, historical cultural and natural beauties help to get a share of tourism (Türkben et. al., 2012: 49). In Turkey, as in the world of 'new tourism' name, called varieties developed in different areas of tourism it has become active. Alternative tourism types include health, faith, hunting, plateau, cave, mountain, culture and river tourism. In addition, alternative tourism includes eco tourism, winter tourism, thermal tourism, health tourism, congress tourism, education tourism, sports tourism and religious tourism. (Aydın, 2017: 20; Albayrak, 2013: 71). Festivals can be counted in event tourism as they can be shown in cultural tourism.

In this context, the culinary culture and habits of the countries have become one of the important instruments of tourism marketing. The presentation and preparation of the food, which is applied in culinary cultures, are all parts of tourism. (Güneli, 2012). Gastronomic activities should be examined from a wider perspective without narrow coverage, such as recipes or quality food and beverage businesses. Gastronomy activities include culinary schools, gastronomy tours and guides, gastronomy-related media (TV, magazines, etc.), wine houses, cheese houses, places to produce drinks, fields and gardens that produce (Çağlı, 2012: 32).

Festivals in general are linked to the culture of the local people. Localizations and legacies can be arranged on topics such as ethnic culture, popular culture, art and artists (Helgadóttir, 2018: 339). For this reason, it is organized in order to reflect the life style of indigenous people in general. It is important that the festivals be reflected without losing the authenticity and meaning of the regional culture (Cohen, 1979: 186). During the organization of the festivals, the roles will be changed and the quality of life will be improved by providing local visitors as visitors and recreational development (Sayarı & Gün, 2018: 379).

Gastronomy festivals organized in recent years have accelerated the tourism sector. According to Litvin & Fetter (2006: 46), the general definition of the festival is explained as follows; The festival is a set of activities that create activities for food and beverage and hospitality businesses and enable the participation of tourists. Festivals are events where local shows, products and services are offered. Themes such as balloons, kites, music and movies can be handled in festivals. About 30% of the festivals worldwide are gastronomy festivals. his ratio is known to increase every year (Doğdubay & İlsay, 2016). Food festivals held around the world include a wide variety of food production, food tastings, food presentations, recipes, food rituals, and traditions of the cuisine (Cohen & Avieli, 2004).

Turkey Gastronomy Festivals

Festivals may vary according to their purpose. These objectives include promotion, social goals and support to various foundations (Yeoman et. al., 2012: 34). In addition to the many influences of the festivals, it cannot be denied that the number of visitors to the region increases, the recognition increases and the image of the region improves (Getz, 2000; Lee et. al., 2008: 56; Kim et. al., 2010: 297). When the economic effects of gastronomy festivals are considered, it can be said that it has an important contribution to the country in itself. An example of this is the 'Charleston Wine and Food Festival' in the United States.

The festival contributed $7.3 million to the national economy in 2011 (Doğdubay & İlsay, 2016). This gain is above the tourism income of many countries. The power of the festival to attract the target audience, the target audience, media promotions, event infrastructure, the area organized and the organizers are among the important criteria (Jayswall, 2008). After the gastronomy festivals became widespread, it became the main purpose of tourists to know the culinary culture of the region they travel. Many of the destinations that have become aware of this have started to use culinary culture as the attraction force (Horng & Tsai, 2012: 47). Recognizing the importance of gastronomy in recent years, tourism in Turkey showed a significant development in the field of gastronomy. One of the important areas of this development is the festivals. Especially festivals on gastronomy attracted the attention of the world. The following table shows held in Turkey in 2019–2020 and held the planned festivities were transferred.

Table 1: Held in Turkey in the Year 2019 to 2020 and Regulation of Planned Events (Festival, 2019)

Festival name	Date	Scope	Place of the Festival
Izmir waffle festival	January	Innovative	İzmir- Konak
Burhaniye international olive and olive oil festival	January	Olives	Balıkesir- Burhaniye
Dikmen anchovy festival	January	Fish	Sakarya- Hendek
Sivrihisar sausage festival	February	Sausage	Eskişehir-Sivrihisar
Ankara breakfast festival	February	Breakfast	Ankara- Altındağ
International Istanbul culinary days	February	General	İstanbul – Beylikdüzü
Dalyan mullet fish festival	February	Fish	Muğla- Ortaca
Bodrum bitter weed festival	February	Wild herbs	Muğla- Bodrum
İrişkit Festival	March	Local product	Kahramanmaraş- Merkez
Çağla and günlük festival	March	Local product	Mersin- Silifke
Alaçatı herb festival	April	Wild herbs	İzmir- Çeşme
Giritlilerherb festivali	April	Wild herbs	İzmir- Tire
Bodrum ınternational cheese festival	April	Cheese	Muğla- Bodrum
İzmir chocolate and dessertfestival	April	Çikolata	İzmir-Bayraklı
İstanbul cheese festival	April	Cheese	İstanbul- Şişli
Mersin lemonade festival	April	Local product	Mersin- Erdemli
Halikarnassos two collar culture festival	April	Local product	Muğla- Bodrum
Kapadokya traditional dishes festival	April	Local product	Nevşehir- Kapadokya
International mesir macunu festival	April	Local product	Manisa
International urla artichoke festival	April	Local product	İzmir- Urla
Elazığ coffee and chocolate festival	April	coffee and chocolate	Elazığ
Antalya coffee and chocolatefestival	April	coffee and chocolate	Antalya- Muratpaşa
International Istanbul tea festival	May	Tea	İstanbul
Halfeti fruit dishes festival	May	Local dishes	Şanlıurfa- Halfeti
International Street delicaciesfestival	May	Street delicacies	Antalya- Muratpaşa
Mut Karacaoğlan apricot festival	May	Local product	Mersin- Mut
Bornova cherry festival	Jun	Local product	İzmir- Bornova
Yeşilli ınternationalcherry festival	Jun	Local product	Mardin- Mazıdağı
Kemalpaşa gold cherry culture and art festival	Jun	Local product	İzmir- Kemalpaşa
Samsun local herb dishesfestival	Jun	Wild herbs	Samsun- Atakum

(continued on next page)

Table 1: (continued)

Festival name	Date	Scope	Place of the Festival
Koçarlı pistachios festival	Jun	Local product	Aydın- Koçarlı
Traditional Iğdır apricot festival	Jun	Local product	Iğdır
Yeşilhisar culture and apricot festival	Jun	Local product	Kayseri- Yeşilhisar
Serinhisar roasted chickpea, knife and culture festival	July	Local product	Denizli- Serinhisar
Çemişgezek mulberry and cheese festival	July	Local product	Tunceli- Çemişgezek
Mihalıççık cherry festival	July	Local product	Eskişehir- Mahmudiye
Hekimhan walnut, mine and culture festival	July	Local product	Malatya- Hekimhan
Cumalıkızık raspberry festival	July	Local product	Bursa- Yıldırım
Geleneksel gravyer ve kaşar festivali	July	Cheese	Kars
International culture and banana festival	July	Local product	Mersin- Anamur
Sultandağı culture, art and cherry festival	July	Local product	Afyonkarahisar- Sultandağı
Traditional Gediz tarhana festival	July	Local product	Kütahya- Gediz
National Gökçebey chickenandculture festival	July	Local product	Zonguldak- Gökçebey
Manisa Adala peach festival	July	Local product	Manisa- Salihli
Silivri yoghurt festival	July	Local product	İstanbul- Silivri
Akseki ayran festival	July	Local product	Antalya- Akseki
Chestnut Honey festival	July	Honey	Kastamonu- Doğanyurt
Ardahan national culture and honeyfestival	July	Honey	Ardahan
Çamoluk honey festival	August	Honey	Giresun- Çamoluk
İskilip stuffed pickles and strawberries festival	August	Local product	Çorum- İskilip
Karacakılavuz keşkek festival	August	Local product	Tekirdağ- Merkez
Refahiye culture and honey festival	August	Honey	Erzincan- Refahiye
Zara honey and culture festival	August	Honey	Sivas- Zara
Çayıralan honey and culture festival	August	Honey	Yozgat- Çayıralan
Slide festival	August	Local product	Nevşehir
Tonya butter and culture festival	August	Local product	Trabzon- Tonya

(continued on next page)

Table 1: (continued)

Festival name	Date	Scope	Place of the Festival
Umurbey traditional peach festival	August	Local product	Çanakkale- Lapseki
Cheese festival	August	Cheese	Kahramanmaraş-Pazarcık
Hayrabolu sunflower festival	August	Local product	Tekirdağ- Hayrabolu
Grapes and pepper festival	August	Local product	Gaziantep- Islahiye
Bolvadin traditional slide festival	August	Local product	Afyonkarahisar-Bolvadin
Pülümür honey festival	August	Honey	Tunceli- Pülümür
Dikili okra festival	August	Local product	İzmir- Dikili
Gülnar çakşır honey and yörük festival	August	Honey	Mersin- Gülnar
Orhangazi gedelek pickle festival	August	Local product	Bursa Orhangazi
Enrez honey festival	August	Honey	Antalya- Finike
Gazyağcı Konya Dishes festival	September	Local product	Konya- Meram
International Germencik fig culture and art festival	September	Local product	Aydın- Germencik
Gaziantep ınternational gastronomy festival	September	Local product	Gaziantep-Şehitkamil
Uluslararası pistachios culture and art festival	September	Local product	Gaziantep- Şahinbey
International kalecik karası grapes festival	September	Local product	Ankara- Kalecik
Kütahya cimcik festival	September	Local product	Kütahya
Ağrı honey festival	September	Honey	Ağrı
İzmir boyoz festival	October	Local product	İzmir- Konak
Appetizer Festival	October	Appetizer	Antalya- Muratpaşa
Akhisar olives harvest festival	October	Olives	Manisa- Akhisar
Local product promotion festival	October	Local product	Bursa-Mustafakemalpaşa
Yalova subaşı kiwi festival	October	Local product	Yalova-Altınova
Pamukova local product and Quince festival	October	Local product	Sakarya- Pamukova
Local product festival	October	Local product	Antalya- Aksu
Beydağ chestnut festival	October	Local product	İzmir- Beydağ
Pepeçura festival	October	Local product	Rize
Ödemiş chestnut festival	November	Local product	İzmir- Ödemiş
Antalya coffee festival	November	Coffee	Antalya- Kepez

(continued on next page)

Table 1: (continued)

Festival name	Date	Scope	Place of the Festival
Seferihisar mandarin festival	November	Local product	İzmir- Seferihisar
International bread festival	November	Local product	Ankara- Çankaya
Bodrum mandarin harvest festival	December	Local product	Muğla- Bodrum
Çıntar festival	December	Local product	Muğla

Source: (Festival, 2019).

When we examine the table gastronomy festival in Turkey trails and features can be better explained. Turkey is known to be close to about 100 known gastronomic festival. These festivals can be heard in national or international media depending on their promotional status. It is seen that the festivals organized especially in the cities such as Istanbul, İzmir, Antalya and Ankara attract more attention and take place in the media.

The biggest reason for this situation can be said to be the presence of media organizations in these cities and the close relationship of the regulators with the media. The recognition of the city in terms of gastronomy is also an important issue. For example, with the participation of united nations, education, science and culture organization (UNESCO) in the creative cities network of Gaziantep, the city's culinary culture, food types and table manners became prominent and became known in the world. Therefore held in Gaziantep 'Gaziantep international gastronomic festival' shows how every year from all corners of the world and Turkey tens of thousands in attendance. International networks to understand the importance of the participation of Turkey after the Opium Hatay and the city has been included in the UNESCO's hometown creative gastronomy in the network. In the future, these two cities are expected to come to the forefront with festivals such as Gaziantep. Today, there are not enough gastronomic activities and festivals in these cities.

When the table is examined in terms of cities, it is seen that many gastronomy festivals are organized in İzmir, İstanbul, Muğla and Antalya. This is another side of the city from the tourist point of each of the partners is among the leading cities of Turkey.

All of these cities outside Istanbul's main sea forms the backbone of tourism in Turkey, sand, tourism has become the center of the sun. Istanbul, on the other hand, has come to the forefront with its historical and cultural appeal. At the same time, Istanbul is one of the important gastronomic cities in the world known for its Ottoman palace cuisine. İzmir, Muğla and Antalya have increased the number of gastronomy festivals in order to spread tourism activities to twelve months apart from sea, sand and sun tourism. In these cities, in addition to local product festivals (Honey, cheese, weeds, etc.), it is seen that festivals are organized among products that are widely consumed in the world such as chocolate, coffee and waffles to attract the attention of world cuisine. Among the Anatolian cities, Ankara, Mersin and Bursa have more gastronomy festivals than other provinces. Ankara Turkey is more than the number of the festival due to the capital of the republic is expected. It is known that in Mersin and Bursa provinces festivals are organized as a result of active work of local institutions and organizations. In these cities, it is seen that more festivals are organized especially on local products. Apart from these cities, Konya, Nevşehir, Kayseri, Eskişehir, Kutahya, one or two gastronomy festivals are held in cities such as. This shows that Anatolian cities are not well organized and do not have enough experience in organizing the festival. Almost all of the festivals organized in these cities are about local products and dishes. When the cities in the Black Sea region are examined, Samsun, Zonguldak, Giresun and Rize provinces are among the cities where festivals are organized, but only one gastronomy festival arrangement shows that the region has not developed much in terms of festivals. The Black Sea region is particularly prominent with its natural beauties and lakes. After receiving abundant rainfall, especially in the region is the most important region with Turkey's Aegean region in the diversity of edible wild herbs. For this reason, it is important to organize gastronomic festivals about weeds and meals made from weeds in the region. It is obvious that there are many festivals that are not included in the table but are organized in the region. It is of great importance to announce and promote these festivals through the media.

If the festivals are handled according to the dates, the number of festivals is quite high in April, June, July, August, September and October. It is thought that there is an agglomeration on these dates due to the fact that the tourism season continues in these months and it is easy to reach domestic and foreign tourists. In addition, the participation of local and foreign tourists coming to visit a region to the festivals means that the tourists contribute financially to the local people. The reported dates are generally expected to be high in aggregates due to the harvest period. In January, February, March, May, November and December, the number of festivals decreased. Since these dates are not holiday periods, the number of festivals is low. At the same time, the low number of harvested products during these periods also contributed to the low number of festivals. Increasing the number of gastronomy festivals organized during these periods will contribute to the development of the tourism area in particular.

When we look at the scope of the festivals, it is seen that most of them are festivals related to local products. This situation shows the richness of Turkish geography in terms of local products. Among the local products, especially vegetables, fruits, cheese and honey are brought to the fore by the peoples of the region. In the foreground of these products, it is effective to consume a lot in the region and have different characteristics from other regions.

The festive period of these products, which are held in festivals, is usually the period in which they are harvested. It is known that festivals are organized in different periods if the harvest period is not suitable for the festival. However, the subsequent festivals are not as effective as the festivals held during the harvest period. It was also found that many activities were actively organized on products such as coffee and chocolate. In particular, although longer grown coffee raw materials in Turkey is of great importance in Turkish culture. Therefore, it is one of the most important products on which festivals are held. Chocolate festivals are mostly organized in the west and presented to consumers and people who participate in the festival with different presentations. But in recent years in Turkey's eastern province is seen that some chocolate festivals held. It is thought that chocolate and chocolate sector is effective in organizing chocolate festivals in Western provinces.

Result

Gastronomy is basically combining culture and food under one heading.

Food alone does not make sense, nor does it make sense in culture without an object to complement i.e. region's natural beauty, sea, sand, sun and history in terms of attracting attention, may not be enough alone. Exercises that can reflect extra food and culture will contribute to the regions getting more shares from tourism income and financial development of the people of the region.

It is important to evaluate the values of a region in the context of tourism, to exclude tourism from being a seasonal activity and to turn the values obtained in terms of sustainability into advantage. The perception of a great experience for tourists visiting the region can be provided by gastronomic festivals. Gastronomy festivals are the visual festivals that can keep the regions, cities and countries in the tourism field for a long time.

When the gastronomy festivals are taken into consideration, it will be seen how much the value and advantages it adds to the destination are great. Organizing festivals by going beyond the usual tourism diversity will make a significant contribution to the image and sustainability of the region. Apart from bringing different destination areas to tourism with gastronomy festivals, it is thought that the places that already have the characteristics of tourism regions will increase the attraction element. Considering all the factors, it is thought that gastronomy festivals strengthen the food and beverage businesses serving at the local level, contribute to the development and preservation of the region's products and biological diversity, and at the same time contribute to the continuation of local identities.

Gastronomy festivals held in Turkey has contributed to the country in terms of both social and economic. These contributions can be examined under the following headings.

- Regulation of gastronomic festivals, Turkey will serve subtracting alternative tourism activities have become standard in the tourism activities will contribute to the formation of new destinations.
- The sowing period usually starts in April, ending tourism activities in Turkey will provide 12 months. This situation will contribute to the development of the welfare level of the local people and increase the socioeconomic level.
- Produced in the region and that indigenous food, drinks and products in Turkey and will contribute to the recognition in the world. After this recognition, local job opportunities will increase and migration from village to city will decrease.
- Forgotten or forgotten meals and customs, one of the topics discussed in the local festivals, will be brought back to the day by remembering and will have a chance to be recorded.
- Gastronomy festivals will not only provide food and beverages, but also cultural development and transfer of culture to new generations.
- Organizing gastronomy festivals in an area that is not well known will contribute to the increasing national and international recognition of many unknown destinations and to attract more visitors.
- It is thought that due to the necessity of landscaping and infrastructure arrangements before the festival to be held in a region, arrangements will be made due to the festival.
- It will attract media attention after the festival is comprehensive or interesting and will increase the awareness level of the destination.
- Considering the importance of the brand in tourism, the process of branding of the festival area or product will begin and the added value of the product will increase.
- It will contribute to the tourists visiting a region once to visit the same region in the following period.
- As a result of the increase in the interaction of tourists and local people, a sincere atmosphere will be created and this will provide confidence in product marketing.

However, the high level of participation after the festival may bring some negativities. At the beginning of these negativities;
- The high number of people coming to the festival can cause confusion.
- Tourists coming to the region after the festivals to take place and behave in a way that will affect the life of the local people.
- The small amount of products or products introduced during the festival and production takes time may cause discontent to the tourists who come.
- Lack of sufficient infrastructure of the festival area may disturb both tourists and local people.
- Although the products introduced at the festival are unique to the region, they can be imitated after they attract attention.
- After the festival, people coming to the region to settle in the region from time to time can be met with discontent by the local people.

References

Albayrak, A. (2013). *Alternatif Turizm*. Ankara: Detay Yayıncılık.

Atalay, R. (2017). Alternatif Turizmin Bölgesel Kalkınmaya Etkisi: Beyşehir Yöresi Örneği. Unpublished Master's Thesis, Selçuk Üniversitesi Sosyal Bilimler Enstitüsü, Konya.

Aydın, Ö. (2017). Turistlerin Sinop İlinin Alternatif Turizm Potansiyeline Yönelik Algılarının İncelenmesi. Unpublished Master's Thesis, İskenderun Teknik Üniversitesi Sosyal Bilimler Enstitüsü, Hatay.

Cohen, E. (1979). *A Phenomenology of Tourist Experiences*. Sociology, 13(2): 179–201.

Cohen, E. & Avieli, N. (2004). *Food in Tourism: Attraction and Impediment*, Annals of Tourism Research, 31(4): 755–778.

Çağlı, I. B. (2012). Türkiye'de Yerel Kültürün Turizm Odaklı Kalkınmadaki Rolü: Gastronomi Turizmi Örneği. İstanbul Teknik Üniversitesi, Fen Bilimleri Enstitüsü, Unpublished Master's Thesis, İstanbul.

Doğdubay, M. & İlsay, S. (2016). *Bir İletişim Biçimi Olarak Gastronomi Konulu Festivaller*, HakanDoğdubay ve İlsay (Ed.), Bir İletişim Biçimi Olarak Gastronomi içinde (s. 169–193). Ankara: Detay Yayıncılık.

Festival, (2019). https://festivall.com.tr/fest-kategori/13/yeme-icme/ date of access: 12.10.2019.

Getz, D. (2000). Developing a research agenda for the event management field. Events Beyond 2000: Setting the Agenda, Proceedings of Conference on Event Evaluation, Research and Education. Editors J. Allen, R. Harris, L.

K. Jago and A. J. Veal. Sydney: Australian Centre for Event Management, University of Technology Sydney.

Güneli, M. (2012). *Gastronomi ve İmaj Devri* (Date of access: 15.10.2019). http://www.turizmdebusabah.com/yazarlar/gastronomi-ve-imaj-devri-mehmetguneli-62638.html.

Helgadóttir, G. (2018). Herding Livestock and Managing People: The Cultural Sustainability of a Harvest Festival, (Editor) Judith Mair: The Routledge Handbook of Festivals 1st Edition (335-343), Publisher: Routledge Handbook

Horng, J. S. ve Chen-Tsang, T.(2012). *Culinary Tourism Strategic Development: An Asia-Pacific Perspective*, International Tourism of Journal Research, 14, 40-55.

Jayswal, T. (2008). *Events Tourism: Potential to Build a Brand Destination*, Conference on Tourism in India – Challenges Ahead, 15-17 May, IIMK.

Kim, S. S., Prideaux, B. & Chon, K. (2010). *A Comparison of Results of Three Statistical Methods to Understand the Determinants of Festival Participants* Expenditures. International Journal of Hospitality Management, 29(2): 297-307.

Lee, Y. K., Lee, C. K., Lee, S. K. & Babin, B. J. (2008). *Festival scapes and Patrons Emotions, Satisfaction, and Loyalty*. Journal of Business Research, 61(1): 56-64.

Litvin, S. W. & Fetter, E. (2006). *Can a Festival Be Too Successful a Review of Spoleto, Usa*. International Journal of Contemporary Hospitality Management, 18(1): 41-49.

Sayarı, B. K. & Gün, T. (2018). *Placemaking Betwixt and Between Festivals and Daily Life*, (Editor) J. Mair: The Routledge Handbook of Festivals 1st Edition içinde (ss. 373-383), Routledge.

Türkben, C., Fulya, G. & Yılmaz, U. (2012). *Türkiye'de Bağcılığın Tarım Turizmi (Agro-Turizm) İçinde Yeri ve Önemi*, KMÜ Sosyal ve Ekonomik Araştırmalar Dergisi, 14 (23): 47-50.

Yeoman, I., Robertson, M., Ali-Knight, J., Drummond, S. & Mcmahon-Beattie, U. (2012). *Festival and Events Management*, Publisher: Routledge Handbook

List of Figures

Hotel Managers' Metaphoric Perceptions for Smart Hotel
Ebru GÖZEN
Figure 1: Henn-na Hotel Reception Desk. Source: Yamak, 2017 47
Figure 2: Word Cloud 57

Diaspora Tourism
Emre AYKAÇ and Ömer Ceyhun APAK
Figure 1: Conceptual Framework of Diaspora Tourism. Source: Li,
McKercher & Eric TakHin Chan, 2016 76
Figure 2: Intersection of diaspora tourism and culture-based special
interest tourism types 77

Gastronomy Tourism and Geographical Indications in Tokat
Emin ARSLAN
Map 1: Location of the Study Area (Coğrafya Harita, 2019) 100
Figure 1: Some Gastronomic Flavors of Tokat (Tokat Provincial
Directorate of Culture and Tourism, 2019) 101

Digital Transformation & Marketing in Tourism Industry

Samet GÖKKAYA
Figure 1: Historical Development of Industry 4.0. Source: (RSA
Solutions. 2015) 153

Cultural Heritage Tourism Inventory in Tokat Province
Hakan KENDİR
Map 1: Study Area Location (Coğrafya Harita, 2019) 166
Figure 1: Some Touristic Cultural Heritage Assets of Tokat (Turkey's
Culture Portal, 2019) 172

Effects of Digitalization in Tourism
Serdar SÜNNETÇİOĞLU
Figure 1: Information Flow in a Smart Accommodation Network.
Resource: Buhalis and Leung, 2018: 47 183
Figure 2: Technological Ecosystems of Restaurants. Resource: www.
forbes.com, 15.11.2019 184

Figure 3: Tourists Face to Face and Online Interaction at Destination.
Resource: Fan, Buhalis and Lin (2019) 187
Figure 4: Social Interaction Model in Digital Tourism.
Resource: Sünnetçioğlu (2019: 614) 188

Overtourism

Irem BOZKURT and Enes YILDIRIM
Figure1: The Conceptual Model of Overtourism Source: (IPOL, 2019) 210

Qualitative Approaches for Tourism Research
Kansu GENÇER
Figure 1: Hermeneutic Cycle. Source: Sfu.ca (2019) 230
Figure 2: Basic Phenomenology Concepts. Source: Boeree (1998) 233

Event Tourism
Mehmet CAN and Çağla ÜST CAN
Figure 1: Events According to Their Contents: (Getz, 2008: 404) 245

The Effect of Nepotism
Mustafa Cüneyt ŞAPCILAR and Ahmet BÜYÜKŞALVARCI
Figure 1: Research Mode 258

List of Tables

Hotel Managers' Metaphoric Perceptions for Smart Hotel
Ebru GÖZEN
Table 1: Smart Applications Used in Smart Hotels 45
Table 2: Metaphors Created by Managers for Smart Hotel Concept 52
Table 3: Categories of Metaphors Created by Managers for Smart Hotel
Concept 53
Table 4: Category Distribution by Demographic Data 56

Ecoturism and Geographical Information Systems Applications
Yasin DÖNMEZ and Sevgi ÖZTÜRK
Table 1: Definitions of GIS (Jovanović, 2016) 92
Table 2: GIS data elements (Jovanović, 2016) 92

Gastronomy Tourism and Geographical Indications in Tokat
Emin ARSLAN
Table 1: Tokat's Geographical Indication Registration and Gastronomic
Products in Registration Process 102
Table 2: Erbaa Narince Vineyard Leaf 103
Table 3: Niksar Walnut 104
Table 4: Tokat Kebab 104
Table 5: Tokat Narince Pickled Vine Leaves 105
Table 6: Turhal Yoghurtmach 106
Table 7: Zile Churchkhela 106
Table 8: Zile Molasse 107
Table 9: SWOT Matrix for Tokat's Geographically Indicated Products
and Gastronomy Tourism 108

To Determine the Recreational Potential of Trabzon
Mehmet Mert PASLI and Evren GÜÇER
Table 1: The National Parks in Turkey 136
Table 2: Population, Sampling and Representative Number 139
Table 3: Personality Types According to Gender 139
Table 4: Frequency of Activities in the National Park According to
Gender Variable 140
Table 5: Distribution of Personality Types According to Age Variable 141
Table 6: Frequency of Activities in the National Park According to Age
Variable 142

Table 7: Distribution of Personality Types According to Education Variable 143
Table 8: Frequency of Activities in the National Park According to Education Variable 144
Table 9: Correlation Analysis of the Relationship between the Personality Types of the Participants and the Activity Frequency in the National Park 145

Digital Transformation & Marketing in Tourism Industry
Samet GÖKKAYA
Table 1: First, Second and Third İndustrial Transformation Technologies 154

Cultural Heritage Tourism Inventory in Tokat Province
Hakan KENDİR
Table 1: Tokat's Tangible Cultural Heritage Inventory 169

Effects of Digitalization in Tourism
Serdar SÜNNETÇİOĞLU
Table 1: Smart Application Examples from Different Tourism Businesses 181
Table 2: Transformation in Tourism as a Result of Digitalization 182

Overtourism
Irem BOZKURT and Enes YILDIRIM
Table 1: Overtourism Living Destinations and Developed Solution Strategies 214

Qualitative Approaches for Tourism Research
Kansu GENÇER
Table 1: Hermeneutic Phenomenology Preliminary Guide for Tourism Research 236

The Effect of Nepotism
Mustafa Cüneyt ŞAPCILAR and Ahmet BÜYÜKŞALVARCI
Table 1: Participant Opinions on Nepotism Perception 260
Table 2: Participant Opinions on Intention to Leave 261
Table 3: Participant Opinions on Job Satisfaction Perception 261
Table 4: Correlation Analysis of Nepotism and Job Satisfaction Dimensions (Resort Hotels) 263

Table 5: Correlation Analysis of Nepotism and Job Satisfaction
Dimensions (City Hotels) 264
Table 6: Correlation Analysis of Nepotism and Intention to Leave
(Resort Hotels) 265
Table 7: Correlation Analysis of Nepotism and Intention to Leave
(City Hotels) 265
Table 8: Results of Regression Analysis on Nepotism and Job
Satisfaction Dimensions (Resort Hotels) 266
Table 9: Results of Regression Analysis on Nepotism and Job
Satisfaction Dimensions (City Hotels) 267
Table 10: Results of Regression Analysis on Nepotism and Intention to
Leave (Resort Hotels) 268
Table 11: Results of Regression Analysis on Nepotism and Intention to
Leave (City Hotels) 268

City Tourism and Kütahya
Uğur CEYLAN
Table1: Overnight Statistics of Tourism and Municipality Certified
Enterprises 298
Table 2: Number of Hotels and Beds in Kütahya (June, 2019) 298
Table 3: Number of Visitors of Museums in Kütahya City Center
(January–June 2019) 299

An Overview of Creative Tourism Concept
Yeliz PEKERŞEN
Table 1: Activities That Can Be Used for Creative Tourism 315

Gastronomy Festivals in Turkey
Yılmaz SEÇİM
Table 1: Held in Turkey in the Year 2019 to 2020 and Regulation of
Planned Events (Festival, 2019) 325

www.ingramcontent.com/pod-product-compliance
Ingram Content Group UK Ltd.
Pitfield, Milton Keynes, MK11 3LW, UK
UKHW022236230426
12048UKWH00018BA/1295